DNA—Ligand Interactions
From Drugs to Proteins

NATO ASI Series

Advanced Science Institutes Series

A series presenting the results of activities sponsored by the NATO Science Committee, which aims at the dissemination of advanced scientific and technological knowledge, with a view to strengthening links between scientific communities.

The series is published by an international board of publishers in conjunction with the NATO Scientific Affairs Division

A	**Life Sciences**	Plenum Publishing Corporation
B	**Physics**	New York and London
C	**Mathematical**	D. Reidel Publishing Company
	and Physical Sciences	Dordrecht, Boston, and Lancaster
D	**Behavioral and Social Sciences**	Martinus Nijhoff Publishers
E	**Engineering and**	The Hague, Boston, Dordrecht, and Lancaster
	Materials Sciences	
F	**Computer and Systems Sciences**	Springer-Verlag
G	**Ecological Sciences**	Berlin, Heidelberg, New York, London,
H	**Cell Biology**	Paris, and Tokyo

Recent Volumes in this Series

Volume 131—Fat Production and Consumption:
Technologies and Nutritional
Implications
edited by C. Galli and E. Fedeli

Volume 132—Biomechanics of Cell Division
edited by Nuri Akkas

Volume 133—Membrane Receptors, Dynamics, and Energetics
edited by K. W. A. Wirtz

Volume 134—Plant Vacuoles: Their Importance in Solute Compartmentation in
Cells and Their Applications in Plant Biotechnology
edited by B. Marin

Volume 135—Signal Transduction and Protein Phosphorylation
edited by L. M. G. Heilmeyer

Volume 136—The Molecular Basis of Viral Replication
edited by R. Perez Bercoff

Volume 137—DNA– Ligand Interactions: From Drugs to Proteins
edited by Wilhelm Guschlbauer and Wolfram Saenger

Series A: Life Sciences

DNA—Ligand Interactions

From Drugs to Proteins

Edited by

Wilhelm Guschlbauer

Centre d'Etudes Nucléaires de Saclay
Gif-sur-Yvette, France

and

Wolfram Saenger

Freie Universität Berlin
Berlin, Federal Republic of Germany

Plenum Press
New York and London
Published in cooperation with NATO Scientific Affairs Division

Proceedings of a NATO ASI/FEBS Course on
DNA-Ligand Interactions: From Drugs to Proteins,
held August 30–September 11, 1986,
in Abbey of Fontevraud, France

Library of Congress Cataloging in Publication Data

NATO ASI/FEBS Course on DNA-Ligand Interactions: From Drugs to Proteins
    (1986: Abbey of Fontevraud)
    DNA-Ligand interactions.

    (NATO ASI series. Series A, Life sciences; v. 137)
    "Proceedings of NATO ASI/FEBS Course on DNA-Ligand Interactions: From
Drugs to Proteins, held August 30–September 11, 1986, in Abbey of Fontevraud,
France"—T.p. verso.
    "Published in cooperation with NATO Scientific Affairs Division."
    Includes bibliographies and index.
    1. Deoxyribonucleic acid—Congresses. 2. Protein binding—Congresses. 3.
Drugs—Metabolism—Congresses. I. Guschlbauer, Wilhelm. II. Saenger,
Wolfram. III. NATO Advanced Study Institute. IV. Federation of European
Biochemical Societies. V. North Atlantic Treaty Organization. Scientific Affairs
Division. IV. Title. VII. Series. [DNLM: /. Molecular Biology—congresses. 2. Poly-
nucleotide Synthetases—metabolism—congresses. QU 58 N27935d 1986]
QP624.N38   1986   574.87'3282   87-20339
ISBN 978-1-4684-5385-0   ISBN 978-1-4684-5383-6 (eBook)
DOI 10.1007/978-1-4684-5383-6

© 1987 Plenum Press, New York
Softcover reprint of the hardcover 1st edition 1987
A Division of Plenum Publishing Corporation
233 Spring Street, New York, N.Y. 10013

NATO Advanced Study Institute 20/86
FEBS Advanced Course 08/86

Organizing Committee

Wilhelm GUSCHLBAUER, Director
Service de Biochimie, Bat. 142, Département de Biologie
Centre d'Etudes Nucléaires de Saclay
F-91191 Gif-sur-Yvette Cedex, France

Wolfram SAENGER, Codirector
Institut für Kristallographie, Freie Universität
Takustrasse 6
D-1000 Berlin 33, German Federal Republic

Michael WARING   Marc LENG
Department of Pharmacology   Centre de Biophysique Moléculaire du CNRS
University of Cambridge, Hills Road   1A, Ave. de la Recherche Scientifique
Cambridge CB2 2QQ, U.K.   45045 Orléans Cédex, France

ACKNOWLEDGEMENTS

This Advanced Study Institute was held in the Centre Culturel de l'Ouest (Royal Abbey of
Fontevraud) near Saumur (France) from August 30th to September 11th, 1986 under the
auspices of the North Atlantic Treaty Organization (NATO) and the Federation of European
Biochemical Societies (FEBS).

The following institutions and companies have sponsored the participation of students at this
ASI:
    National Science Foundation, Washington, D.C., U.S.A.
    ORIS-Industrie, Paris, France
    Département de Biologie, Centre d'Etudes Nucléaires de Saclay, Gif-sur-Yvette, France

    Schering A.G., Berlin, G.F.R.
    Farmitalia - Carlo Erba, S.p.A., Milano, Italy
    Hoffmann-Laroche & Co. Ltd., Basle, Switzerland
    Institut Scientifique Roussel - Roussel-Uclaf, Paris, France
    Institut de Récherches Internationales Servier, Neuilly-sur-Seine, France
    Elf-Biorecherches, Castanet Tolosan, France
    Ligue Française contre le Cancer, Paris, France
    L'Oréal, Clichy, France

Support received from the following companies and institutions is also acknowledged:
    Boehringer-Mannheim France, Meylan, France
    Bruker - Spectrospin, Wissembourg, France
    Jobin-Yvon ISA, Longjumeau, France
    Perkin Elmer France, Bois d'Arcy, France
    Siemens A.G., Darmstadt, G.F.R.
    Region des Pays de la Loire, Nantes, France
    Centre Culturel de l'Ouest, Fontevraud, France

Free advertising was offered by the following Journals:

NUCLEIC ACIDS RESEARCH and EMBO JOURNAL (IRL Press Ltd., Eynsham, Oxford, U.K.)

JOURNAL OF BIOMOLECULAR STRUCTURE AND DYNAMICS (Adenine Press, Guilderland NY, USA)

BIOCHIMIE (Société de Chimie Biologique, Paris, France)

PREFACE

This volume contains the texts of the nineteen lectures presented at the NATO-ASI - FEBS Course on "DNA - ligand interactions: from drugs to proteins." The Advanced Study Institute (ASI) was held from August 30th to September 11th, 1986 in the Abbey of Fontevraud (France). The ASI was attended by 112 participants from a wide scientific horizon and from twentyone different countries. It was in some way a follow-up of the ASI held in Maratea, Italy in May 1981 and which was published in the NATO ASI Life Science series as volume 45.

While much has been learned about the way the cellular machinery maintains and transmits the genetic heritage, as well as how these processes are regulated, little is known about how the interactions between the various partners involved are taking place. The interactions of drugs and proteins with nucleic acids are of evident importance in the understanding of these problems.

The spectacular advances in recombinant DNA technology and the increased sophistication of biophysical techniques, in particular x-ray diffraction and nuclear magnetic resonance, have created a scientific environment which is highly promising for the future of research in molecular biology. These advances permit the serious hope that biology on the molecular level may become a reality. Some of the contributions at the ASI presented the most recent advances in this exciting field.

To understand fully the posssibilities of these techniques, structural biologists have to learn about the problems of modern biology and molecular biologists have to become familiar with the new structural techniques. We realized that only such an interdisciplinary approach can be crowned by success and this ASI was planned and organized around this idea.

The choice of the subjects and lecturers was directed by the desire to present few, but particularly interesting biological problems in a concise way and then to offer various aspects of the way how physico-chemical and biochemical methodology can attack them. For reasons of time limitation only few subjects could be selected. Therefore several highly interesting, but voluminous subjects were excluded from the programme, like chromatin structure or RNA - protein interactions. We preferred to concentrate on a few precise problems which are actively advancing. The scientific content of the interdisciplinary Study Institute centered around the following subjects:

* DNA structure, its role in gene regulation, DNA hydration and interaction with drugs.
* Repressors: interactions with operators, structure and function.
* Methylases and restriction nucleases: role in regulatory processes, structure and mode of action.
* Finally, an outlook at a much more complicated system: DNA - antibody interactions.

From the various contributions in this volume it can be seen that even this volontarily limited content demands interest in such varied subjects as genetics, enzymology, spectroscopy, physical chemistry, pharmacology and crystallography.

In addition to the 19 lectures, three round table discussions on the three main themes were held. The active involvement of the participants was manifested by 18 research seminars and 48 posters, presented in three poster sessions.

The choice of Fontevraud Abbey was also a deliberate one: a quiet place at the crossroad of cultural influences, one of the spiritual centers of Western Europe for six centuries, it contains such significant relics as the tombs (with their colored reclining effigies) of Richard Lionheart and his parents, Henry II, King of England and Aliénor d'Aquitaine. This grandiose Abbey, rich in history and architecture in one of the most beautiful regions of Europe has now been converted, after 150 years as a prison, into a Seminar Center, the Centre Culturel de l'Ouest. Within the first days several participants noticed a gothic angel (left) on one of the vaults of the refectory holding an ominous attribute: a braided ring, a left-handed helical circle. Some unknown stone mason five hundred years ago must have forseen that some day plasmids will be discovered.

In conclusion the organizers wish to thank their colleagues, Michael Waring and Marc Leng, for their help in the early stages of the organization of the ASI. We are also grateful to all participants for contributing to stimulating discussions and to the excellent atmosphere. We thank, in particular, the lecturers for their brilliant contributions and for producing their manuscripts in due time. The lecturers also kindly furnished the reading material which was made available to the participants in four ring binders in the entrance of the lecture hall. Our thanks go to the staff of the Centre Culturel de l'Ouest in the Abbey of Fontevraud for their constant attention and, in particular, to **Miss Annie Quillet**, responsible of the Meeting Department of the Centre. Last, but not least, our thanks go to my faithful collaborator, **Miss Anne Woisard**, who tirelessly was troubleshooting, taxiing people around, making phone calls, accompanying buses and helping whenever necessary, before, during and after the meeting. Merci, Anne.

The editors hope that this volume will find its way on the desks and into the laboratories of researchers and students alike as an interdisciplinary text and important source of reference material.

View of the Abbey of Fontevraud from the South.

Wilhelm Guschlbauer
Wolfram Saenger

December 1986

CONTENTS

A) DNA Structure and Interaction with Small Molecules

DNA structure: Current results from single crystal x-ray diffraction studies   1
    O. Kennard

One- and two- dimensional NMR studies of the structures of simple sequence DNAs . .   23
    D.R. Kearns

The major and minor grooves of the DNA helix as conduits for information transfer . . .   45
    R.E. Dickerson*, M.L. Kopka and P.E. Pjura

Unusual DNA structures and gene regulation   63
    R.D. Wells

Specificity and dynamics of protein – nucleic acid interactions   85
    D. Porschke

DNA and its counterions: metal ions, amino acids, histones and protamines   105
    J. Subiraná

Dynamic aspects of antibiotic – DNA interaction   113
    M. Waring

Specific gene regulation of oligodeoxynucleotides covalently
        linked to intercalating agents   127
    C. Hélène

Pharmacology of DNA binding drugs   141
    B. Lambert and J.-B. LePecq

B) DNA – Protein Interactions: Repressor – Operator Systems

On the nature and specificity of DNA-protein interactions
        in the regulation of gene expression   159
    P.H. von Hippel and O.G. Berg

Searching for the code of ideal protein-DNA recognition   173
    N. Lehming, J. Sartorius, B. von Wilcken-Bergmann and B. Mueller-Hill

trp Repressor, a crystallographic study of allostery in genetic regulation   183
    P.B. Sigler, A. Joachimiak, R.W. Schevitz, C.L. Lawson, R.-G. Zhang, Z. Otwinowski
    and R. Marmorstein

Structural studies of three DNA binding proteins: catabolite gene activator protein,
    resolvase and the Klenow fragment of DNA polymerase I 185
    T.A. Steitz, L. Beese, B. Engelman, P. Fremont, J. Friedman, M. Sanderson,
    S. Schultz, G. Shields and J. Warwicker

A two-dimensional NMR study of the complex of lac repressor headpiece
    with a 14 base pair lac operator fragment 191
    R. Boelens, R.M. Scheek, R.M.J.N. Lamerichs, J. de Vlieg, J.H. van Boom
    and R. Kaptein

C) DNA - Protein Interactions: Methylation and Restriction Enzymes.

DNA methylation and mismatch repair: molecular specificities 217
    M. Radman

Recognition of DNA sequences by restriction endonucleases 225
    G. Maass

Mechanism and specificity of two restriction enzymes, CauI and CauII,
    that recognize asymmetrical DNA sequences 239
    S.P. Bennett and S.E. Halford

Structure of the DNA-EcoRI endonuclease recognition complex 251
    J.M. Rosenberg, J.A. McClarin, C.A. Frederick, B.-C. Wang, J. Grable,
    H.W. Boyer and P. Greene

D) Outlook: DNA-Protein Interactions of Complex Systems

Antibodies to nucleic acids 257
    M. Leng

Summary and Conclusions 269
    W. Saenger

Research Seminars 275

Posters 277

Participants 281

Index 285

* In multi-author articles the speaker's name is in bold face print.

DNA STRUCTURE

CURRENT RESULTS FROM SINGLE CRYSTAL X-RAY DIFFRACTION STUDIES

Olga Kennard

University Chemical Laboratory
Lensfield Road
Cambridge, U.K.

INTRODUCTION

These lecture-notes aim to summarise the results of single crystal
X-ray determinations of deoxyoligonucleotides carried out in the period
1978-1986. Also included are diagrams illustrating some of the geometrical
features which are fundamental to an understanding of the differentiation
between the various DNA conformations. Finally a selected list of
references is given to individual structure determinations, to a textbook,
and to some of the most important reviews. It is hoped that these notes
will prove to be of help to participants of the NATO Study Institute and
remain useful to the general reader when the inevitable progress of science
modifies and perhaps makes obsolete the ideas and speculations of relating
to the mechanism of DNA-ligand interactions which was the substance of the
lecture given at the meeting.

SINGLE CRYSTAL X-RAY STRUCTURE DETERMINATIONS OF OLIGONUCLEOTIDES
CONTAINING 4 OR MORE BASES.

Table 1 lists selected references to structure determinations published
between 1979 and 1986. This list is not comprehensive since in most cases
several structure determinations have been reported for each sequence at
different temperatures with different counterions, bromination or methylation
of some of the bases, and other modifications.

Note that currently only 8 different structure types have been fully
analysed. There are two types of B-DNA, a dodecamer and a hexamer
crystallising in different orthorhombic space groups and utilising different
packing motifs. There are two types of A-DNA crystals, one tetramer, several
octamers, a nonamer, and a DNA-RNA hybrid decamer, which crystallise in
either a tetragonal or a hexagonal space group. In both space groups the
packing interactions involve hydrogen bonding and van der Waals contacts
between the minor grooves or adjacent molecules. Z-DNA tetramers crystallise
in two orthorhombic space groups and Z-DNA hexamers are variations of the
same principles of molecular packing in an orthorhombic space group. Compare
this with around 300 protein structures reported in the literature.

Table 2 lists selected references to deoxyoligonucleotides containing
mismatched base pairs. These structures are variants on the structure types
listed in Table 1.

Table 3 gives selected reference to deoxyoligonucleotide drug complexes where the resolution was sufficient to locate the oligonucleotide component. Preliminary reports of other complexes have been published but are not listed.

Figure 1 illustrates some geometric features of Watson-Crick complementary base pairs and the mismatched base pairs G.A(syn), C.A and T.G as determined by single crystal X-ray diffraction studies. The distances between the C1'...C1' atoms are given in Å and the angles between these vectors and the glycosyl bonds in degrees.

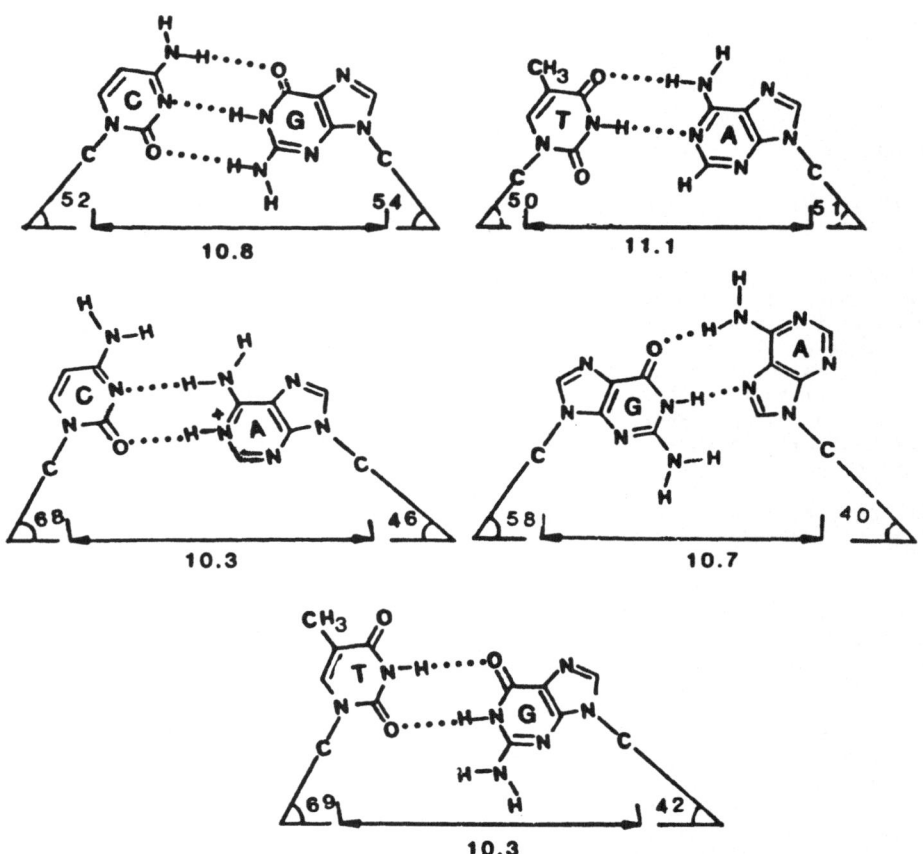

FIGURE 1: Watson-Crick base pairs and mismatched base pairs: C.A,G.A,T.G

TABLE 1   Single-crystal X-Ray structure of DNA fragments (containing four
or more bases) 1979-1986

Right-handed DNA

d(pATAT)	Alternating	B-DNA	Viswamitra et al.	(1982)
d(^{I}CCGC)		A-DNA	Conner et al.	(1982)
d(GGTATACC)		A-DNA	Shakked et al.	(1983)
d(GGCCGGCC)		A-DNA	Wang et al.	(1982a)
d(GGGGCCCC)		A-DNA	McCall et al.	(1985)
d(GGATGGGAG)		A-DNA	McCall et al.	(1986)
r(GCG)d(TATAGGC)		A-RNA/DNA	Wang et al.	(1982b)
d(CGCGAATTCGCG)		B-DNA	Drew et al.	(1981)
d(Gp(S)CpGp(S)CpGp(S)C)		B-DNA	Cruse et al.	(1986)
d(GGCCGGCC)		A-DNA	Haran et al.	(1986)
d(CCCCGGGG)		A-DNA	Haran et al.	(1986)

Left-handed DNA

d(CGCG)	Crawford et al.	(1980)
d(CGCG)	Drew et al.	(1980)
d(CGCGCG)	Wang et al.	(1979)
d(^{me}CGTA^{me}CG)	Wang et al.	(1984b)
d(^{me}CGAT^{me}CG)	Wang et al.	(1985)

Note that many variations on d(CGCGCG) containing methylated cytosines have
been omitted from this table.

TABLE 2

Mismatched base-pairs in DNA

d(GGGGCTCC)	T.G in A-DNA	Brown et al.	(1985)
d(GGGGTCCC)	T.G in A-DNA	Kneale et al.	(1985)
d(CGCGAATTGCG)	G.T in B-DNA	Kennard	(1985)
d(TGCGCG)	T.G in Z-DNA	Kennard	(1985)
d(^{Br}UGCGCG)	^{Br}U.G in Z-DNA	Brown et al.	(1986a)
d(CGCGTG)	T.G in Z-DNA	Ho et al.	(1985)
d(CGCGAATTAGCG)	G.A in B-DNA	Brown et al.	(1986b)
d(CGCAAATTCGCG)	C.A in B-DNA	Hunter et al.	(1986)
d(CGCIAATTAGCG)	I.T in B-DNA	Corfield et al.	(1986)

TABLE 3

DNA complexed with drugs

d(CGTACG) with daunomycin	Quigley et al.	(1980)
d(CGTACG) with triostin A	Ughetto et al.	(1985)
d(CGTACG) with echinomycin	Ughetto et al.	(1985)
d(GCGTACGC) with triostin A	Wang et al.	(1984a)
d(CGCGAATTCGCG) with cisplatin	Wing et al.	(1984)
d(CGCGAATTCGCG) with netropsin	Kopka et al.	(1985)

TABLE 4

DNA complexed with proteins

d(TCGCGAATTCGCG) with Eco RI	Frederick et al.	(1984)
Nucleosome core	Richmond et al.	(1984)
d(ACAATATATATTGT) with phage 434 repressor	Anderson et al.	(1985)

TECHNIQUES OF X-RAY ANALYSIS

Synthesis, crystallisation and data collection

Oligonucleotides for X-ray analysis are usually synthesised by the triester method (Gait et al. 1982, Arentzen et al. 1979). Stringent purification is essential for successful crystallisation. Crystallisation is usually by vapour diffusion of a precipitating agent into a solution containing DNA, various counterions, and in many cases, spermine. The resulting crystals contain around 50% DNA, and are always highly hydrated. Crystals of fragments with only Watson-Crick base pairs are generally stable at room temperature, while those with mismatched base pairs are less stable. Intensity measurements are frequently carried out at 4°C or below. Crystal perfection and size determine the degree of resolution. The Z-DNA hexamers form the most perfect crystals, often diffracting to $1\overset{\circ}{A}$ resolution. Other sequences diffract to around $1.8-2.5\overset{\circ}{A}$, although the resolutions may be improved somewhat by the use of synchrotron radiation (Kennard et al. 1986). A typical data collection on a diffractometer with a crystal diffracting to about $2.0\overset{\circ}{A}$ results in about 2500 intensity measurements, which are used for the analysis.

Methods of structure solution and refinement

A variety of methods are used for structure solution: heavy atoms, molecular replacement, direct space search, etc. The aim is to find a starting model for the fundamental structural motif, the asymmetric unit, which can then be refined by comparing structure factors calculated from the model with experimentally determined values. The asymmetric unit consists, usually, of two independent DNA strands, which form a double helix. Sometimes the motif is only one strand and the double helix is formed by a two-fold symmetry operation.

4

Resolution limits the degree of structural detail which can be
accurately determined. Some of the Z-DNA hexamers diffract to around 1Å
and thus it is possible to derive individual atomic coordinates. For other
structures constrained methods of refinement are used in an attempt to use
the experimental data optimally for the accurate determination of the
principal torsion angles and the mutual orientation of the bases [for details
consult the original references].

Refinement always includes solvent molecules which are located from
electron density maps, often with the aid of computer graphics. Usually
around 80-100 solvent molecules per DNA are sufficiently ordered to be
detected by these techniques. Counterions have only been identified with
confidence in the Z-DNA hexamers. In the other structures solvent positions
were assigned as water oxygen atoms. The refinement is terminated when there
is no further improvement in the agreement factor R, defined as $\Sigma\,|Fo\text{-}Fc|\,/\Sigma|Fc|$
where Fo= observed structure factor, Fc= calculated structure factor.

A variant of the above technique was used for the structures with base
pair mismatches. In these structures the contributions of atoms of the
mismatched bases were initially omitted from the calculation. At an
intermediate stage of the refinement the positions of atoms in the base
pair mismatch were located from electron density maps and subsequently
included in the calculations. In this way the analysis was not biased by
a preconceived model of the nature of the base pair mismatch. Similar
techniques were used for some of the oligonucleotide-drug complexes and
have led to the discovery of novel features such as Hoogsteen base pairing
in DNA fragments complexed with bifunctional intercalators (Ughetto et al.
1985a, Quigley et al. 1986).

CONVENTIONS USED IN INTERPRETING STRUCTURAL INFORMATION

PROPERTIES OF A SINGLE OLIGONUCLEOTIDE STRAND

Figure 2 illustrates a fragment of DNA with the sequence A, G, C, T.
linked by 3'5' phosphodiester bonds.

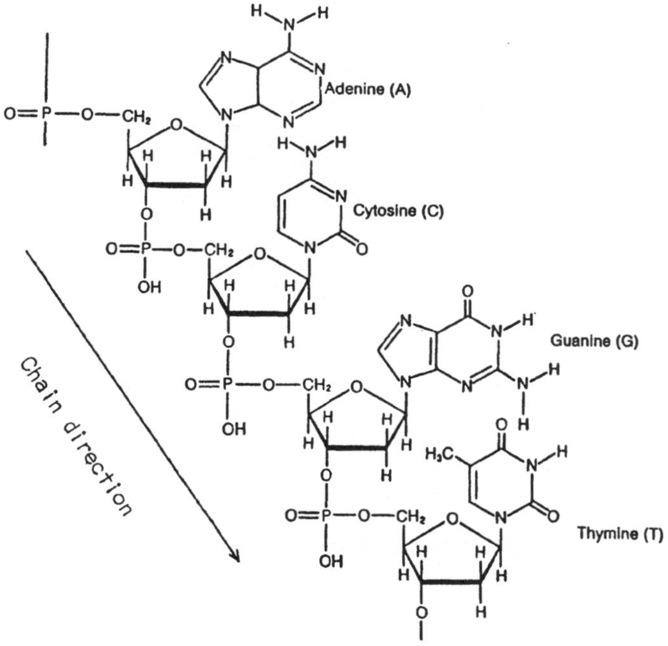

FIGURE 2

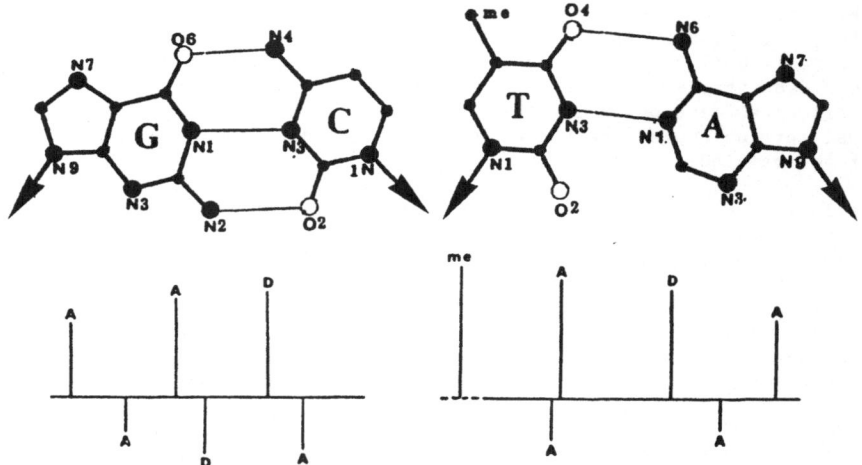

FIGURE 3: Atomic numbering scheme for Watson-Crick base pairs and schematic drawing of the distribution of donor and acceptor groups; after Woodbury and von Hippel (1981).

PROPERTIES OF INDIVIDUAL BASE PAIRS

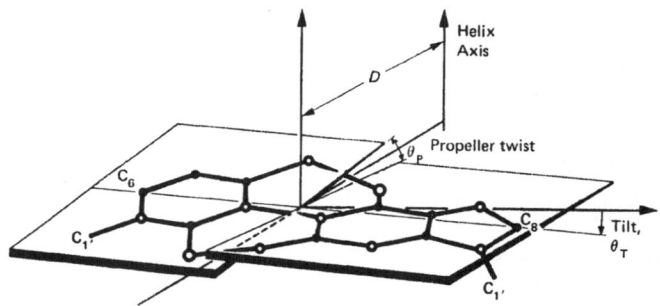

FIGURE 4: A view of the mutual orientation of two bases of a base pair and their relation to the helix axis.

D: Displacement of base pairs from the helix axis.

Propeller twist: The dihedral angle between individual base planes viewed along the long axis of the base pair. The angle is positive for the clockwise rotation of the nearer base.

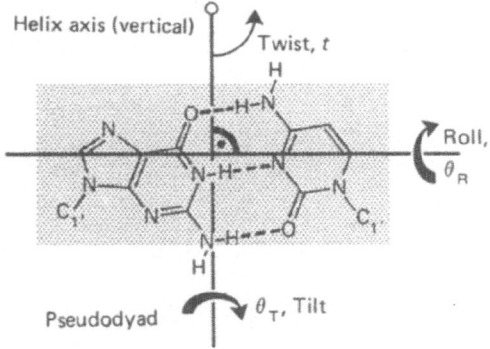

FIGURE 5: View of a base pair perpendicular to the mean plane of the bases.

Helix twist t: defines the orientation of a base pair with respect to the helix axis. Local values are estimated by the angle between successive intrastrand C1'...C1' vectors projected down a plane normal to the global helix axis.

Base pair tilt: θT defines the orientation of a base pair relative to the helix axis, and is measured by the rotation about the pseudodyad axis which lies in the mean plane of the base pair and relates to the C1'-N glycosyl bonds.

Base pair roll: θR defines the orientation of a base pair relative to the helix axis and is measured by the rotation about a line perpendicular to the helix axis and the pseudodyad. This line is closely approximated by the C6 (pyrimidine)-C8 (purine) vector.

ATOMIC NUMBERING SCHEME AND DEFINITION OF TORSION ANGLES

Figure 6 illustrates the sugar-phosphate backbone, the atomic numbering scheme and the definition of torsion angles.

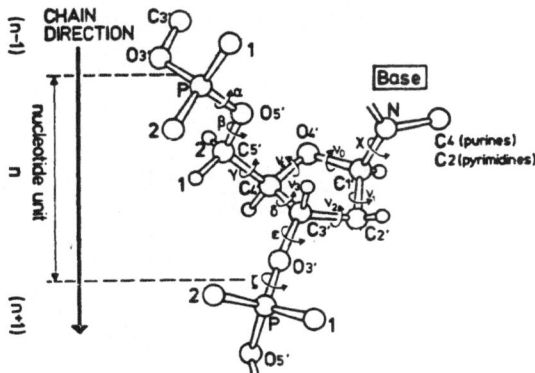

FIGURE 6: The seven conformational angles (α to χ) defining the sugar phosphate backbone. See Table 10 for data.

Figure 7 illustrates the P....P distances associated with the two principal sugar conformations and Figure 8 the _anti_ and _syn_ orientations of the sugar and base moieties.

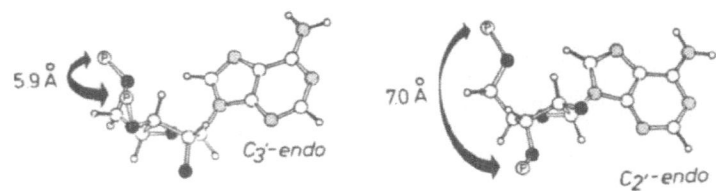

FIGURE 7

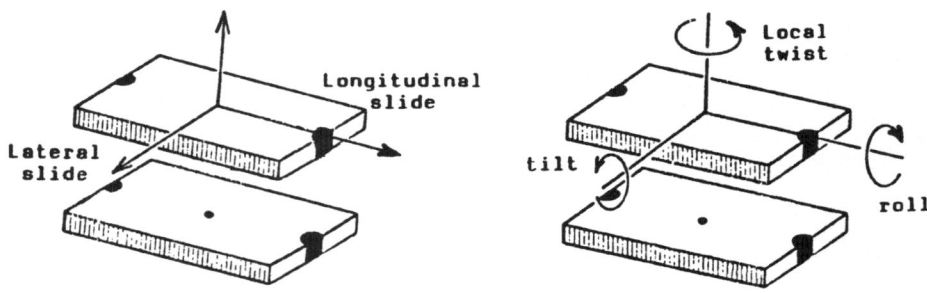

FIGURE 8

PROPERTIES OF SUCCESSIVE BASE PAIRS (BASE STEPS)

Figure 9 illustrates the various parameters which define the mutual orientation of successive base pairs.

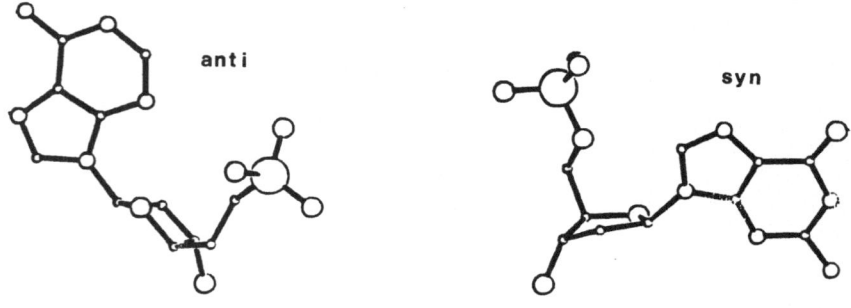

FIGURE 9

Local twist: orientation of successive base pairs with respect to a local helix axis.

Roll: θR the angle between successive base pairs in the direction of the short axis of the base pair. Positive if successive base pairs open towards the minor groove.

Tilt: θT the angle between successive base pairs in the direction of the long axis of the base pair.

Lateral slide: relative displacement of successive base pairs in the direction perpendicular to the long axis.

Longitudinal slide: relative displacement of successive base pairs in the direction of the long axis.

Rise per base pair: distance between successive base pairs along the helix axis.

For analytical expressions see Dickerson and Drew (1981) and Dickerson (1983). For calculation of the overall helix axis see Shakked et al. (1983).

PROPERTIES OF GROOVES: (For method of calculation see Arnott (1976).)

Groove width: estimated as the shortest P-P vector across a groove less 5.8Å (twice the van der Waals radius of a phosphate group).

Minor groove depth: distance between P and N2 of guanine or O2 of thymine less the sum of the van der Waals radii.

Major groove depth: distance between P and O6 of guanine or N6 of thymine.

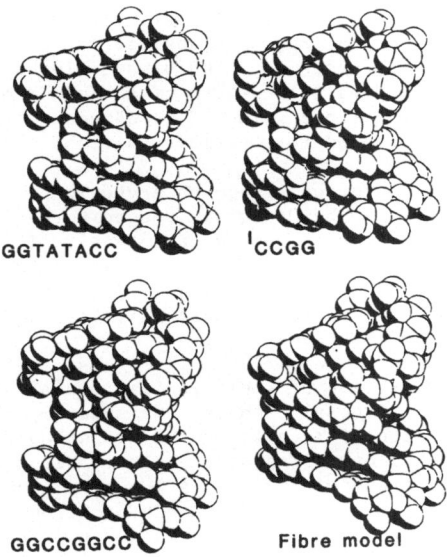

GGTATACC   ᴵCCGG

GGCCGGCC   Fibre model

FIGURE 10: Space-filling drawings of three A-DNA octamers (see Table 5) and an octamer based on the fibre model of A-DNA (Arnott and Hukins, 1972). Note the variation in the groove shapes. The major groove (left) becomes wider as the rise per base pair increases and the base tilt decreases.

NUMERICAL DATA

   Tables 5-11 give numerical information derived from X-ray diffraction
analyses of oligonucleotide fragments, published between 1980-1986.

   Tables 5, 6, and 7 give data for DNA fragments crystallising as A, B,
and Z helices. For each crystal structure the following information is
quoted: temperature of X-ray measurement, space group, cell dimensions and
Z, the number of oligomers in the unit cell. In all these crystals two
strands of the oligomer form a double helix, in most of the structures
(depending on the space group and the value of Z) the double helix has no
imposed crystallographic symmetry. In addition the tables indicate the
resolution of the analysis. Note that whereas most of the Z-type crystals
diffract to near atomic resolution the A and B type generally only diffract
to a resolution of between 2.0 - 2.5Å.

   Table 6 gives similar data for DNA fragments which form double helices
containing either T.G, ^{Br}U.G, G.A or C.A mismatched base pairs. A comparison
of this Table with Tables 5-7 shows clearly that within each conformational
type (A, B or Z) the structures containing mismatches crystallise
isomorphously with the native structures containing Watson-Crick complementary
base pairs only. Further details of all these structures can be obtained
from the original papers referenced by the Tables.

   Table 9 gives the average (or global) helical parameter for A and B
type helices observed in the major single crystal X-ray diffraction analyses
and, for comparison, in fibres. The Table is restricted to helices without
mismatched base pairs. The parameters include the helix twist, rise per
base pair, base pair tilt, propeller twist, groove width, distance of the
P atoms from the helix axis and the displacement of the base pairs (mid points
of the C6-C8 vectors) from the helix axis. These parameters can best be
understood by reference to Figures 4 and 5 and the definitions given in the
text.

   Note that there are quite substantial variations for each of these
parameters in the different DNA fragments analysed. These variations arise
from the differences in base sequence. The groove widths are particularly
affected. In the A-type crystals the width of the minor groove ranges from
10.2 to 9.4Å compared to the fibre value of 11.0Å and the width of the major
groove from 8.6 to 3.2Å compared to the fibre value of 2.3Å.

   The B-DNA helix has so far only been observed in two types of oligomers:
the dodecamer, first described by Dickerson and co-workers (Drew et al. 1981),
and the phosphorothioate hexamer (Cruse et al. 1986). In both structures
the helix has a wide major groove ranging from 10.2 to 12.2Å (c.f. 11.4Å for
fibre) and a narrow minor groove 4.6 to 6.4Å (c.f. 6.0Å for fibre) (Table 9).
Note also the large range of values for base pair tilt and propeller twist
which again are related to base sequences.

   Table 10 gives a comparison of the principal global helical parameters,
the orientation about the glycosyl bond and the sugar conformation for A,
B, and Z helices. The values were obtained by averaging the global parameters
of the corresponding double helix fragments. The Table gives a rapid and
numeric overview of the parameters which characterise each of the conformational types. Note the two types of B-DNA conformation. In one of these,
the hexamer, marked B' in the Table, the backbone conformation alternates
between two conformational states which leads to two values of the helix
twist and rise per base pair.

Table 11 lists the average backbone torsion angles for the individual single crystal structure determinations as well as the values based on fibre data. The definition of the torsion angles is best understood by reference to Figure 6. There are quite substantial variations even in these average torsion angles for the individual structures. The variations become more pronounced when one considers the local rather than the global (averaged) torsion angles. For details see the individual structure determinations referenced in Table 1.

For a recent and comprehensive discussion of the effects of base sequences on the fine structure of the DNA double helix see a review by Shakked and Rabinovich (1986). I am most grateful to these authors for permission to use the tables from their paper in these notes.

SUMMARY

These lecture notes attempt to summarise three aspects of single crystal X-ray analyses of deoxyoligonucleotide containing double helical DNA fragments.

1) Comprehensive reference to single crystal X-ray structure analyses of DNA fragments containing only Watson-Crick complementary base pairs and fragments with Watson-Crick and mismatched base pairs.

2) Definitions and illustrations of the most commonly used terms and parameters which define the detailed conformation of single stranded and double helical DNA.

3) Numerical tables of the results of single crystal X-ray structure determinations which highlight the variations in the conformational parameters found in DNA fragments with different base sequences.

Possibly the most striking result of the X-ray diffraction analyses carried out to date is the influence of base sequences on conformations. This can result in such gross effects as the formation, under certain physical conditions, of left-handed Z helices. It is of interest that so far all attempts at crystallising the same base sequences in two different DNA conformations have been unsuccessful.

Within each conformational type there are large fluctuations in the parameters, including groove width, base orientation, etc. which are related to the base sequences (Calladine, 1982; Shakked and Rabinovich, 1986). A great deal more experimental information is needed before a clear-cut theory about the correlation between sequence and conformation emerges and many laboratories are actively working in this area. These notes will be updated from time to time to include new material as it is published.

ACKNOWLEDGEMENTS

I thank Dr Z Shakked and Professor D Rabinovich for permission to use some of the tables from their review in "Progress in Biophysics and Molecular Biology" (1986, in press) and Dr W B T Cruse for help with the manuscript. Figures 2, 4, 6 and 7 were reproduced from "Principles of Nucleic Acid Structure", W Saenger, Springer-Verlag, New York, 1984, by permission and Figure 1 was redrawn from a paper by C R Calladine and H R Drew (1984), J. Molec. Biol., 178, 773-272. I am grateful to Mrs H A Belchamber for her careful preparation of the camera ready copy of the manuscript.

TABLE 5: CRYSTAL DATA OF A-DNA STRUCTURES

COMPOUND	TEMPERATURE (°C)	SPACE GROUP	CELL DIMENSIONS (Å) a	b	c	Z^{b}	RESOLUTION (Å)	REFERENCE
d(GGTATACC)	r.t.[a]	$P6_1$	45.01	45.01	41.55	12	1.8	(1,2)
d(GGBr^{5}UABr^{5}UACC)	r.t.	$P6_1$	45.08	45.08	41.72	12	2.25	(1,2)
d(GGGGCCC)	r.t.	$P6_1$	45.32	45.32	42.25	12	2.5	(3)
d(^{I}CCGG)	-2	$P4_3 2_1 2$	41.10	41.10	26.70	16	2.25	(4,5)
d(GGCCGGCC)	-8	$P4_3 2_1 2$	42.06	42.06	25.17	8	2.25	(6)
d(GGCCGGCC)	-18	$P4_3 2_1 2$	40.51	40.51	24.67	8	2.25	(6)
d(GGCCGGCC)	r.t.	$P4_3 2_1 2$	42.04	42.04	25.09	8	2.25	(7)
d(CCCCGGGG)	r.t.	$P4_3 2_1 2$	43.36	43.36	24.83	8	2.25	(7)
r(GCG)d(TATACGC)	0	$P2_1 2_1 2_1$	24.20	43.46	49.40	8	2.0	(8)
d(GGATGGGAG)	r.t.	$P4_3$	45.29	45.29	24.73	8	3.0	(9)

(a) Room temperature ranging from 15 to 20°C
(b) Number of oligomers in the unit cell

(1) Shakked et al., 1981
(2) Shakked et al., 1983
(3) McCall et al., 1985
(4) Conner et al., 1982
(5) Conner et al., 1984
(6) Wang et al., 1982a
(7) Haran et al., 1986
(8) Wang et al., 1982b
(9) McCall et al., 1986

TABLE 6: CRYSTAL DATA OF B-DNA STRUCTURES

COMPOUND	TEMPERATURE ($^\circ$C)	SPACE GROUP	CELL DIMENSIONS (Å)			Z	RESOLUTION (Å)	REFERENCE
			a	b	c			
d(CGCGAATTCGCG)	20	$P2_12_12_1$	24.87	40.39	66.20	8	1.9	(1,2)
d(CGCGAATTCGCG)	-257	$P2_12_12_1$	23.44	29.31	65.26	8	2.7	(2)
d(CGCGAATTCGCG)	20	$P2_12_12_1$	24.16	39.93	66.12	8	2.6	(3)
d(CGCGAATTBr^{5}CGCG)	20	$P2_12_12_1$	24.71	40.56	65.56	8	3.0	(4)
d(CGCGAATTBr^{5}CGCG)	7	$P2_12_12_1$	24.20	40.09	63.95	8	2.3	(4)
d(Gp(S)CpGp(S)CpGp(S)C)	14	$P2_12_12_1$	34.90	39.15	20.64	8	2.2	(5)

(1) Dickerson and Drew, 1981
(2) Drew et al., 1982
(3) Wing et al., 1984
(4) Fratini et al., 1982
(5) Cruse et al., 1986

TABLE 7: CRYSTAL DATA OF Z-DNA STRUCTURES

COMPOUND	SPACE GROUP	CELL DIMENSIONS (Å)			Z	RESOLUTION (Å)	REFERENCE
		a	b	c			
d(CGCGCG)(Mg^{2+} spermine)	$P2_12_12_1$	17.88	31.55	44.58	8	0.9	(1)
d(CGCGCG)(Mg^{2+} spermidine)	$P2_12_12_1$	17.96	31.19	44.73	8	1.3	(1)
d(CGCGCG)(Ba^{2+} spermine)	$P2_12_12_1$	17.94	31.54	44.68	8	1.1	(1)
d(CGCGCG)(Mg^{2+} and/or Na^+)	$P2_12_12_1$	17.98	30.94	44.81	8	1.8	(1)
d(m^5CGm^5CGm^5CG)(Mg^{2+} spermine)[a]	$P2_12_12_1$	17.76	30.57	45.42	8	1.3	(2)
d(Br^5CGTABr^5CG)(Mg^{2+}/Na^+)[b]	$P2_12_12_1$	17.85	30.74	44.78	8	1.5	(3)
d(m^5CGTAm^5CG)(Mg^{2+}/Na^+)	$P2_12_12_1$	17.91	30.43	44.96	8	1.2	(3)
d(Br^5CGATBr^5CG)(Mg^{2+} cobalt hexamine)	$P2_12_12_1$	18.30	31.11	44.10	8	1.5	(7)
d(CGCG)(Mg^{2+} and/or Na^+)	$C222_1$	19.50	31.27	64.67	16	1.5	(4)
d(CGCGCGGC)(Mg^{2+} and/or Na^+)	$P6_5$	31.27	31.27	43.56	9	1.6	(6)
d(CGCATGCG)(Mg^{2+} and/or Na^+)	$P6_5$	30.90	30.90	43.14	9	2.5	(5)

(a) m^5C = 5-methyl cytosine
(b) Br^5C = 5-bromo cytosine
(1) Wang et al., 1979
(2) Fujii et al., 1982b
(3) Wang et al., 1984
(4) Drew et al., 1980
(5) Fujii et al., 1985

(6) Crawford et al., 1980
(7) Wang et al., 1985

TABLE 8: CRYSTAL DATA OF MISMATCHED DNA STRUCTURES

MISMATCH	SEQUENCE	DNA TYPE	SPACE GROUP	a	b (Å)	c	RESOLUTION (Å)	R FACTOR %
T.G (1)	d(GGGGTTCC)	A	$P6_1$	44.71	44.71	42.36	2.1	14
T.G (2)	d(GGGGTCCC)	A	$P6_1$	45.20	45.20	42.97	2.1	14
G.T (3)	d(CCCGAATTTGCG)	B	$P2_12_12_1$	25.53	41.22	65.63	2.5	18
T.G (4)	d(TGCGCG)	Z	$P2_12_12_1$	17.97	30.73	45.11	1.5	18
^{Br}U.G(5)	d(^{Br}UGCGCG)	Z	$P2_12_12_1$	17.94	30.85	49.94	2.2	16
T.G (6)	d(CCCGTG)	Z	$P2_12_12_1$	17.45	31.63	45.56	1.0	19
G.A (7)	d(CCCGAATTAGCG)	B	$P2_12_12_1$	25.69	41.96	65.19	2.5	17
C.A (8)	d(CGCAAATTCGCG)	B	$P2_12_12_1$	25.37	41.44	65.20	2.5	19

(1) Brown et al., 1985
(2) Kneale et al., 1985
(3) Kennard, 1985
(4) Kennard, 1985
(5) Brown et al., 1986b
(6) Ho et al., 1985
(7) Brown et al., 1986a
(8) Hunter et al., 1986

TABLE 9 : AVERAGE HELIX PARAMETERS*

	HELIX[a] TWIST (°)	RISE PER[a] BASE-PAIR (Å)	BASE-PAIR TILT (°)	PROP.[b] TWIST (°)	GROOVE WIDTH Minor	GROOVE WIDTH Major	R_p(Å)[c]	D(Å)[d]
A helix:								
d(GGTATACC)	31.9	2.92	13.2	10.2	10.2	6.3	9.3	4.0
d(GGGGCCCC)	31.6	2.94	13.3	10.5	10.1	6.8	9.3	4.1
d(^{I}CCGG)/d(^{I}CCGG)	34.0	2.84	14.0	16.2	9.5	4.8	9.4	3.5
d(GGCCGGCC)(-8°C)	33.2	3.00	11.9	8.6	9.6	7.9	9.4	3.6
d(GGCCGGCC)	32.9	3.08	11.6	12.7	9.8	7.5	9.5	3.5
d(CCCCGGGG)	33.5	3.10	9.0	10.8	9.4	8.6	9.8	3.9
r(GCG)d(TATAGCG)	33.3	2.52	19.3	11.8	10.2	3.2	9.3	4.5
A-DNA[f]	32.7	2.56	22.0	6.0	11.0	2.3	8.6	4.4
B helix:								
d(GCGCAATTCGCG)(Native)	35.9	3.34	1.5	13.0	5.3	11.7	9.3	-0.2
d(GCGAATTBr^{5}CGCG)(MPD7)	36.0	3.37	-2.0	17.7	4.6	12.2	9.3	-0.2
B-DNA[f]	36.0	3.38	2.4	13.3	6.0	11.4	9.3	-0.2
d(Gp(S)CpGp(S)CpGp(S)C)[e]	34.1	3.51	0.2	9.4	6.4	10.2	9.1	-0.6
	36.8	3.30						

(a) Helix twist is calculated from the angles between
    successive interstrand C1'-C1' vectors projected
    down a plane normal to the overall helix axis; rise
    per base pair, from intrastrand C1'-C1' vectors
    projected onto the helix axis.
(b) Propeller twist is the dihedral angle between normals
    to the two bases of a pair viewed along its long axis.
(c) Distance of P atoms from the helix axis.
(d) Distance of the midpoints of the C6-C8 vectors from
    helix axis.
(e) Two conformational states
(f) Based on fiber data (Arnott et al., 1982)
(*) See figure 4.

TABLE 10: COMPARISON OF A- B- AND Z-DNA

	A-DNA(a)	B-DNA[a]	B'-DNA[b]	Z-DNA[c]
Helix sense	Right-handed	Right-handed	Right-handed	Left-handed
Base pairs/turn	11	10	10	12 (6 dimers)
Helix twist (^{o})	32.7	36.0	34.1,36.8	-10, -50
Rise per base-pair (Å)	2.9	3.4	3.51,3.30	3.7
Helix pitch (Å)	32	34	34	45
Base pair tilt (^{o})	13	0	0	-7
Distance of P from axis (Å)	9.5	9.3	9.1	6.9, 8.0
Glycosidic orientation	anti	anti	anti	anti syn[d]
Sugar pucker	C3'-endo	wide range	C2'-endo normal range	C2'-endo, C3'-endo[d]

(a) The numerical values for each form were obtained by averaging the global parameters of the corresponding double-helix fragments.

(b) B'-DNA values are for a double helix with backbone conformation alternating between conformational states I and II (Cruse et al. 1986).

(c) Two values correspond to CpG and GpC steps for the helix twist and P distance values, to cytidine and guanosine for the others.

(d) Two values correspond to the two conformational states.

17

TABLE 11: AVERAGE TORSION ANGLES (°)*

	α (P-O5')	β(O5'-C5')	γ(C5'-C4')[d]	δ(C4'-C3')	ε(C3'-O3')	ζ(O3'-P)	χ(glycosyl)
A helix:							
d(GGTATACC)[a]	-62(12)	173(8)	52(14)	88(3)	-152(8)	-78(7)	-160(8)
d(GGGGCCCC)	-76(14)	178(10)	63(14)	84(11)	-156(13)	-70(10)	-158(10)
d(ᴵCCGG)	-73(4)	180(8)	64(11)	89(25)	-157(14)	-66(5)	-157(14)
d(GGCCGGCC)(-8°C)	-75(38)	185(14)	60(26)	98(23)	-166(20)	-75(21)	-149(11)
d(GGCCGGCC)[b]	-67(10)	170(19)	58(7)	80(3)	-142(21)	-76(12)	-162(11)
d(CCCCGGGG)[b]	-87(11)	182(13)	72(7)	79(2)	-153(13)	-66(7)	-162(9)
r(GCG)d(TATACGC)	-69(33)	175(15)	58(25)	82(9)	-151(15)	-75(17)	-162(10)
A-DNA[c]	-50	172	42	79	-146	-78	-159
A-RNA[c]	-68	178	54	82	-153	-71	-158
B helix:							
d(CGCGAATTCGCG)(Native)	-63(8)	171(14)	54(8)	122(21)	-169(23)	-108(36)	-117(14)
d(CGCGAATTBr⁵CGCG)(MPD7)	-62(10)	170(18)	55(10)	120(27)	-169(32)	-109(42)	-120(18)
d(Gp(S)CpGp(S)CpGp(S)C)	-53	154	46	138	-164,-114[e]	-102,-192[e]	-109
B-DNA[c]	-33	138	33	142	-141	-157	-139

(a) Average values of d(GGTATACC) and d(GGBr⁵UABr⁵UACC).

(b) Excluding α and γ of fifth nucleotide.

(c) Based on fiber data of Arnott et al. (1982).

(d) Omitting 5'-terminal values.

(e) Corresponds to the conformational states I and II.

(*) See figure 6.

REFERENCES

Anderson, J.E., Ptashne, M. & Harrison, S.C. (1985), A phagerepressor
    operator at 7A resolution. Nature, 316, 596-601
Arnott, S. (1976) in "Organisation and Expression of Chromosomes"
    V.G. Allfrey, E.K.F. Bautz, B.J. McCarthy, R.T. Schimke, A. Tissieres,
    eds., 209-222, Dahlem Conference, Berlin
Arnott, S. & Hukins, D.W.J. (1972), Optimised parameters for A-DNA and
    B-DNA. Biochem. Res. Comm., 47, 1504-1509
Arnott, S., Chanrdrasekaran, R., Hall, I.H., Puigjaner, L.C., Walker, J.K.
    & Wang, M. (1982) Cold Spring Harbour Symp., Quantum Biology, 47, 53-65
Arentzen, R., van Boekel, C.A.A., van der Marel, G. & van Boom, J.H. (1979)
    Synthesis 137
Brown, T., Kennard, O., Kneale, G. & Rabinovich, D. (1985), High-
    resolution structure of a DNA helix containing mismatched base pairs.
    Nature, 315, 604-606
Brown, T., Kneale, G., Hunter, W.N. & Kennard, O. (1986a), Structural
    characterisation of the bromouracil.guanine base pair mismatch in a
    Z-DNA fragment. Nucleic Acids Res., 14, 4, 1801-1809
Brown, T., Hunter, W.N., Kneale, G. & Kennard, O. (1986b), Molecular structure
    of the G.A base-pair and its implications for the mechanism of
    transversion mutations. Proc. Nat. Acad. Sci. USA, 83, 2402-2406
Calladine, C.R. (1982), Mechanics of sequence-dependent stacking of bases
    in B-DNA. J. Mol. Biol., 161, 343-352
Conner, B.N., Takano, T., Tanaka, S., Itakura, K. & Dickerson, R.E. (1982),
    The molecular structure of d(CCGG): a fragment of right-handed
    double helical A-DNA. Nature, 295, 294-299
Conner, B.N., Yoon, C., Dickerson, J.L. & Dickerson, R.E. (1984), Helix
    geometry and hydration in A-DNA tetramer: C-C-G-C, J. Mol. Biol.,
    174, 664-695
Corfield, P., Brown, T., Hunter, W.N. & Kennard, O. (1986), Private
    Communication
Crawford, J.L., Kolpak, F.J., Wang, A.H.-J., Quigley, G.J., van Boom, J.H.,
    van der Marel, G.A. & Rich, A. (1980), The tetramer d(CpGpCpG)
    crystallises as a left-handed double helix. Proc. Nat. Acad. Sci.
    USA, 77, 4016-4020
Cruse, W.B.T., Salisbury, S.A., Brown, T., Eckstein, F., Costick, R. &
    Kennard, O. (1986), Chiral phosphorothioate analogues of B-DNA: The
    crystal structure of Rp-d(G-p(S)C-pG-p(S)C-pG-p(S)C). J. Mol. Biol.
    in press
Dickerson, R.E. & Drew, H.R. (1981), Structure of a B-DNA dodecamer II.
    Influence of base sequence on helix structure. J. Mol. Biol., 149,
    761-786
Dickerson, R.E. (1983), Base sequence and helix structure variations in B
    and A DNA. J. Mol. Biol., 166, 419-441
Drew, H., Takano, T., Tanaka, S., Itakura, K. & Dickerson, R.E. (1980),
    High salt d(GpGpCpG): a left-handed Z-DNA double helix. Nature,
    286, 567-573
Drew, H.R., Wing, R.M., Takano, T., Broka, C., Tanaka, S., Itakura, K. &
    Dickerson, R.E. (1981), Structure of a B-DNA dodecamer: conformation
    and dynamics. Proc. Nat. Acad. Sci. USA, 78, 2179-2183
Drew, H.R., Samsom, S. & Dickerson, R.E. (1982), Structure of a B-DNA
    dodecamer at 16K. Proc. Nat. Acad. Sci. USA, 79, 4040-4022
Frederick, C.A., Grable, J., Melia, M., Samudzi, C., Jen-Jacobsen, L.,
    Wang, B.-C., Greene, P., Boyer, H.W. & Rosenberg, J.M. (1984), Kinked
    DNA in crystalline complex with EcoRI endonuclease. Nature, 309,
    327-331
Frattini, A.V., Kopka, M.L., Drew, H.R. & Dickerson, R.E. (1982), Reversible
    bending and helix geometry in a B-DNA dodecamer: CGCGAATT[Br]CGCG.
    J. Biol. Chem., 257, 14686-14707

Fujii, S., Wang, A.H.-J., van der Marel, J.H., van Boom, J.H. & Rich, A. (1982), Molecular structure of (m^{5}dC-dG): the role of the methyl group on 5-methyl cytosine in stabilising Z-DNA. Nucleic Acids Res., 10, 7879-7892

Fujii, S., Wang, A.H.-J., van der Marel. G.A., van Boom, J.H. & Rich, A. (1985), The octamers d(CGCGCGCG) and d(CGCATGCG) both crystallise with Z-DNA in the same hexagonal lattice. Biopolymers, 24, 243-250

Gait, M.J., Matthes, H.W.D., Singh, M., Sproat, B.S. & Titmus, R.C. (1982), Rapid synthesis of oligodeoxyribonucleotides VII. Solid phase synthesis of oligodeoxyribonucleotides by a continuous flow phosphotriester method on a kieselguhr-polyamide support. Nucleic Acids Res., 10, 6243-6248

Haran, T.E., Shakked, Z., Wang, A.H.-J. & Rich, A. (1986) J. Mol. Biol., submitted

Ho, P.S., Frederick, C.A., Quigley, G., van der Marel, G.A., van Boom, J.H., Wang, A.H.-J. & Rich, A. (1985), G.T wobble pairing in Z-DNA at 1.0Å atomic resolution; the crystal structure of d(CGCGTG).

Hunter, W.N., Brown, T., Anand, N.N. & Kennard, O. (1986), Structure of an adenine-cytosine base pair in DNA and its implications for mismatch repair. Nature, 320, 552-555

Kennard, O. (1985), Structural studies of DNA fragments: The G.T wobble base pair in A, B and Z DNA; The G.A base pair in B-DNA. J. Biomol. Struct. Dynam., 3, 205-225

Kennard, O., Cruse, W.B.T., Nachman, J., Prange, T., Shakked, S. & Rabinovich, D. (1986), Ordered water structure in an A-DNA octamer at 1.7Å resolution. J. Biomol. Struct. Dynam., 4, 623-647

Kneale, G., Brown, T., Kennard, O. & Rabinovich, D. (1985), G.T base-pairs in a DNA helix: The crystal structure of d(GGGGTCCC). J. Mol. Biol., 186, 805-814

Kopka, M.L., Yoon, C., Goodsell, D., Pjura, P. & Dickerson, R.E. (1985), Binding of an antitumour drug to DNA. Netropsin and C-G-C-G-A-A-T-T-C-G-C-G. J. Mol. Biol., 183, 553-563

McCall, M., Brown, T. & Kennard, O. (1985), The crystal structure of d(GGGGCCCC) - a model for poly (dG).poly (dC). J. Mol. Biol., 183, 385-396

McCall, M., Brown, T., Hunter, W.N. & Kennard, O. (1986), The crystal structure of d(GGATGGGAG) forms an essential part of the binding site for transcription factor IIIA. Nature, 322, 661-664

Quigley, G.J., Wang, A.H.-J., Ughetto, G., van der Marel, G., van Boom, J.H. & Rich, A. (1980), Molecular structure of an anti-cancer drug DNA complex: daunomycin plus d(CpGpTpApCpG). Proc. Nat. Acad. Sci. USA, 77, 7204-7208

Quigley, G.J., Ughetto, G., van der Marel, G.A., van Boom, J.H., Wang, A.H.-J. & Rich, A. (1986), Non-Watson-Crick G.C and A.T base-pairs in a DNA antibiotic complex. Science, 232, 1255-1258

Richmond, T.J., Finch, J.T., Rushton, B., Rhodes, D. & Klug, A. (1984), Structure of the nucleosome core particle at 7Å resolution. Nature, 311, 532-537

Shakked, Z., Rabinovich, D., Cruse, W.B.T., Egert, E., Kennard, O., Sala, G., Salisbury, S.A. & Viswamitra, M.A. (1981), Crystalline A-DNA: The X-ray analysis of the fragment d(GGTATACC). Proc. R. Soc., Lond., B213, 479-487

Shakked, Z., Rabinovich, D., Kennard, O., Cruse, W.B.T., Salisbury, S.A. & Viswamitra, M.A. (1983), Sequence dependent conformation of an A-DNA double helix: the crystal structure of the octamer d(GCTATACC). J. Mol. Biol., 166, 183-201

Shakked, Z. & Rabinovich, D. (1986), The effect of the base sequence on the fine structure of the DNA double helix in "Progress in Biophysics and Molecular Biology", in press.

Ughetto, G., Wang, A.H.-J., Quigley, G.J., van der Marel, G.A., van Boom, J.H. & Rich, A. (1985), A comparison of the structure of echinomycin and triostin A complexed to a DNA fragment. *Nucleic Acids Res.*, 13, 2305-2323

Viswamitra, M.A., Shakked, Z., Jones, P.G., Sheldrick, G.M., Salisbury, S.A. & Kennard, O. (1982), Structure of the deoytetranucleotide d-pApTpApT and a sequence-dependent model for poly (dA-dT). *Biopolymers*, 21, 513-532

Wang, A.H.-J., Quigley, G.J., Kolpack, F.J., Crawford, L., van Boom, J.H., van der Marel, G. & Rich, A. (1979), Molecular structure of a left-handed double helical DNA fragment at atomic resolution. *Nature*, 282, 680-686

Wang, A.H.-J., Fujii, S., van Boom, J. & Rich, A. (1982a), Molecular structure of the octamer d(GGCCGGCC): modified A-DNA. *Proc. Nat. Acad. Sci.* USA, 79, 3968-1972

Wang, A.H.-J., Fujii, S., van Boom, J.H., van der Marel, G.A., van Boeckel, S.A.A. & Rich, A. (1982b), Molecular structure of r(GCG)d(TATACGC): A DNA-RNA hybrid helix joined to double helical DNA. *Nature*, 299, 601-604

Wang, A.H.-J., Ughetto, G., Quigley, G.J., Hakoshima, T., van der Marel. G.A., van Boom, J.H. & Rich, A. (1984a), The molecular structure of a DNA-triostin A complex. *Science*, 252, 1115-1121

Wang, A.H.-J., Hakoshima, T., van der Marel, G., van Boom, J.H. & Rich, A. (1984b), A.T base-pairs are less stable that G.C base-pairs in Z-DNA: The crystal structure of d(m^{5}CGTAm^{5}CG). *Cell*, 37, 321-331

Wang, A.H.-J., Gessner, R.V., van der Marel, G.A., van Boom, J.H. & Rich, A. (1985), Crystal structure of a Z-DNA without an alternating purine-pyrimidine sequence. *Proc. Nat. Acad. Sci.* USA, 82, 3611-3615

Wing, R.M., Pjura, P., Drew, H.R. & Dickerson, R.E. (1984), The primary mode of binding of cisplatin to a B-DNA dodecamer: C-G-C-G-A-A-T-T-C-G-C-G. *EMBO J.*, 3, 1201-1206

Woodbury, C.P., Jr, & von Hippel, P.H. (1981) *in* "Gene Amplification and Analysis", J.G. Chirikjian ed, Vol 1., 181-207, Elsevier/North Holland, New York

TEXTBOOK AND REVIEWS

Saenger, W. (1984) "Principles of Nucleic Acid Structure". Springer-Verlag, New York

"Biological Macromolecules and Assemblies: Volume 2 - Nucleic Acids and Interactive Proteins", A. MacPherson and F. Jurnak eds., J. Wiley & Sons, New York

Dickerson, R.E. (1983), The DNA helix and how it is read. *Scientific American*, (Dec.), 94-111

Rich, A., Nordheim, A. & Wang, A.H.-J. (1984), Chemistry and biology of left-handed Z-DNA. *Ann. Rev. Biochem.* 53, 791-846

Kennard, O. (1986), The molecular structure of base-pair mismatches *in* "Trends in Nucleic Acids Research", D. Lilley & F. Eckstein eds., Springer-Verlag, Heidelberg

ONE- AND TWO-DIMENSIONAL NMR STUDIES OF THE STRUCTURES OF SIMPLE SEQUENCE DNAS

David R. Kearns

Department of Chemistry, B-014
University of California, San Diego
La Jolla, California 92093

INTRODUCTION

Progress towards understanding the biological functions of proteins and DNA at the molecular level is moving at a very rapid pace as a result of the development of DNA sequencing and footprinting methods, techniques for cloning and overproduction of proteins, and the crystallization of DNA, proteins and their complexes. Many of the original problems have been solved but important questions remain unanswered. Perhaps one of the most important questions still to be answered is, how do proteins recognize specific DNA sequences? Specific interactions (hydrogen bonding, electrostatic, hydrophobic) between the DNA bases and the amino acid residues will clearly play a major role (Seeman et al., 1976) but sequence dependent variations in conformation heterogeneity may be important. Several different DNA conformations have been identified including left-handed Z-DNA (see Fig. 1) and there is evidence that sequence can have a pronounced effect on the formation of certain conformations (Arnott, 1970; Arnott & Hukins, 1972; Pohl & Jovin, 1972; Wang et al., 1979, 1981; Dickerson, 1983).

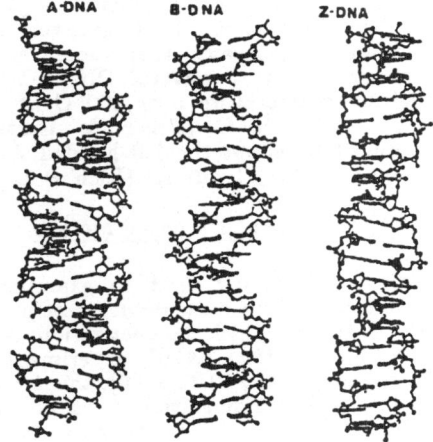

Fig. 1.   The double helical structures of the three principal forms of DNA A, B and Z are shown (adapted from Dickerson et al. (1983)).

DNAs with alternating GC sequences convert to the Z-form in high salt solutions and some substituted GC polymers are stabilized in the Z-form even under mild conditions (Pohl & Jovin, 1972; Rich et al., 1984), whereas most (linear) DNAs do not adopt a Z-conformation even under favorable conditions (high salt and added organic solvent (Wells et al., 1977, 1980; Saenger, 1984). Circular dichroism measurements suggest that there may be other examples of sequence-dependent conformational transitions (Mitsui et al., 1970; Vorlickova et al., 1982, 1983; Chen et al., 1983). In addition to sequence effects on major conformational rearrangements in DNA (eg. B, A, Z) more subtle (Dickerson & Kopka, 1983; Dickerson et al., 1983) sequence-dependent variations in the B-DNA structure (pitch, roll, propellering, glycosidic torsional angle, sugar pucker, backbone angles) might be large enough to be recognized by different proteins (Dickerson & Kopka, 1983; Dickerson et al., 1983).

Numerous protein sequences and associated DNA binding site sequences have been determined, but so far no simple rules have evolved that allow us to make any strong predictions about the major factors involved in the recognition. X-ray diffraction studies have recently provided atomic resolution structures of several short DNA duplexes (Viswamitra et al., 1978, 1982; Wang et al., 1979, 1981, 1985; Crawford et al., 1980; Wing et al., 1980; Drew et al., 1981) and the structures of several DNA binding proteins have been determined (Anderson et al., 1981; McKay & Steitz, 1981; Brayer & McPherson, 1983; Steitz et al., 1983; Tanaka et al., 1984). The structure of the complex between the Eco RI restriction enzyme and a 13 base pair DNA duplex has been solved at 3 Å resolution (Frederick et al., 1984) and the analysis of higher resolution data will be reported shortly (McClarin et al., 1986). These x-ray diffraction studies will undoubtedly answer many questions regarding the role of structure in the recognition process and, in favorable cases, even provide information about the amplitudes of conformational fluctuations (Holbrook & Kim, 1984). However, because of the difficulties in obtaining suitable crystals, the effort involved in solving one structure, and the possibility of crystal effects on structure it is important to develop solution-state techniques for studying the structure and dynamics of proteins and DNA in solution. Fortunately, new developments in one- and two-dimensional (1D and 2D) NMR relaxation techniques have provided the spectroscopic tools for accomplishing this. The first 2D NMR spectra of a DNA duplex were reported in 1982 (Feigon et al., 1982) and since then a multitude of 2D NMR techniques have been applied to a variety of DNA molecules. In our discussion here we consider only applications of ^{1}H 2D NMR spectroscopy to studies of the structure of simple sequence DNAs. Simple sequence DNA are important because they have been extensively used to obtain physical parameters (stacking energies, CD, nearest neighbor, interaction energies, persistence length) for predicting the behavior of natural DNA (Bloomfield et al., 1974; Saenger, 1984). The synthetic simple-sequence DNAs are interesting because they should exhibit the extremes of behavior that might be encountered with natural DNA. For example, poly(dG-dC)·poly(dG-dC) has a melting temperature over 100° C at 0.1 M NaCl whereas poly(dA-dT)·poly(dA-dT) melts at 60° C (Sober & Harte, 1970). Poly(dG-dC)·poly(dG-dC) can be induced to adopt the left-handed Z-conformation in high salt solution but poly(dA-dT)·poly(dA-dT) will not (Rich et al., 1984).

Proton Relaxation, NOEs and Structure and Internal Motion

The three NMR methods commonly used to examine the structural and dynamic properties of DNA all depend on measurements of proton spin-lattice cross relaxation rates, that is the rates at which magnetization is transferred between spins. In the 2D nuclear Overhauser effect (NOE) experiment (Freeman & Morris, 1979; Jeener et al., 1979; Nagayama et al., 1979, 1980), this transfer is manifested by a modulation of resonance intensities arising from cross relaxation (magnetization transfer) processes. In truncated-driven NOE (Wagner & Wuthrich, 1979) or presteady-state NOE (Bothner-By & Noggle, 1979) experiments, individual resonances in the spectrum are irradiated (saturated) for varying periods of time and the rate of transfer of magnetization to other resonances is monitored. Finally, various direct measurements of spin-lattice relaxation rates (Early et al., 1980; Assa-Munt et al., 1981, 1984; Broido & Kearns, 1982) can be used to obtain the same information as in the presteady-state NOE or 2D NOE experiments.

The cross-relaxation rates, σ_{is}, extracted from the experimental measurements are related to the interproton distance, r_{is}, as follows (Abragam, 1978):

$$\sigma_{is} = \frac{1}{2}\,\gamma_i^2\gamma_s^2\hbar^2 r_{is}^{-6}[J_0(O) - 6J_2(2\omega)] \tag{1}$$

where γ_i and γ_s are the gyromagnetic ratios for spins i and s, $\hbar$ is Planck's constant over 2π, and ω_i and ω_s are the Larmor frequencies. The $J_n(\omega)$, n=0,1,2, are the spectral densities representing the relative amount of radio frequency power at different frequencies generated by the *overall* and *internal* molecular motions. A slowly tumbling, rigid molecule generates only low frequency power, whereas a molecule with fast internal motions generates higher frequencies. For DNA we expect $J_0(O) \gg J_1(\omega_H)$, $J_2(2\omega_H)$ and therefore

$$\sigma_{is} = \frac{1}{2}\gamma_H^4\hbar^2 r_{is}^{-6}J_0(O) \quad . \tag{2}$$

Therefore, by measuring σ_{is}, r_{is} can be determined provided $J_0(O)$ is known or can be determined (from measurements of σ_{is} for interacting protons with fixed internuclear separation).

Which technique is best depends upon the size of the molecule, number of interacting protons to be studied, the amount of sample available and various experimental factors. With short DNA, a judicious set of presteady-state NOE measurements might be preferable to a 2D NOE experiment because of problems in quantitatively extracting cross relaxation rates from 2D NOE measurements. In experiments that we have carried out with DNA decamers (Feigon & Kearns, 1979; Feigon et al., 1983b) 2D NOE (350 ms mixing time) and 2D correlated spectroscopy (COSY) spectra were collected in ~15 hrs on samples containing 3-4 mg of DNA in 0.2 ml. With higher molecular weight (e.g. 60 base pair) simple sequence DNAs (poly(dA-dT), poly(dG-dC), poly(dA)·poly(dT)) useful 2D spectra were obtained in 10-20 hrs (Assa-Munt & Kearns, 1984; Mirau & Kearns, 1984).

A variety of NMR relaxation techniques have been used to assess the amplitudes and correlation times for internal motions in DNA (Bolton & James, 1979, 1980; Hogan & Jardetzky, 1980a,b; Shindo, 1980; Bendel et al., 1983; Clore & Gronenborn, 1984; James, 1984). In our work (Early et al., 1980; Assa-Munt et al., 1984; Mirau et al., 1985; Behling & Kearns, 1986) we studied the cross relaxation of protons by neighbors with known fixed internuclear separations to evaluate internal motion in different simple-sequence DNAs ~60-70 base pairs long. These results demonstrated that the bases in these DNAs exhibit considerable internal librational motion with large amplitudes ($\pm 20°$-$30°$) and short correlation times (~1-2 ns) for the torsional motions but longer (10-100 ns) correlation times for out-of-plane motions. ^{13}C and ^{31}P NMR relaxation measurements on high molecular weight DNA also indicate there are large amplitude ($\pm 35°$) high frequency (1-2 ns) motions in the DNA backbone (Hogan & Jardetzky, 1979, 1980a; Rill et al., 1980, 1981; Shindo, 1980; Bendel et al., 1983; James, 1984). The combined results of these various NMR experiments, summarized in Table 1, demonstrate there are substantial internal motions present in all parts of the DNA molecule. Because the amplitudes are large and the correlation times are short the effect of internal molecular motion must be considered in discussing the structural features of the DNA. In particular, it means that internucleotide distances determined by NMR will have a necessary uncertainty connected with them, and that when we refer to the "structure" of DNA we mean a time-averaged structure.

Table 1. A summary of internal motions in DNA detected by NMR

NMR Experiment	Group	Type of Motion	Amplitude	Correlation Time
^{13}C	Sugar	Repuckering	±25°	1-2 ns
^{31}P	Phosphate	Rotation	±25°	1-2 ns
^{1}H	Sugar	Repucker	±35°	1-2 ns
^{1}H	Base	Out-of-Plane	±25°	10-50 ns
^{1}H	Base	Torsional	±30°	1-3 ns

DNA STRUCTURAL FEATURES

The three major forms of DNA (A-, B- and Z-) are shown in Fig. 1 and several of the intra- and inter-nucleotide proton-proton distances that are associated with each structure are summarized in Table 2. Some of the shorter interproton separations (less than about 2.7 Å) are indicated in bold type since these are useful in qualitatively distinguishing between the different forms. The chemical composition of a double-stranded DNA and a schematic diagram depicting the important structural features that characterize the DNA secondary structure are shown in Fig. 2.

Table 2. Approximate interproton distances (in Å) for A, B and Z-DNA.[a]

Intranucleotide[b] Interactions	A	Model B	Z
BH→H1′	3.8	3.8	3.7 (Pyr) 2.6 (Pur)
BH→H2′	3.8	**2.0**	3.1 (Pyr) 4.1 (Pur)
BH→H2′′	4.6	3.5	4.3 (Pyr) 4.7 (Pur)
BH→H3′	**2.6**	4.0	4.5 (Pyr) 5.3 (Pur)

Inter-Nucleotide Interactions[b,c]	A	Model B	Z
BH→BH	4.7	5.0	6.1 (Pyr-Pur) 5.3 (Pur-Pyr)
BH→H1′	4.0	2.9	3.7 (Pyr) 6.5 (Pur)
BH→H2′	**2.0**	3.9	3.1 (Pyr) 7.1 (Pur)
BH→H2′′	3.0	**2.5**	4.3 (Pyr) >7.5 (Pur)
BH→H3′	3.4	5.1	4.5 (Pyr) 6.6 (Pur)
BH→CH5		>6	
BH→TCH3		>6	

[a]Taken from Westerink et al.,(1984).
[b]BH refers to purine H8 or pyrimidine H6 proton.
[b]Between nucleotide and its 5′ neighbor.

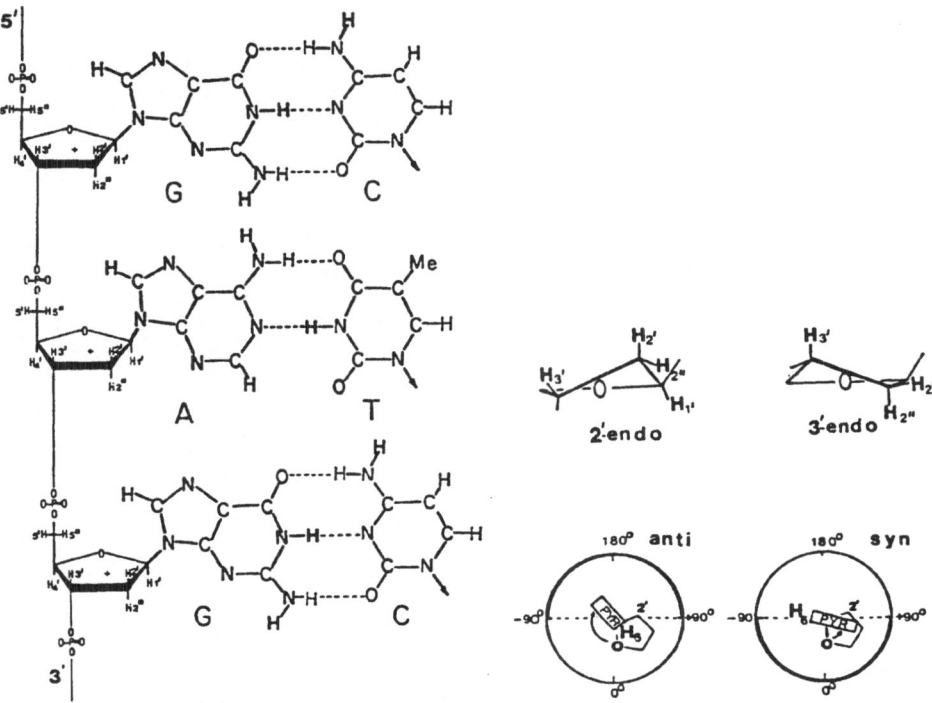

Fig. 2. (left) A schematic diagram illustrating the chemical composition of the DNA. In this diagram, the backbone from only one strand is indicated. (right) A schematic diagram illustrating the two principal sugar puckers (top) and the two major ranges of torsional angles (*anti* and *syn*) adopted by nucleotides (bottom).

Key structural features that can be determined from semi-quantitative NMR relaxation measurements are (1) sugar pucker (2) glycosidic torsional angle, (3) base pairing, and (4) helical sense.

Sugar Pucker. The ribose group in DNA can adopt many different puckers but the two main conformations, according to x-ray diffraction studies of small molecules (nucleosides, nucleotides) are the 2′-endo and the 3′-endo conformation (Altona & Sundaralingam, 1973; Altona, 1975; Altona et al., 1976; Haasnoot et al., 1981). While other sugar conformations are accessible, spectroscopic studies of DNA have often been interpreted in terms of mixtures of just these two conformations (Cheng et al., 1982; Kan et al., 1982; Wartell & Harrell, 1986). In A-DNA the sugar pucker is 3′-endo, but in the standard B-DNA, the pucker is 2′-endo.

Glycosidic Torsional Angle. The glycosidic torsional angle, χ, is the angle between the C1′-O4′ bond of the sugar relative to the N1(N9)-C2(C4) bond of the pyrimidine (purine) of the attached base (see Fig. 2). The range $\pm90°$ about $\chi=180°$ is referred to as the *anti*-conformation and the range $\pm90°$ about $\chi=0°$ is referred to as *syn*-conformation (Saenger, 1984). Note that other conventions for defining χ exist so that care must be exercised in comparing results by different authors. In B-DNA and A-DNA all nucleotides have an *anti*-conformation, but in the left-handed Z-DNA they alternate between *anti*- and *syn*-.

Base Pairing. The standard Watson-Crick pairings are illustrated in Fig. 2. The hydrogen bonded exchangeable protons are useful in monitoring the base pairing structure and the dynamics of base pairing (Kallenbach et al., 1981; Mirau & Kearns, 1983; Assa-Munt et al., 1984). In addition to Watson-Crick base pairs, other types of hydrogen bonding schemes have been observed (Hoogsteen, 1963).

Helical Sense. Under most experimental conditions, natural DNA and most synthetic DNAs are believed to be in a right-handed helical form (Wells et al., 1977, 1980; Saenger, 1984). However, special DNA sequences (notably $d(C-G)_n$ and $d(m^5C-G)_n$) can be converted to a left-handed Z-form in the presence of certain cations (Rich et al., 1984) and even natural DNA contains sequences that convert to the Z-form when subjected to sufficient torsional stress by supercoiling (Nordheim et al., 1982; Stirdivant et al., 1982). There is evidence to suggest that the left-handed Z-DNA is involved in the process of recombination (Kmiec & Holloman, 1984) and it seems likely that Z-DNA will play a role in other biochemical processes (Zarling et al., 1984). Because left-handed forms of DNA have been discovered, it can no longer be assumed that all DNA helices are right-handed. Once the base pairing, the approximate sugar pucker glycosidic torsional angles, and the helical sense have been determined, many of the gross secondary structural features of the DNA are established.

A Strategy for Structure Determination

The determination of DNA structure by NMR relaxation techniques will usually involve the following steps (Feigon et al., 1983b; Hare et al., 1983; Clore et al., 1984; Weiss et al., 1984; Wemmer et al., 1984): (1) Classification of resonances in the 1D spectrum according to type. (2) Identification of resonances from protons within individual nucleotide units using 2D NOE and 2D COSY spectra. (3) Sequential assignment of resonances to specific nucleotides in the sequence using 1- or 2D NOE methods. (4) Semi-quantitative evaluation of cross relaxation rates to determine sugar pucker, *syn-* or *anti*-orientation of the bases, base pairing and helical sense. (5) Quantitative measurement of cross relaxation rates by 1- or 2D methods and conversion of rates to interproton distances. (6) Comparison of experimental interproton distances with those calculated for models of DNA helices and interactive adjustment of the structure to fit the experimental distances.

Classification of Resonances in the 1D Spectra. The spectrum, in D_2O, of the DNA dodecamer d(CCGCCACTGATGG)·d(CCATCAGTCGG) is shown in Fig. 3 along with the assignment to type of resonance (Feigon et al., 1983a). The positions of the resonances observed for the different protons in this molecule are typical of DNA and it is therefore easy to classify many resonances in the spectrum according to type of protons (i.e. aromatic base proton, H1′ sugar, methyl etc.) (Kearns, 1977; Patel, 1978). The aromatic base proton resonances are located between 8.5 and 7.0 ppm, the H1′ sugar protons and the cytidine H5 protons resonate between 6.5-5.5 ppm, the 3′ resonances are located around 5.0 ppm, the remaining sugar protons resonate between 4.6-1.5 ppm and the thymine methyl protons have the highest field resonances, located between 2.0 and 1.5 ppm.

The AH8 and GH8 protons can be distinguished from other resonances by the fact that both can be exchanged for deuterium by heating the DNA in D_2O (Benevides et al., 1984). The AH2 resonances are usually the sharpest ones in the spectrum because the AH2 protons are not scalar coupled to any other protons in the molecule and they are physically distant from most other *non*-exchangeable protons in the molecule (except the AH2 of adjacent adenines) (Behling & Kearns, 1985). The AH2 proton, however, is close (~2.8 Å) to the exchangeable imino proton of the complementary thymine base and therefore in H_2O (but not in D_2O) will exhibit strong NOEs with these protons. An illustration of this interaction is shown in the 1D NOE spectrum of poly(dI-dC)·poly(dI-dC) presented in Fig. 4.

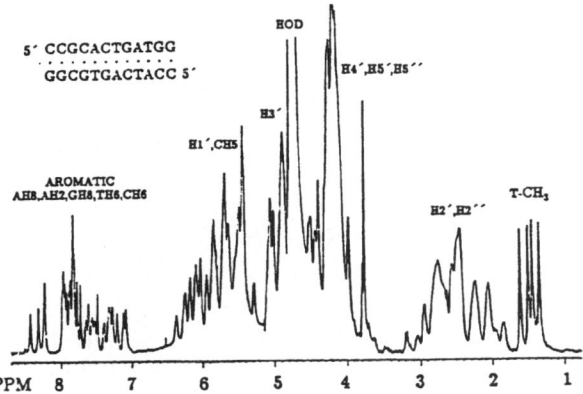

Fig. 3. The 500 MHz ^{1}H spectrum of the non-exchangeable protons of 5'd(CCGCACTGATGG)·(CCATCAGTGCGG) is shown along with classification of the various resonances according to type (T=35° C).

The strongest peak in the difference spectrum, at ~7.4 ppm, is assigned to the neighboring IH2 proton in the Watson-Crick base pair. The absence of an effect on the IH8 resonance eliminates the possibility of Hoogsteen pairing (Kearns et al., 1983).

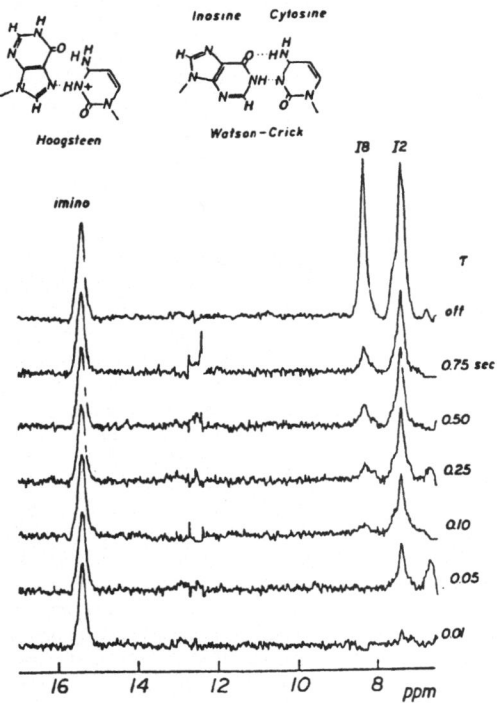

Fig. 4. A presteady-state NOE experiment on poly(dI-dC)·poly(dI-dC) obtained by irradiating the inosine imino proton resonance at 15.3 ppm for the times indicated at the right side of the figure. These spectra demonstrate that there is a transfer of magnetization from the inosine imino proton to the IH2 proton that resonates at 7.4 ppm, confirming that the base pairing is Watson-Crick.

29

To proceed further in the assignment of the spectra requires 2D NOE and COSY measurements.

Identification of Resonances from Individual Nucleotide Units. Protons within a single sugar group can be identified through their scalar coupling by a COSY spectrum. The through-bond scalar connectivities are indicated below.

$$H1 \underset{H2''}{\overset{H2'}{\lessgtr}} \updownarrow \gtrdot H3' \rightarrow H4' \underset{H5''}{\overset{H5'}{\lessgtr}} \updownarrow$$

Therefore in the COSY spectrum each H1′ resonance is coupled to an H2′ and H2′′ resonance, which in turn are both coupled to a single H3′ and so forth. By connecting the appropriate crosspeaks in the COSY spectrum it is usually possible to identify most of the resonances associated with each sugar residue in the molecule, subject to limitations imposed by spectral resolution. Because the H1′ proton is scalar coupled to *both* the H2′ and H2′′ resonances, the COSY spectrum alone is usually not adequate to permit distinction between these two protons, but this can be done by a 2D NOE experiment. This method depends on the fact that within a sugar residue the H1′ proton is closer to the H2′′ than to the H2′ proton (2.45 vs 3.0 Å) regardless of the sugar pucker. The H2′ proton in turn is closer to the 3′ proton than is the H2′′ and so forth. Therefore, in a 2D NOE spectrum, obtained at short mixing times to prevent spin diffusion, the following relative intensity patterns for the sugar crosspeaks are expected: (H1′-H2′′)>(H1′-H2′), (H2′-H3′)>(H2′′-H3′). The overlap of the 5′ and 5′′ resonances is usually so severe that little effort is made to make these assignments. In addition to these intra-sugar connectivities, strong NOE crosspeaks between the TH6 and TMe resonances of each thymine and between the CH5 and CH6 resonances of each cytidine are expected. The COSY spectra, in conjunction with 2D NOE data, should provide redundant information on the identification of many resonances in the DNA spectra.

Sequential Assignment of Resonances. The next step in the assignment procedure is to use 1D or 2D NOE techniques to assign resonances to protons of specific nucleotides in the sequence. The basis for sequential assignment of resonances from NOE measurements is indicated schematically in Fig. 5. For a right-handed helix with all bases in the *anti*-conformation, a base proton of Base-N (Pyr H6 or Pur H8) can exhibit an NOE interaction with the H1′ proton of its own sugar group and the H1′ sugar of the 5′ neighboring nucleotide (N-1). These interactions may be the result of a direct transfer of magnetization (if the pair of interacting protons are physically close together) or as a result of secondary transfer of magnetization. In B-DNA a direct transfer between the base proton and the H1′ proton of the 5′ neighboring sugar proton is expected, but the distance between the base proton (Pur H8 or Pyr H6) and its own H1′ sugar proton is large (3.5 Å). Consequently, the NOE arising from direct magnetization transfer from a base proton to its own H1′ proton is expected to be weak. However, by allowing sufficient time for secondary transfer of magnetization (base→H2′→H2′′→H1′) a crosspeak between a base proton and its own H1′ proton can be observed. Regardless of the precise conformation of the DNA, each base proton in the interior of a right-handed double helix should exhibit crosspeaks with two H1′ resonances in the spectrum, one from its own sugar and one from the 5′ neighboring nucleotide. The resonances from Base N+1 then can be used to link the H1′ resonances from nucleotide N and nucleotide N+1, and so forth. In this way, all the base and H1′ protons in the molecule can be sequentially assigned to specific nucleotides by "walking" through the molecule. The analysis can start with any base and proceed in both directions until all base and H1′ sugar resonances within a given strand have been assigned. In addition to sequence uniqueness, there are additional checks on the assignments because the 5′ terminal bases on each strand can have a crosspeak with only one H1′ sugar resonance. An example of this method of assignment is presented in Fig. 5.

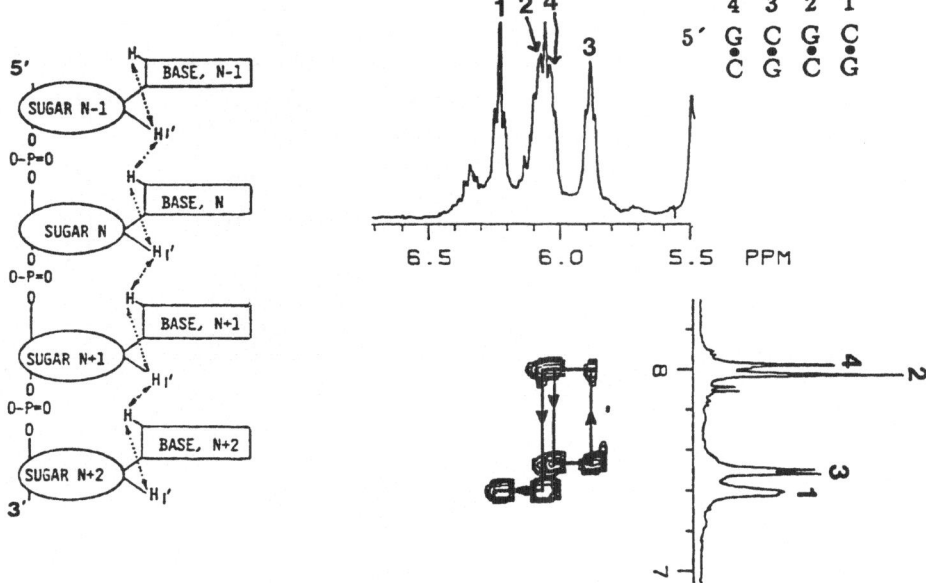

Fig. 5.   (Left) This diagram illustrates that the base proton (Pur H8 or Pyr H6) in a regular right-handed DNA helix exhibits an NOE with the H1′ sugar proton of the same nucleotide and with the H1′ of the 5′ neighboring (N-1) nucleotide. This forms the basis for sequentially identifying the base and sugar protons resonances in a DNA. (Right) A 2D NOE contour plot illustrating the sequential assignment of base (GH8 and CH6) and H1′ sugar resonances in d(GCGC)·d(GCGC). (S. Ikuta & D.R. Kearns, unpublished)

Once the H1′ resonances have been assigned to specific sugar residues in the sequence, the COSY and NOE data discussed above may be used to assign the other (H2′,H2′′,H3′,H4′) resonances in the same sugar. This combination of procedures completes the assignment of the resonances in the spectrum. The next step is to carry out a semi-quantitative or quantitative study of the structure of the DNA.

NMR STUDIES OF SIMPLE SEQUENCE DNAS

We now describe the application of 1D and 2D NMR techniques to determine many of the qualitative structural features (sugar pucker, glycosidic torsional angle, base pairing structure and helical sense) of simple-sequence DNAs.

Poly(dA−dT)·Poly(dA−dT)

Poly(dA-dT)·poly(dA-dT) provides a good example of the application of 1D and 2D NMR techniques because a number of models have been proposed to account for the structure of this molecule in solution and in fibers. Most models are based on a right-handed helix with Watson-Crick base pairing (Arnott & Hukins, 1973; Arnott & Selsing, 1974; Klug et al., 1979), but left-handed structures have also been proposed (Drew & Dickerson, 1982; Gupta et al., 1983). In view of the considerable evidence for environmental effects on DNA conformation and the discovery of Z-DNA, none of the proposed structures can be totally discounted. A summary of some of the important features of these models is given in Table 3. We now consider the various steps involved in the analysis of the poly(dA−dT)·poly(dA−dT) structure since the same ones are used in the analysis of other DNA structures.

Table 3. A summary of the key structural features of models proposed for poly(dA−dT)·poly(dA−dT).

Model	Helix Sense	Base Pairs	Sugar Pucker	Glycosidic Torsion	Reference
1	Right	Watson-Crick	3′-endo(A) 2′-endo(T)	Anti	(a)
2	Right	Watson-Crick	2′-endo	Anti	(b)
3	Right	Watson-Crick	2′-endo	Anti	(c)
4	Left	Hoogsteen		A(Syn) T(Anti)	(d)
5	Left	Watson-Crick		A(Syn) T(Anti)	(e)

a) (Klug et al., 1979)
b) (Arnott & Hukins, 1973)
c) (Arnott & Selsing, 1974)
d) (Drew & Dickerson, 1982)
e) (Gupta et al., 1983)

Determination of the Base Pairing Structure. The base pairing structure is unequivocally established as Watson-Crick by a simple 1D NOE experiment (analogous to that shown in Fig. 4) in which the lowfield thymine imino resonance was saturated and changes elsewhere in the spectrum were monitored (Assa-Munt & Kearns, 1984). At short irradiation times the AH2 resonance is the only one that exhibits an NOE. This shows that the AH2 proton is located close to the thymine imino proton and definitively establishes that the base pairing is Watson-Crick (Assa-Munt & Kearns, 1984).

Identification of H1′, H2′, H2′′ Resonances from the A and T Sugar Groups. With small oligonucleotides both COSY and 2D NOE spectra can be used to identify the sugar proton resonance (Feigon et al., 1982, 1983b,c; Hare et al., 1983; Scheek et al., 1984; Weiss et al., 1984; Wemmer et al., 1984). With larger DNA the COSY method does not work because of more rapid T_2 relaxation and one has to rely on the 2D NOE and the fact that the H1′ is always closer to the H2′′ than to the H2′ proton of the same nucleoside unit, but remote from other neighboring sugar protons. A 2D NOE spectrum of poly(dA−dT)·poly(dA−dT) (see Fig. 6) shows that the H1′ resonance at 6.18 ppm interacts strongly with a H2′′ resonance at 2.86 ppm and weakly with the H2′ resonance at 2.58 ppm (Assa-Munt & Kearns, 1984). The corresponding resonances from the other sugar groups are located at 5.66 (H1′), 2.50 (H2′′), and 2.05 (H2′) ppm. From these experiments the H1′, H2′ and H2′′ resonances from each sugar can be identified but they do not establish which set belongs to the A or T nucleotides.

Determination of the Syn/Anti-Conformation of the Nucleotide Units and Sugar Pucker. In the anti-range of conformations, the TH6 and AH8 base protons are located close to the H2′ (or H3′) sugar protons on the same nucleotide unit, whereas in the syn-conformation the base protons are located close to the H1′ proton and far from the H2′ proton (see Fig. 2, Table 2) and therefore the NOE effects can be used to distinguish between these two families of conformations. Experimentally the TH6 resonance exhibits a strong crosspeak with a H2′ resonance at 2.06 ppm, but weak crosspeaks with the two H1′ resonances. The magnitude of the TH6-TMe crosspeak provides an internal calibration that proves that the TH6-TH1′ crosspeak is weak and inconsistent with a syn-conformation. These observations immediately establish that the T nucleotide is in the anti-conformation and that the H2′ resonance at 2.06 ppm belongs to the T nucleotide. From step (2), this also permits the remaining sugar protons to be assigned. An NMR analysis of the interactions

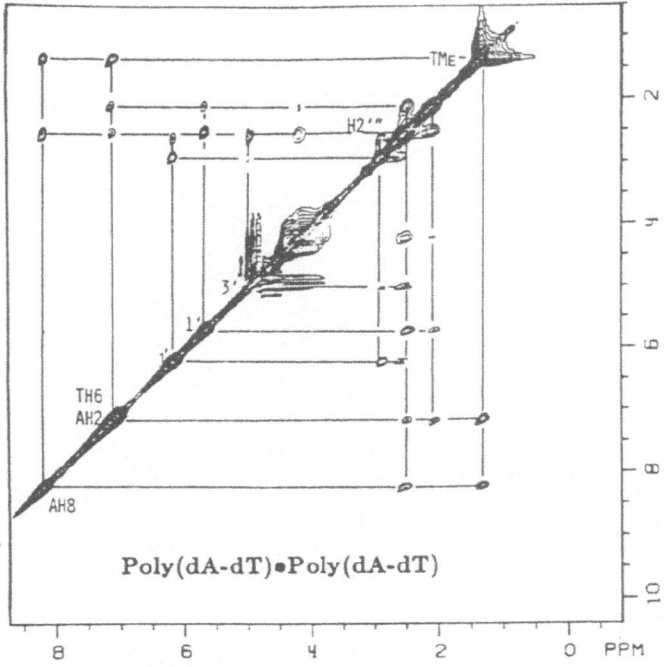

Fig. 6. (A) A 360 MHz pure absorption phase 2D NOE spectrum of poly(dA−dT)·poly(dA−dT) at 21° C obtained with a 50 ms mixing time. A contour plot is shown, and the assignments of the various resonances are indicated on the spectrum.

between the AH8 resonance (8.15) ppm and the sugar protons establishes that the A nucleotide unit also has an *anti*-conformation. These same experiments also establish the approximate sugar pucker of the A and T nucleotides. In the *anti*-conformation of the nucleotide unit the base protons (TH6 and AH8) are close to the H2' (~2.0-2.2 Å) and remote (4.0 Å) from the H3' proton *if* the sugar pucker is approximately 2'-endo. For a 3'-endo conformation (see Table 2) the converse is true. Because the intra-nucleotide crosspeaks between the A and T base protons and the H2' resonances are strong, whereas base-H3' crosspeaks are weak or absent, this is consistent with a structure in which *both* the A and T nucleotide units have sugar puckers close to 2'-endo.

Determination of the Helix Sense. Once the nucleotide configuration has been established (i.e. *anti* or *syn*, 2'-endo or 3'-endo), the internucleotide interactions can be used to rigorously distinguish between left- and right-handed helices from general topological considerations, without making any assumptions regarding exact structural details. This follows from a consideration of the differences between the general features of a right- and a left-handed stack of bases, all with an *anti*-conformation of the nucleotide units. From examination of Fig. 7 it is apparent when the bases are stacked in a right-handed helix, the H6 or H8 base proton *can* be relatively close to the H2'' (and possibly the H1' and H2') protons of the 5' neighboring sugar. However, in a left-handed helix both the H2' and H2'' protons of one nucleotide must be remote from the base protons of neighboring nucleotide. Therefore, the observation of a moderately strong interaction between TH6 and the H2'' proton of the A sugar but weaker interaction with the H1' proton of A proves that poly(dA-dT)·poly(dA-dT) has a right-handed helical structure (Assa-Munt & Kearns, 1984).

33

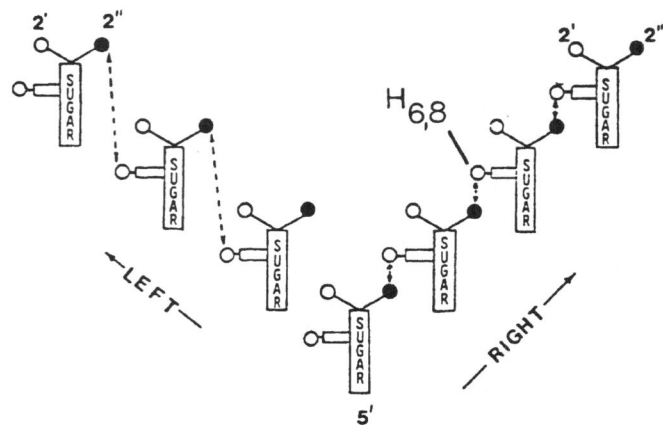

Fig. 7. A diagram comparing the stacking of bases in a right-handed and left-handed helix and illustrating the differences in the base-H2′ internucleotide interactions in the right- and left-handed helices. All bases are in an *anti*-conformation, but no other assumptions regarding the sugar pucker or precise alignment are made.

A comparison of the relative magnitudes of the crosspeaks with the interproton distance calculated for different DNA structures (Table 2) suggests that the structure is close to that expected for a B-DNA structure (Assa-Munt & Kearns, 1984). Specifically, both the AH8 and TH6 protons are close to the H2′ proton of the same nucleotide, slightly further from the H2′′ proton of the 5′ neighboring nucleotides, and distant from the H1′ proton of either sugar. Klug et al. (1979) have described a model for poly(dA-dT) in which the sugar attached to adenine is in a 3′-endo conformation and the sugar attached to thymine is 2′-endo, but this is not consistent with the 2D NOE data. Moreover, this model predicts that the AH8 proton is relatively far from its own H2′ proton (3.1 Å) and close to the thymine H2′′ sugar proton (2.2 Å) whereas the reversed pattern of interactions is observed. Moreover, the base (AH8 or TH6) sugar interactions are so similar for the A and T nucleotides we conclude that there is little evidence in the 2D NOE spectra for a structure in which the A and T nucleotides have different conformations. While poly(dA-dT)·poly(dA-dT) may adopt an alternating structure under some conditions, the solution-state NMR data are much more consistent with a regular B-form structure.

It should be noted that the above experiments refer to the conformation of poly(dA−dT)·poly(dA−dT) in 0.1 M NaCl. In solutions containing high concentrations of CsF there are changes in the CD spectrum that have been interpreted as indicating a transition to a left-handed Z-form (Vorlickova et al., 1983). This possibility has been explored by   Borah et al.,(1985a)   using 2D NMR and they find that even in 6 M CsF and in ethanol the molecule remains in a B-form, although there are changes in the ^{31}PNMR spectra and CD circular dichroism spectra that indicate an apparent transition to a new state (Vorlickova et al., 1983; Borah et al., 1985b). The CD changes might be due to some changes in base stacking without large structural changes and the ^{31}P changes could be due in part to a salt effect on the structure of water around the DNA and an alteration of the ^{31}P chemical shift (Lerner et al., 1984). In summary, the qualitative 2D NOE measurements demonstrate that poly(dA−dT)·poly(dA−dT) has a structure that is close to B-form over a wide range of experimental conditions.

Poly(dA)·Poly(dT)

Several models have been proposed to account for x-ray diffraction (Arnott & Selsing, 1974; Arnott et al., 1983) of poly(dA)·poly(dT). In most previous analyses of the fiber x-ray diffraction data from simple-sequence DNA the two strands of DNA were constrained to have identical backbone conformations. However Arnott et al. (1983) obtained a better fit of the data for poly(dA)·poly(dT) by allowing the two strands to adopt different conformations. In their refined structure (referred to as the heteronomous model) the poly(dA) strand has C3′-endo sugar pucker (C3′-endo/anti) whereas the sugars on the poly(dT) strand has a C2′-endo pucker. From the pattern of crosspeak intensities observed in the 2D NOE spectrum of poly(dA)·poly(dT), it is evident that both the A and T nucleotides have a more or less standard C2′-endo sugar conformation with an *anti*-arrangement of the bases relative to the sugar (Behling & Kearns, 1986). Within experimental error (±10%), the corresponding cross relaxation rates for interactions on the A and T strands are identical indicating that the distances between the base protons (AH8 or TH6) to the sugar protons (H1′ or H2′) are identical in each strand (Behling & Kearns, 1986). A quantitative evaluation of the poly(dA)·poly(dT) structure, based on combining the NMR data with molecular mechanics and dynamics calculations is discussed below.

Poly(dI-dC)·Poly(dI-dC)

A number of different structures have been suggested for poly(dI-dC)·poly(dI-dC). To account for the fiber x-ray diffraction data, a right-handed D-form (Ramaswamy et al., 1982), a left-handed helical structure with Watson-Crick base pairing (Mitsui et al., 1970), and a left-handed helical structure with Hoogsteen pairing (Drew & Dickerson, 1982) have been proposed. Some solution-state spectroscopic and enzymatic digestion studies indicate a more normal B-form (Grant et al., 1972), but the long wavelength CD spectrum is inverted from normal B-DNA (Mitsui et al., 1970), and this led several groups to propose left-handed helical structure. However, more recent studies show that vacuum UV CD is characteristic of B-DNA (Sutherland & Griffin, 1983). The 2D NOE spectrum of this molecule is essentially the same as that obtained for poly(dA-dT) and the same structural conclusions are to be drawn (Mirau & Kearns, 1984). Both bases are in an *anti*-configuration with sugar puckers that appear to be primarily 2′-endo.

Poly(dI-dbr⁵C)

The 2D NOE spectrum for this molecule in 0.1 NaCl is sufficiently similar to the poly(dI-dC) spectrum (Mirau & Kearns, 1984) to conclude that both molecules have similar right-handed helical structures.

The B- and Z-Forms of Poly(dG-dC)

Poly(dG-dC) is of considerable historical interest because it was the first DNA that was found to convert to a left-handed helical form (Z-DNA). Earlier CD measurements by Pohl and Jovin (1972) provided strong evidence for a conversion form of normal B-form DNA to some new form, but it was not until 1979 that Wang et al. (1979) solved the structure of d(C-G)$_3$·d(C-G)$_3$ and unequivocally showed it formed a left-handed helical structure. Since then numerous other sequences have been found that can be converted to the Z-form under appropriate conditions (Rich et al., 1984). In low salt, the 2D NOE spectrum for poly(dG-dC) is consistent with a right-handed B-form structure (Mirau & Kearns, 1984; Borah et al., 1985b) but in high salt solutions where this polymer is expected to be in the left-handed Z-form (Pohl & Jovin, 1972; Patel et al., 1979; Behe & Felsenfeld, 1981; Chen et al., 1983), there are significant changes in the 2D NOE spectra (Mirau & Kearns, 1984; Borah et al., 1985b). A major change is that the crosspeak between the GH8 base proton and GH1′ sugar proton resonances is one of the strongest peaks in the 2D NOE spectrum whereas the corresponding peak in the B-form 2D spectrum was weak. A strong GH8-GH1′ NOE, first observed in 1D NOE measurements (Patel et al., 1982), indicates the G nucleotides is in a *syn* conformation, whereas the moderately strong cross-peak between the CH6 proton and the CH2′ proton indicates that C-nucleotide

35

is in the *anti*-conformation. The *syn/anti* alternation observed in the polymer is characteristic of Z-DNA (Wang et al., 1979). These 2D-NOE studies on the B- and Z-forms of poly(dG-dC) provide a foundation for recognizing the characteristics of the Z-form of other polymers.

Poly(dm^{5}C-dG)

After the initial discovery of the Z-form of poly(dG-dC), a search for other sequences that might convert to Z-DNA was carried out. Behe and Felsenfeld (1981) showed that methylation of cytidine facilitates conversion to the Z-form and they discovered a number of polyvalent ions including [Co(NH$_3$)$_6^{3+}$] that were especially effective in indicating the Z-form of poly(dG-dm^{5}C). Borah and Cohen (1985) have used 2D NMR to examine the conformation of this polymer in 50 mM NaCl and in 3 mM MgCl$_2$. The 2D NOE spectrum in 50 mM NaCl is characteristic of a B-form DNA: the crosspeaks between the CH6 and GH8 protons and their respective 2′ sugar protons are moderately strong, the crosspeak between the GH8 and CMe protons is strong (the analogous crosspeak is seen in poly(dA−dT)·poly(dA−dT)), and the crosspeak between CH6 and the H1′ proton of the adjacent G sugar is weak. The corresponding GH8-H1′ internucleotide interaction was too weak to be observed. In 3 mM MgCl$_2$ the molecule converts to Z-DNA and this is manifested in the 2D spectrum by a strong crosspeak between the GH8 resonance and the GH1′ sugar peak characteristic of a *syn*-conformation and the disappearance of the GH8-CMe crosspeak. Because inter-nucleotide interactions are weak (see Table 2), these 2D NOE spectra on high molecular weight are mainly useful for identifying major structural features associated with intra-nucleotide conformations. To obtain detailed information about the structure of Z-form double helix requires studies of short oligonucleotides.

2D-NOE Studies on Two Z-Forming Oligonucleotides (dm^{5}C-dG)$_3$·(dm^{5}C-dG)$_3$

Feigon et al. (1984) measured the 1- and 2D NMR spectra of (m^{5}dC-dG)$_3$· (m^{5}dC-dG)$_3$ and found in normal salt solutions, and at slightly elevated temperature, the molecule is primarily in the B-form. At low temperatures the Z-form is stabilized by the addition of methanol and therefore it was possible to find conditions where the B- and Z-forms coexisted in comparable amounts. Using the 2D methods discussed in a previous section, all base resonances and most sugar proton resonances could be assigned in the B-form. However, the same procedure cannot be used to assign the resonances in the Z-form because of the different spatial proximities of the base and H1′ protons in the Z-form (no interaction between a base proton and the H1′ of either neighboring nucleotide is expected). Feigon et al. (1984) ingeniously overcame this limitation by using the 2D NOE experiment in the way that is was originally used to study chemical exchange (Jeener et al., 1979). By choosing conditions where both the B- and Z-conformations were present, and slowly interconverting, they were able to observe crosspeaks between the resonances from the same proton in the two different conformations. Once resonances in the B-conformer spectrum were assigned, resolved base resonances in the Z-form could be assigned and used to analyze the 2D NOE spectra of the pure Z-form. Often there is a question about the relation between the structure of a molecule in the crystal and in solution and in case of 5′d(m^{5}C−G−m^{5}C−G−m^{5}C−G)$_2$ Feigon et al. (1984) were able to test for possible differences. In the Z-form 2D NOE spectrum they observed a crosspeak between the methyl resonances of the cytidines at position 1 and 5 in the sequence 5′d(m^{5}C$_1$-G$_2$-m^5-C$_3$-G$_4$-m^{5}C$_5$-G$_6$) arising from an inter-strand interaction between a methyl group in one strand with the methyl group on the opposite strand. In the crystal the molecule is in the Z-form and the observed carbon-carbon distance was 4.6 Å, corresponding to a distance of closest approach of 3.0 Å between the attached protons (Fujii et al., 1982). Thus, they were able to demonstrate by 2D-NMR that many of the important structural features observed in the crystalline Z-form are also present in solution.

The Z-Form of d(br^{5}C-G-br^{5}C-G-A-T-br^{5}C-G-br^{5}C-G)$_2$

The alternation between *anti* and *syn*-conformation of the bases along each strand of the duplex is one unique characteristic of Z-DNA. Because purines adopt the *syn*-conformation more readily than pyrimidines (Davies, 1978; Saenger, 1984), this might account for the resistance of simple-sequence DNA with nonalternating purine-pyrimidine sequences to convert to the Z-form. There is, however, indirect biochemical evidence that naturally occurring DNA sequences can be driven into the Z-form (Rich et al., 1984) even when there is not perfect alternation of purines and pyrimidines and x-ray diffraction studies show that the oligomer d(br^{5}C-G-A-T-br^{5}C-G) adopts the Z-form (Wang et al., 1985). Again, using mixed solvent systems Feigon et al. (1985) showed by CD, UV, ^{31}P and by 2D NOE (Feigon et al., 1985) that 5'-d(br^{5}C-G-br^{5}C-G-A-T-br^{5}C-G-br^{5}C-G) adopts a Z-form with all of the guanines and the thymine base in the *syn*-conformation (indicated by very strong base-H1' crosspeaks). The cytidine and adenine residues all remained in an *anti*-conformation. These studies conclusively prove that strict pyrimidine-purine alternation is not required to form Z-DNA in solution and suggest that the barrier to the *syn*-conformation of the pyrimidines is not as unfavorable as previously thought. They also provide one more example where features detected in the x-ray diffraction studies of single crystals are seen in solution by 2D NMR.

Poly(dNH$_2$A-dT)

It has been suggested that the presence of the third amino hydrogen bond in the G·C base pair is one reason why poly(dG−dC)·poly(dG−dC) adopts a Z-conformation in high salt, but poly(dA−dT)·poly(dA−dT) does not (Gaffney et al., 1982; Howard & Miles, 1983). Following this line of reasoning one might expect poly(dNH$_2$A−dT) to form Z-DNA. Borah et al. (1985) have recently used 2D NMR to test this possibility. In low salt solution they observe a general 2D NOE crosspeak pattern that is very reminiscent of B-form poly(dA-dT) (strong peaks between the AH8 and TH6 protons and their respective 2'',2' sugar protons but not with the H1' protons, and strong crosspeaks between the TH6 and the AH8 and the TMe resonance). In high salt (4 M NaCl) solution the only major change in the 2D NOE spectrum was the appearance of a strong crosspeak between the H3' proton and base (AH8 and/or TH6) proton characteristic of a C3'-endo sugar conformation. (The AH8 and TH6 resonances overlap in the high salt form, so it was not possible to rigorously establish that both protons interact with H3'.) The authors conclude that the molecule has adopted an A-form in high salt (Borah et al., 1985a) rather than a Z-form. Whether or not the molecule can be converted to the Z-form using more drastic conditions remains to be seen.

Poly(rA)·poly(dT)

This is an interesting hybrid of RNA and DNA that has been studied by Gupta et al. (1985). Using 1D NOE techniques they conclude that the nucleotidyl units in *both* rA and dT strands have equivalent conformations (C2'-endo/C1'-exo, x≃240-260°) consistent with a right-handed B-DNA geometry. This solution-state structure is different from those proposed to account for the fiber form (Zimmerman & Pheiffer, 1981).

r(CGCGCG)·r(CGCGCG)

This molecule provides a test of the NMR method because most RNA molecules are expected to have an A-like conformation in solution (Bloomfield et al., 1974; Saenger, 1984). This was confirmed by comparing the observed 2D NOE peak patterns with those expected for A-RNA and A-, B- and Z-DNA conformations (Westerink et al., 1984). A key distinguishing characteristic of the A-RNA form was the strong inter-nucleotide NOE between the base (GH8 or CH6) resonance and the H2' resonance of the 5' neighboring residue and correspondingly weaker intra-nucleotide interaction between this same pair of resonances.

QUANTITATIVE DETERMINATION OF DNA STRUCTURES

Earlier 2D studies on DNA oligonucleotides have been concerned with assignment of spectra and qualitative structural studies but now the focus has turned toward quantitative determination of structures in solution. The first attempt to *quantitatively* determine the complete structure of DNA in solution was reported by Clore and Gronenborn (1985). From presteady-state NOE experiments they obtained cross relaxation rates for 150 interactions (~14 interactions per base pair) for d(AAGTGTGACAT)·d(ATGTCACACTT), and evaluated the corresponding interproton distances. When these experimental interproton distances were then used in those calculated for a B-DNA structure, the RMS difference was 0.62 Å (Clore & Gronenborn, 1985). However, when the B-DNA structure was relaxed using a restrained least-squares refinement program that makes use solely of distance and planarity restraints, the RMS value of the difference was reduced to 0.2 Å. In this refined structure, they found large variations in the values of certain conformational parameters. For example, the propeller twist of the bases ranged from low values of ~0° up to high values of 20°, with an average value of 10°, base roll varied from -15° to +15° and possibly larger, and base tilt varied from +14° to -6°. These values seem large, but similarly conformational heterogeneity was reported for the dodecamer d(CGCGAATTCGCG)$_2$ in the crystal (Dickerson & Drew, 1981). In more recent work (Nilsson et al., 1986), presteady state NOE values (Clore & Gronenborn, 1983) were combined with restrained molecular dynamics calculations to generate a structure for the hexamer 5'd(CGTACG)$_2$. Independent of the starting conformation (A- or B-DNA) the molecular dynamics calculations converged to the same B-type structure. However, when they only used restrained molecular mechanics calculations they did not obtain the same final structure starting from different initial conformations. This strongly suggests that structures obtained using just molecular mechanics or distance geometry algorithms could be misleading.

We (Behling et al., 1986) recently developed a quantitative model for the poly(dA)·poly(dT) structure by combining the structural constraints derived from our NMR measurements with molecular mechanics and annealed molecular dynamics based on the AMBER program (Weiner et al., 1984). Three different starting structures were used in the calculations: the B-form and A-form derived from x-ray diffraction measurements on DNA fibers (Arnott & Hukins, 1972) and the heteronomous structure proposed by Arnott et al. (1983). We find that starting with either the heteronomous model (Arnott et al., 1983) or the standard B-type helix (Arnott & Hukins, 1972) energy minimization with annealed molecular dynamics (both including NMR constraints), leads to the same structure (see Fig. 8). The same calculations starting with the A-form DNA resulted in a highly perturbed A-form structure (both A- and T-sugars are C2'-endo), with an unacceptably high energy and a poorer fit (factor of ~2) of the NMR data.

In considering the above results there are several points to be kept in mind. In addition to the intrinsic limitations of the current theoretical procedures we note that the NMR data are limited in that they mainly include *local* interactions between the base and sugar protons, and between sugar protons. While these interactions are effective in establishing sugar puckers and glycosidic bond angles, they are less effective in defining the positions of the bases relative to one another. For this reason, longer range structural features (eg. bending) may be more difficult to establish by NMR techniques. A second point is that there are large amplitude motions present in solution state DNA, and even in solid DNA (Shindo et al., 1980; Nall et al., 1981; Opella et al., 1981; Mai et al., 1983; Holbrook & Kim, 1984; Vold et al., 1986). It may, therefore, be unrealistic to talk about *the* structure of DNA, although it is reasonable to discuss the average structure.

CONCLUDING REMARKS

In this article we have described the application of 1D- and 2D-NMR proton relaxation techniques to studies of the structures of simple sequence DNAs in solu-

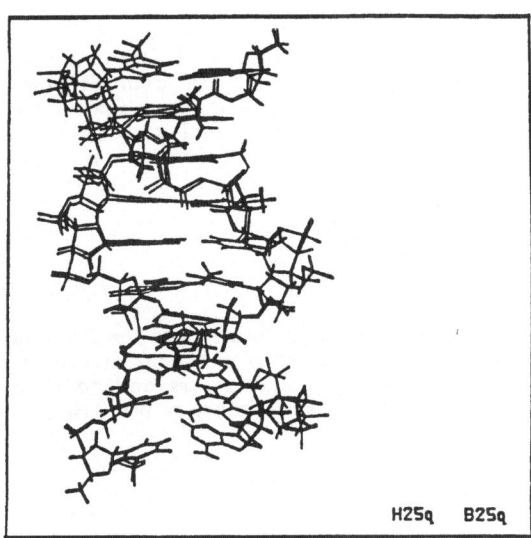

Fig. 8. A comparison of two structures obtained from carrying out a restrained annealed molecular dynamics calculation on $d(A)_{10}\cdot d(T)_{10}$ incorporating the constraints obtained from NMR relaxation measurements on poly(dA)·poly(dT). In this calculation the starting structure was either B-DNA or heteronomous DNA (Arnott et al., 1983) but both converged to the same final structure.

tion. Analysis of relaxation measurements on a number of different DNAs has revealed considerable internal motion in every part of the DNA, and this has two practical consequences. First it causes an uncertainty in distance determinations because of fluctuations in interproton distances. Second, it somewhat diminishes the importance of precise structural details of the DNA in protein recognition of a specific sequence. The 2D NMR techniques have been used to assign nearly all of the resonances in DNA duplexes containing as many as ~20 base pairs, and with improvements in resolution, sensitivity and technique, it should be possible to extend this to higher molecular weight DNA. With simple-sequence DNA, molecules even longer than 100 base pairs can be studied. The 1D and 2D relaxation methods have finally provided a much needed method for determining DNA structures in solution. Once spectra have been assigned, interproton distances can be semi-quantitatively or quantitatively determined from measurements of (i) selective and bi-selective spin-lattice relaxation rates, (ii) the time-dependence of presteady-state NOEs or (iii) the time-dependence of crosspeak intensities in 2D NOE experiments. Using these techniques, examples of all three major DNA conformations (A, B-, Z-form) have been found, and in some cases unusual conformations have been detected (*syn*-conformation for thymine in a Z-DNA). In other cases molecules that were proposed to have unusual structures were found to have normal B-type structures. A number of the short DNA duplexes have been found to have quite regular B-type conformations, although the structure determination carried out by Clore and Gronenborn (1985) on d(AAGTGTGACAT)·d(ATGTCACACTT) indicates that there is substantial conformational microheterogeneity in this molecule.

Complete quantitative analysis of 2D NOE spectra of DNA duplexes is now in progress in various laboratories, and the results of these studies may help answer some of the questions we posed initially. At this point it does not seem overly optimistic to suggest that a major bottleneck in the quantitative determination of DNA structures has been broken and that we shall soon have new insight into the effect of sequence on DNA conformation and the role it plays in DNA-protein interactions.

ACKNOWLEDGMENTS

The support of the American Cancer Society and the National Science Foundation is most gratefully acknowledged. Much of the work described here is due to the collective efforts of my students and collaborators and to them I am indebted. Finally I thank Shelley Hexom for expert assistance in preparing this manuscript.

REFERENCES

Abragam, A. (1978) in *The Principles of Nuclear Magnetism* (Marshall, W.C., & Wilkinson, D.H., Eds.), Ch. 8, Oxford University Press, Oxford, England.

Altona, C. (1975) in *Structure and Conformation of Nucleic Acids and Protein-Nucleic Acid Interactions* (Sundaralingam, M., & Rao, S.T.), p 613-629, University Park Press, Baltimore.

Altona, C., & Sundaralingam, M. (1973) J. Am. Chem. Soc. *95*, 2333-2344.

Altona, C., van Boom, J.H., & Haasnoot, C.A.G. (1976) Eur. J. Biochem. *71*, 557-562.

Anderson, W.F., Ohlendorf, D.H., Takeda, Y., & Matthews, B.W. (1981) Nature *290*, 754-755.

Arnott, S. (1970) Prog. Biophys. Mol. Biol. *21*, 267-319.

Arnott, S., & Hukins, D.W.L. (1972) Biochem. Biophys. Res. Commun. *47*, 1506-1509.

Arnott, S., & Hukins, D.W.L. (1973) J. Mol. Biol. *81*, 93-105.

Arnott, S., & Selsing, E. (1974) J. Mol. Biol. *88*, 509-521.

Arnott, S., Chandrasekaran, R., Hall, I.H., & Puigjaner, L.C. (1983) Nuc. Acids Res. *11*, 4141-4155.

Assa-Munt, N., & Kearns, D.R. (1984) Biochemistry *23*, 791-796.

Assa-Munt, N., Granot, J., & Kearns, D.R. (1981) presented at VII International Biophysics Congress, Aug. 23-28, Abstract Book p 96.

Assa-Munt, N., Granot, J., Behling, R.W., & Kearns, D.R. (1984) Biochemistry *23*, 944-955.

Behe, M., & Felsenfeld, G. (1981) Proc. Natl. Acad. Sci. USA *78*, 1619-1623.

Behling, R.W., & Kearns, D.R. (1985) Biopolymers *24*, 1157-1167.

Behling, R.W., & Kearns, D.R. (1986) Biochemistry *25*, 3335-3346.

Behling, R.W., Rao, S.N., Kollman, P., & Kearns, D.R. (1986) Biochemistry, submitted.

Bendel, P., Laub, O., & James, T.L. (1983) J. Am. Chem. Soc. *105*, 6748.

Benevides, J.M., Lemeur, D., & Thomas, G.J. (1984) Biopolymers *23*, 1011-1024.

Bloomfield, V., Crothers, D.M., & Tinoco, I. (1974) in *The Physical Chemistry of Nucleic Acids*, Harper & Row, New York.

Bolton, P.H., & James, T.L. (1979) J. Phys. Chem. *83*, 3359-3366.

Bolton, P.H., & James, T.L. (1980) Biochemistry *19*, 1388-1392.

Borah, B., Cohen, J.S., & Bax, A. (1985b) Biopolymers *24*, 747-765.

Borah, B., Cohen, J.S., Howard, F.B., & Miles, H.T. (1985a) Biochemistry *24*, 7456.

Bosch, C., Kumar, A., Baumann, R., Ernst, R.R., & Wuthrich, K. (1981) J. Magn. Reson. *42*, 159-163.

Bothner-By, A.A., & Noggle, J.H. (1979) J. Am. Chem. Soc. *101*, 5152-5155.

Brayer, G.D., & McPherson, A. (1983) J. Mol. Biol. *169*, 565-596.

Broido, M.S., & Kearns, D.R. (1982) J. Am. Chem. Soc. *104*, 5207-5216.

Chen, C-W., Cohen, J.S., & Behe, M. (1983) Biochemistry *22*, 2136-2142.

Cheng, D.M., Kan, L.S., Leutzinger, E.E., Jayaraman, K., Miller, P.S., & Ts'o, P.O.P. (1982) Biochemistry *21*, 621-630.

Clore, G.M., & Gronenborn, A.M. (1983) EMBO J. *2*, 2109-2113.

Clore, G.M., & Gronenborn, A.M. (1984) FEBS Lett. *172*, 219-225.

Clore, G.M., & Gronenborn, A.M. (1985) EMBO J. *138*, 447-472.

Clore, G.M., Lauble, H., Frenkiel, T.A., & Gronenborn, A.M. (1984) Eur. J. Biochem. *145*, 629-636.

Crawford, J., Kolpak, F., Wang, A., Quigley, G., van Boom, J., van der Marel, G., & Rich, A. (1980) Proc. Natl. Acad. Sci. USA *77*, 4016-4020.

Davies, D.B. (1978) in *Progress in NMR Spectroscopy, Vol. 12*, p 135-186, Pergamon Press Ltd., Great Britain.

Dickerson, R.E. (1983) in *Nucleic Acids: The Vectors of Life* (Pullman, B., & Jortner, J., Eds.), p 1-15, D. Reidel Pub. Co., Dordrecht, Holland.

Dickerson, R.E., & Drew, H.R. (1981) J. Mol. Biol. *149*, 761-786.

Dickerson, R.E., & Kopka, M.L. (1983) in *Nucleic Acids: The Vectors of Life* (Pullman, B., & Jortner, J., Eds.), D. Reidel Pub. Co., Dordrecht, The Netherlands.

Dickerson, R.E., Kopka, M.L., & Pjura, P. (1983) J. Biomol. Str. Dynam. *1*, 755-771.

Drew, H.R., & Dickerson, R.E. (1982) EMBO J. *1*, 663-667.

Drew, H.R., Wing, R.M., Takano, T., Broka, C., Tanaka, S., Itakura, K., & Dickerson, R.E. (1981) Proc. Natl. Acad. Sci. USA *78*, 2179-2183.

Early, T.A., Kearns, D.R., Hillen, W., & Wells, R.D. (1980) Nuc. Acids Res. *8*, 5795-5812.

Feigon, J., & Kearns, D.R. (1979) Nuc. Acids Res. *6*, 2327-2337.

Feigon, J., Denny, W.A., Leupin, W., & Kearns, D.R. (1983a) Biochemistry *22*, 5930-5942.

Feigon, J., Leupin, W., Denny, W.A., & Kearns, D.R. (1983b) Biochemistry *22*, 5943-5951.

Feigon, J., Wang, A.H.J., van der Marel, G.A., van Boom, J.H., & Rich, A. (1984) Nuc. Acids Res. *12*, 1243-1263.

Feigon, J., Wang, A.H.J., van der Marel, G.A., van Boom, J.H., & Rich, A. (1985) Nuc. Acids Res., in press.

Feigon, J., Wright, J.M., Denny, W.A., Leupin, W., & Kearns, D.R. (1983c) Cold Spring Harbor Symp. on Quant. Biol. *67*, 207-217.

Feigon, J., Wright, J.M., Leupin, W., Denny, W.A., & Kearns, D.R. (1982) J. Am. Chem. Soc. *104*, 5540-5541.

Frederick, C.A., Grable, J., Melia, M., Samudzi, C., Jen-Jacobson, L., Wang, B.C., Greene, P., Boyer, H.W., & Rosenberg, J.M. (1984) Nature *309*, 327-331.

Freeman, R., & Morris, G.A. (1979) Bull. Magn. Reson. *1*, 5.

Fujii, S., Wang, A.H.J., van der Marel, G., van Boom, J.H., & Rich, A. (1982) Nuc. Acids Res. *10*, 7879-7892.

Gaffney, B.L., Marky, L.A., & Jones, R.A. (1982) Nuc. Acids Res. *10*, 4351-4361.

Grant, R., Kodama, M., & Wells, R. (1972) Biochemistry *11*, 805-815.

Gupta, G., Sarma, M.H., & Sarma, R.H. (1985) J. Mol. Biol. *186*, 463-469.

Gupta, G., Sarma, M.H., Dhingra, M.M., Sarma, R.H., Rajagopalan, M., & Sasisekharan, V. (1983) J. Biomol. Str. Dynam. *1*, 395-416.

Haasnoot, C.A.G., de Leeuw, F.A.A.M., de Leeuw, H.P.M., & Altona, C. (1981) Org. Magn. Reson. *15*, 43.

Hare, D.R., Wemmer, D.E., Chou, S.H., Drobny, G., & Reid, B.R. (1983) J. Mol. Biol. *171*, 319-336.

Hogan, M.E., & Jardetzky, O. (1979) Proc. Natl. Acad. Sci. USA *76*, 6341-6345.

Hogan, M.E., & Jardetzky, O. (1980a) Biochemistry *19*, 2079-2085.

Hogan, M.E., & Jardetzky, O. (1980b) Biochemistry *19*, 3460-3468.

Holbrook, S.R., & Kim, S.-H. (1984) J. Mol. Biol. *173*, 361-388.

Hoogsteen, K. (1963) Acta Crystallogr. *16*, 907-916.

Howard, F.B., & Miles, H.T. (1983) in *Nucleic Acids: The Vectors of Life* (Pullman, B., & Jortner, J., Eds.), p. 511-520, D. Reidel Pub. Co., Dordrecht, Holland.

James, T.L. (1984) in *Phosphorus-31 NMR, Principles and Applications* Ch. 12, p 349-400, Academic Press, Inc.

Jeener, J., Meier, B.H., Bachman, P., & Ernst, R.R. (1979) J. Chem. Phys. *71*, 4546-4553.

Kallenbach, N.R., Mandal, C., & Englander, S.W. (1981) in *Biomolecular Stereodynamics* (Sarma, R., Ed.), p 233, Adenine Press, New York.

41

Kan, L.S., Cheng, D.M., Jayaraman, K., Leutzinger, E.E., Miller, P.S., & Ts'o, P.O.P. (1982) Biochemistry 21, 6723-6732.

Kearns, D.R. (1977) Ann. Rev. Biophys. Bioeng. 6, 477-523.

Kearns, D.R., Mirau, P.A., Assa-Munt, N., & Behling, R.W. (1983) in *Nucleic Acids: The Vectors of Life* (Pullman, B., & Jortner, J., Eds.), p 113-125, D. Reidel Pub. Co., Dordrecht, Holland.

Klug, A., Jack, A., Viswamitra, M.A., Kennard, O., Shakked, Z., & Steitz, T.A. (1979) J. Mol. Biol. 131, 669-680.

Kmiec, E.B., & Holloman, W.K. (1984) Cell 36, 593-598.

Lerner, D.B., Becktel, W.J., Everett, R., Goodman, M., & Kearns, D.R. (1984) Biopolymers 23, 2157-2172.

Mai, M.T., Wemmer, D.E., & Jardetzky, O. (1983) J. Am. Chem. Soc. 105, 7149-7152.

McClarin, J.A., Frederick, C.A., Wang, B.C., Greene, P., Boyer, H.W., Grable, J., & Rosenberg, J.M. (1986) Science, in press.

McKay, D.B., & Steitz, T.A. (1981) Nature 290, 744-749.

Mirau, P.A., & Kearns, D.R. (1983) in *Structure and Dynamics: Nucleic Acids and Proteins* (Clementi, E., & Sarma, R.H., Eds.), p 227-239, Adenine Press, New York.

Mirau, P.A., & Kearns, D.R. (1984) Biochemistry 23, 5439-5446.

Mirau, P.A., Behling, R.W., & Kearns, D.R. (1985) Biochemistry 24, 6200-6211.

Mitsui, Y., Langridge, R., Shortle, B., Cantor, C., Grant, R., Kodama, M., & Wells, R.D. (1970) Nature 228, 1166-1169.

Nagayama, K., Kumar, A., Wuthrich, K., & Ernst, R.R. (1980) J. Magn. Reson. 40, 321-334.

Nagayama, K., Wuthrich, K., & Ernst, R.R. (1979) Biochem. Biophys. Res. Commun. 90, 305.

Nall, B.T., Rotwell, W.P., Waugh, J.S., & Rupprecht, A. (1981) Biochemistry 20, 1881-1887.

Nilsson, L., Clore, G.M., Gronenborn, A.M., Brunger, A.T., & Karplus, M. (1986) J. Mol. Biol. 188, 455-475.

Nordheim, A., Lafer, E.M., Peck, L.J., Wang, J.C., Stollar, D., & Rich, A. (1982) Cell 31, 309-318.

Opella, S.J., Wise, W.B., & DiVerdi, J.A. (1981) Biochemistry 20, 284-290.

Patel, D.J. (1978) J. Polymer Sci. 62, 117-141.

Patel, D.J., Canuel, L., & Pohl, F. (1979) Proc. Natl. Acad. Sci. USA 76, 2508-2511.

Patel, D.J., Kozlowski, S.A., Nordheim, A., & Rich, A. (1982) Proc. Natl. Acad. Sci. USA 79, 1413-1417.

Pohl, F.M., & Jovin, T.M. (1972) J. Mol. Biol. 67, 375-396.

Ramaswamy, N., Bansal, M., Gupta, G., & Sasisekharan, V. (1982) Proc. Natl. Acad. Sci. USA 79, 6109-6113.

Rich, A., Nordheim, A., & Wang, H.-J. (1984) Ann. Rev. Biochem. 53, 791-846.

Rill, R.L., Hilliard, P.R., Bailey, J.T., & Levy, G.C. (1980) J. Am. Chem. Soc. 102, 418.

Rill, R.L., Hilliard, P.R., Levy, L.F., & Levy, G.C. (1981) in *Biomolecular Stereodynamics Vol. I* (Sarma, R.H., Ed.), p 383, Adenine Press, New York.

Saenger, W. (1984) in *Principles of Nucleic Acid Structure* (Cantor, C.R., Ed.), Springer-Verlag, New York.

Scheek, R.M., Boelens, R., Russo, N., van Boom, J.H., & Kaptein, R. (1984) Biochemistry 23, 1371-1376.

Seeman, N.C., Rosenberg, J.M., & Rich, A. (1976) Proc. Natl. Acad. Sci. USA 73, 804-808.

Shindo, H. (1980) Biopolymers 19, 509-522.

Shindo, H., Wooten, J.B., Pheiffer, B.H., & Zimmerman, S.B. (1980) Biochemistry 19, 518-526.

Sober, H.A., & Harte, R.A. (1970) in *CRC Handbook of Biochemistry*, p H-18, H-19, The Chemical Rubber Co., Cleveland, Ohio.

Steitz, T.A., Weber, I.T., & Matthew, J.B. (1983) Cold Spring Harbor Symp. Quant. Biol. 67, 419-426.

Stirdivant, S.M., Klysik, J., & Wells, R.D. (1982) J. Biol. Chem. 257, 10159-10165.

Sutherland, J.C, & Griffin, K.P (1983) Biopolymers 22, 1445-1448.

Tanaka, I., Appelt, K., Dijk, J., White, S.W., & Wilson, K.S. (1984) Nature *310*, 376-381.

Viswamitra, M.A., Kennard, O., Jones, P.G., Sheldrick, G.M., Salisbury, S., Falvello, L., & Shakked, Z. (1978) Nature *273*, 687-688.

Viswamitra, M.A., Shakked, Z., Jones, P.G., Sheldrick, G.M., Salisbury, S.A., & Kennard, O. (1982) Biopolymers *21*, 513.

Vold, R.R., Brandes, R., Tsang, P., Kearns, D.R., Vold, R.L., & Rupprecht, A. (1986) J. Am. Chem. Soc. *108*, 302-303.

Vorlickova, M., Kypr, J., & Sklenar, V. (1983) J. Mol. Biol. *166*, 85-92.

Vorlickova, M., Kypr, J., Stokrova, S., & Sponar, J. (1982) Nuc. Acids Res. *10*, 1071.

Wagner, G., & Wuthrich, K. (1979) J. Magn. Reson. *33*, 675-680.

Wang, A.H.J., Gessner, R., van der Mrel, G.A., van Boom, J.H., & Rich, A. (1985) Proc. Natl. Acad. Sci. USA *82*, 3611.

Wang, A.H.J., Quigley, G., Kolpak, F., van der Marel, G., van Boom, J., & Rich, A. (1981) Science *211*, 171-176.

Wang, A.H.J., Quigley, G.J., Kolpak, F.J., Crawford, J.L., van Boom, J.H., van der Marel, G., & Rich, A. (1979) Nature *282*, 680-686.

Wartell, R.M., & Harrell, J.T. (1986) Biochemistry, in press.

Weiner, S.J., Kollman, P.A., Case, D.A., Singh, U.C., Ghio, C., Alagona, G., Profeta, S., & Weiner, P. (1984) J. Am. Chem. Soc. *106*, 765-784.

Weiss, M.A., Patel, D.J., Sauer, R.T., & Karplus, M. (1984) Proc. Natl. Acad. Sci. USA *81*, 130-134.

Wells, R.D., Blakesley, R.W., Hardies, S.C., Horn, G.T., Larson, J.E., Selsing, E., Burd, J.F., Chan, H.W., Dodgson, J.B., Jensen, F.F., Nes, I.F., & Wartell, R.M. (1977) in *CRC Critical Reviews in Biochemistry*, p 305.

Wells, R.D., Goodman, T.C., Hillen, W., Horn, G.T., Klein, R.D., Larson, J.E., Muller, U.R., Neuendorf, S.K., Panayotatos, N., & Stirdivant, S.M. (1980) Prog. Nuc. Acid Res. Mol. Biol. *24*, 167-267.

Wemmer, D.E., Chou, S.H., Hare, D.R., & Reid, B.R. (1984) Biochemistry *23*, 2262-2268.

Westerink, H.P., van der Marel, G.A., van Boom, J.H., & Haasnoot, C.A.G. (1984) Nuc. Acids Res. *12*, 4323-4338.

Wing, R., Drew, H., Takano, T., Broka, C., Tanaka, S., Itakura, K., & Dickerson, R.E. (1980) Nature *287*, 755-758.

Zarling, D.A., Arndt-Jovin, D.J., Robert-Nicoud, M., McIntosh, L.P., Thomae, R., & Jovin, T.M. (1984) J. Mol. Biol. *176* 369-415.

Zimmerman, S.B., & Pheiffer, B.H. (1981) Proc. Natl. Acad. Sci. USA *78*, 78-82.

THE MAJOR AND MINOR GROOVES OF THE DNA HELIX

AS CONDUITS FOR INFORMATION TRANSFER

Richard E. Dickerson, Mary L. Kopka and Philip E. Pjura

Molecular Biology Institute; University of California at

Los Angeles; Los Angeles, CA 90024

The statement that DNA is an informational macromolecule is one that has moved from being an insight, to a commonplace, to dogma, and like so many dogmas has ended as a cliché. But this picture has been given more concrete form in the last six years, with the advent of x-ray crystal structure determinations of synthetic DNA oligomers of defined sequence, both alone and complexed with control proteins such as repressors and with antitumor drugs.[1] It is an attractive idea to a structurally-minded molecular biologist to ask precisely how the information inherent in DNA gets out, in a way that allows it to be used by the host organism.

Intrinsic and Extrinsic Information Readout

DNA actually contains two levels of information. The first and more familiar level is that embodied by the genetic code, by which successions of base triplet codons contain information for successions of amino acids along the polypeptide chain of a protein. The machinery for readout of this genetic information is complex and highly evolved, with many enzymes and intermediate molecules. It is represented schematically in Figure 1. For purposes of the present discussion, one should note that the final product, a polypeptide chain, need bear no intrinsic resemblance to the triplets of DNA bases that code for the chain. The "translation", or the association of one particular codon triple with one special amino acid, is performed by the amino acid-tRNA synthetase enzyme. It is difficult to see how such a complex mechanism could have evolved from simpler precursor methods of readout of the genetic information. But it is equally certain that so complex a mechanism must have been the product of a long evolutionary history. The information contained in this triplet genetic message is used primarily for coding--i.e., for storing information as to the amino acid sequence of the derived polypeptide chains.

There is a far simpler readout of a different type of information in the DNA, that is the subject of this discussion. Many control proteins such as repressors bind specifically to certain DNA base sequences, and control the expression of contiguous genes. Polymerases and other enzymes bind at consensus sites having a particular pattern of bases. Some drug molecules, likewise, show preferential binding at certain sequences. In such cases the readout of information from DNA is simple and immediate: the control protein or drug must exhibit a complementarity to the particular base sequence that it recognizes, either in shape, charge, polarity, or pattern

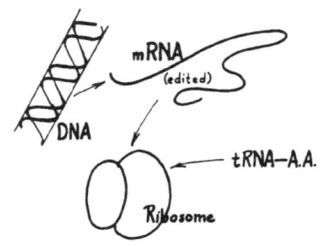

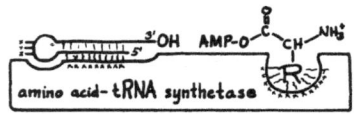

Figure 1. Extrinsic information readout from DNA. The ultimate product, a polymer of amino acids, need have no intrinsic similarity in chemical or physical properties with the succession of base triplets that encode the polymer sequence.

Table I. Modes of Readout of Base Sequence Information from DNA

1. Extrinsic Information Readout

    Multi-step process: DNA, RNA, ribosomes and enzymes.
    No intrinsic resemblance, in chemical or physical properties,
        between amino acids and their triplet codons.
    Critical translation step occurs at matching of tRNA to amino
        acid on the surface of the aminoacyl-tRNA sythetase.
    Elaborate, highly evolved mechanism.
    Base sequence information used for CODING.

2. Intrinsic Information Readout

    Single-step process: Binding directly to DNA.
    Protein or other recognition molecule must have an intrinsic com-
        plementarity to the DNA, in shape, orientation, charge, or
        hydrogen-bonding pattern.
    Critical translation step occurs at binding of molecule to DNA.
    Simple, direct mechanism, possibly primitive.
    Base sequence information used for CONTROL.

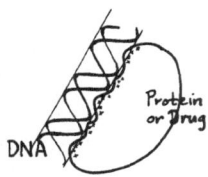

Figure 2. Intrinsic information readout from DNA. The recognition molecule must have an intrinsic complementarity to the particular DNA sequence being recognized and "read", in order to bind preferentially to just that sequence.

of hydrogen bonding (Figure 2). The simplicity of this intrinsic information readout channel suggests that it may well be primitive, and may possibly have been an evolutionary precursor to the more complex extrinsic readout mechanism of the genetic code. The information in this intrinsic readout process is used mainly for control.

Where is information stored in DNA? Neither the phosphate backbone nor the connecting deoxyribose sugars contains information; this resides instead in the order of base pairs along the helix. In extrinsic information readout of the genetic code, the base pairs come apart, and the sequence is "read" by making complementary Watson-Crick pairings with the freed bases. As far as is known, no such unpairing occurs during intrinsic information readout, and base sequence can be sensed only by probes that fit within the major and minor grooves, sensing the chemical features of the edges of Watson-Crick base pairs (Figure 3). These features include the placement of hydrogen bonding donors and acceptors,[2] possibly their spatial orientation,[1,3] and the physical bulk of thymine methyl group and guanine N2 amine.[4]

The conduits for intrinsic information readout from DNA hence are the major and minor grooves of the double helix. It will become significant later that the major groove has twice the information content of the minor groove. A.T base pairs can be differentiated from G.C base pairs via either groove. But reversals of one and the same base pair--A.T vs. T.A, or G.C vs. C.G--are a more subtle problem. As Figure 4 shows, base pair reversal produces distinctive changes in the pattern of hydrogen bond donors and acceptors in the major groove, with the methyl of thymine as an added steric "flag". But base pair reversal leaves the hydrogen bonding pattern in the minor groove unchanged with respect both to type and to location. A control molecule that sensed hydrogen bonding patterns in the minor groove could not discriminate between A.T and T.A base pairs, or G.C and C.G. Hence the major groove has a four-symbol code, whereas the minor groove has only a two-symbol code, and only half the information content.

Intrinsic Information Readout by Proteins

It is interesting that in every known crystal structure analysis of a repressor or other sequence-specific DNA-binding protein, the structure of the protein by itself, when combined with all the available chemical evidence, has suggested that the protein interacts mainly with the major groove

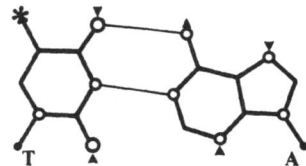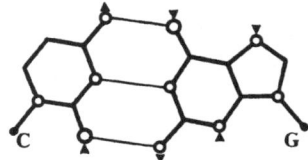

Figure 3. T.A and C.G base pairs, showing potential hydrogen bond donors (arrowhead pointing away from base pair) and acceptors (arrowhead pointing toward base pair). Black dots are C1' atoms of deoxyribose. Small open circles are nitrogens, and large open circles are oxygens. * = methyl group of thymine. The upper edge of each base pair is the major groove edge, and the lower is the minor groove edge.

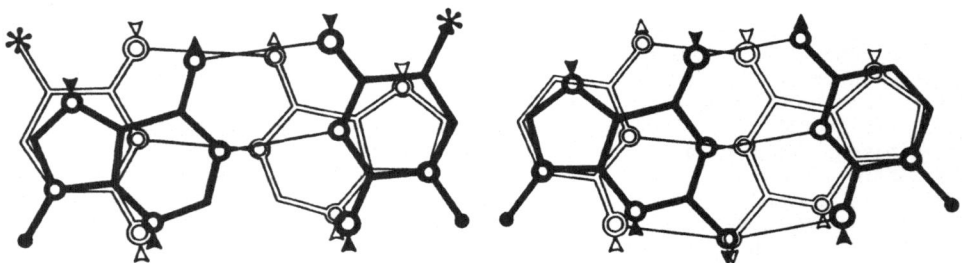

Figure 4. Reversals of base pairs, showing similarities or differences in hydrogen bonding positions. Left: T.A (open bonds) over A.T (solid bonds). Right: C.G (open) over G.C (solid). Note that hydrogen bonding positions nearly overlap in the minor groove, meaning that outside molecules cannot recognize base pair reversals.

of DNA: cro repressor,[5] CAP,[6] lambda repressor headpiece,[7] and trp repressor.[8] Similarly, in the only two structures yet solved of complexes of such proteins with their specific DNA--phage 434 repressor at low resolution[9] and EcoRI restriction enzyme at high resolution[10, 11]--the primary association is seen to involve the major groove of the double helix and not the minor. This is illustrated in Figure 5 for the EcoRI restriction enzyme complexed with a dodecamer B-DNA double helix having initial 5' unpaired thymine: T-C-G-C-G-A-A-T-T-C-G-C-G. The low resolution drawing at left shows that the central EcoRI recognition sequence, G-A-A-T-T-C, has its major groove turned to face the enzyme, and that the two enzyme subunits wrap "arms" around the helix, following the major groove around to the other side of the DNA cylinder. The minor groove is almost totally ignored.

The details of interaction between EcoRI protein side chains and the recognition sequence, diagrammed at right in Figure 5, show a mechanism of recognition that is nearly identical to that proposed by Seeman, Rosenberg and Rich in 1976[2] and outlined earlier in this paper. Hydrogen bond donating arginines, and hydrogen bond accepting

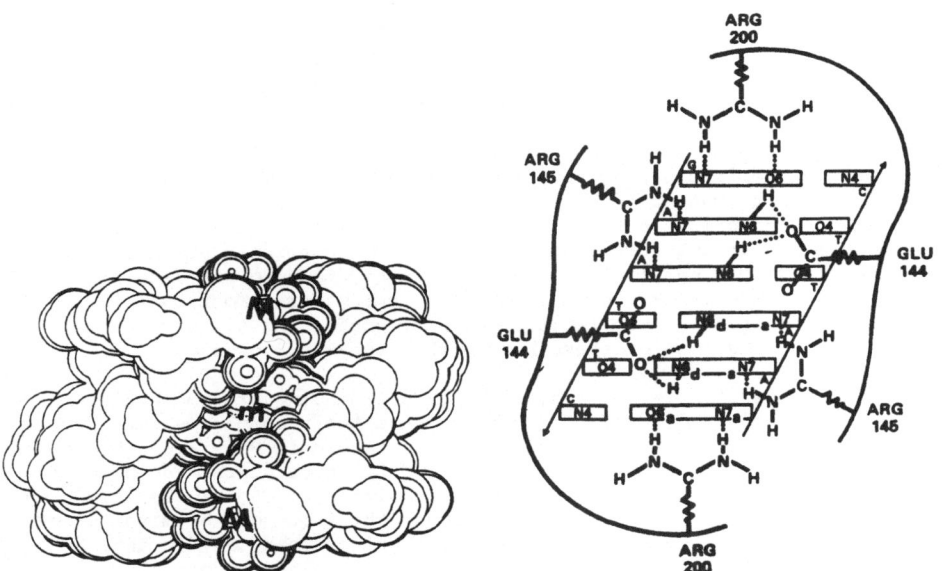

Figure 5. Recognition of the G-A-A-T-T-C site in DNA by the
EcoRI restriction endonuclease, as determined by Rosenberg and
collaborators.[10, 11] Left: The two subunits of the enzyme
(light lines) wrapped around a DNA double helix (dark lines).
Interaction is almost entirely within the major groove (M),
with the minor groove (m) effectively ignored. Right: Details
of hydrogen bonding between EcoRI subunit side chains and base
edge atoms. Black curved lines connect amino acids on the same
enzyme subunit. Notice that in this particular example, the
amino acid side chains ignore the pyrimidines and read only the
edges of the purines. Note also that any change in base sequence
would eliminate one or more of the hydrogen bonds now formed.

glutamic acids, are arranged so that they make complementary interactions
with acceptors and donors on the minor groove edges of the six central
base pairs. Any substitution of one base for another within the recognition sequence, G-A-A-T-T-C, would upset the pattern of hydrogen bonding
and weaken the interaction between DNA and enzyme. It may only be a curi-
osity of this particular EcoRI example that the enzyme completely ignores
pyridines, and focuses entirely on purines.

     It is understandable that a large protein molecule should interact
specifically with DNA only, or primarily, in the major groove. In order
to read and recognize a particular base sequence, the amino acid side chains
must be oriented or pointed toward their hydrogen bonding donors or acceptors on the DNA. This requires a rigid scaffolding to which the side chains
can be attached; any flexibility in the support framework could only lessen
the fidelity of the reading process. A rigid scaffolding, in globular
proteins, usually means an alpha helix. Indeed, in every DNA-reading protein yet solved, one or more alpha helices are either observed or proposed
to be inserted within the major groove. An extended chain, run through the
major groove, would not be able to orient its side groups in a sufficiently
determined way to discriminate between different base sequences. By this
criterion the minor groove of DNA is a poor channel of communicating for
proteins: it is too narrow to admit an alpha helix, and the isolated extended polypeptide chain that could be threaded through the minor groove
cannot orient its side chains sufficiently strongly to be discriminating.

In addition, as was mentioned in connection with Figure 4, the minor groove has only half the information content of the major groove.

Intrinsic Information Readout by Drug Molecules

Antitumor drugs, and other small organic molecules that bind to DNA in a sequence-specific manner, present quite a different picture. Some drugs that bind to DNA do so by intercalating a flat aromatic ring between two adjacent base pairs in the helix stack; others slip into the narrow minor groove. Even intercalating drugs that have nonplanar chains, such as daunomycin,[12] actinomycin,[13] and triostin A[14] tend to bury their non-planar parts in the minor groove rather than the major. Most DNA-binding drugs that exhibit base specificity, do so by preferring A.T base pair regions to G.C, and most of these bind within the minor groove.[15]

Two examples of A.T-favoring, minor groove-binding drug molecules that have been examined crystallographically in their complexes with double-helical B-DNA are Hoechst 33258 and netropsin, shown in Figure 6. Netropsin requires a site consisting of four successive A.T base pairs, and Hoechst has a similar but somewhat less stringent requirement. Both molecules have a crescent shape that curves naturally around the floor of the minor groove, and both consist of planar amide or aromatic groups that can fit into an opening only 3.5 Å wide. Hence they can slip into a narrow minor groove that would be forbidden to the bulkier alpha helix of a protein. Furthermore, on their concave edge (the side that would face the floor of the groove), each molecule has potential hydrogen bond donors or acceptors, and the overall positive charge on each drug molecule facilitates binding to the anionic DNA double helix. In retrospect the two drug molecules in Figure 6 seem admirably engineered to bind within the minor groove of B-DNA.

Single-crystal x-ray structure analyses have been carried out of the complex of netropsin with C-G-C-G-A-A-T-T-brC-G-C-G,[16-18] and of Hoechst 33258 with C-G-C-G-A-A-T-T-C-G-C-G.[18, 19] Figure 7 shows that Hoechst

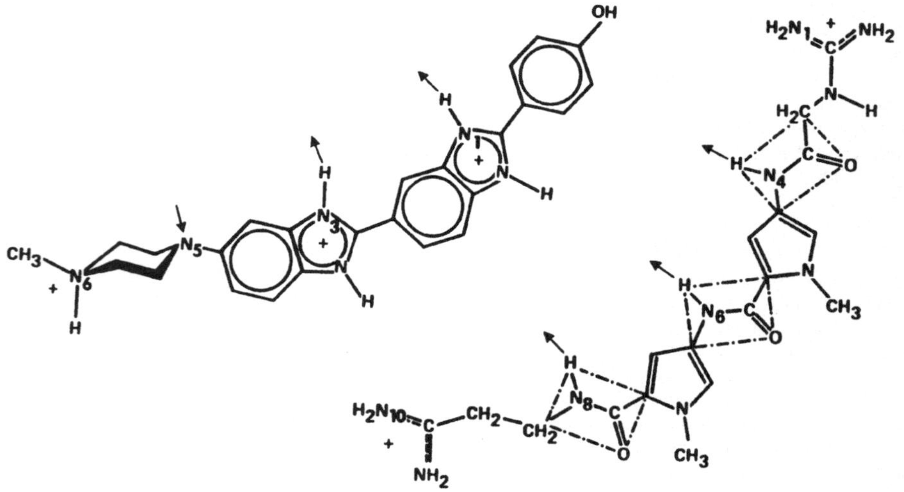

Figure 6. Two base sequence-specific drugs that bind within the minor groove of double-helical B-DNA: Hoechst 33258 at left, and netropsin at right. Arrows indicate hydrogen bonding donors and acceptors on the concave edge of each crescent-shaped molecule.

does indeed bind within the minor groove. When it does so, it displaces
a zig-zag string of water molecules or spine of hydration that runs down
the A-A-T-T center of the drug-free DNA helix.[20, 21]

This minor groove spine of hydration is one of two hydration structures
that help to stabilize the B form of the double helix. It has been known
for many years that the B helix is the high-humidity form, and that certain
sequences can be dehydrated in the absence of salts and converted into the
A helix (or for certain special sequences, the left-handed Z helix).
Saenger and coll.[22] have proposed that conversion to the A or Z helices
under conditions of low water activity occurs because these helix forms
have a greater economy of hydration: the shorter phosphate-phosphate
separations along each strand of helix in the A and Z forms permit bridging of adjacent phosphates by water molecules, whereas the greater inter-
phosphate separation in the B form requires that each phosphate be sepa-
rately hydrated. But this, while undoubtedly a contributing factor, cannot
be the whole story. This drying interconversion from B to A is known experimentally to be easier for sequences with high G.C content, and can be
difficult or even impossible for high-A.T polymers. Moreover, the B-to-A
interconversion is stopped if small cations are present. Neither the sequence dependence nor the ion dependence of the B-to-A transition can be
accounted for by phosphate hydration alone.

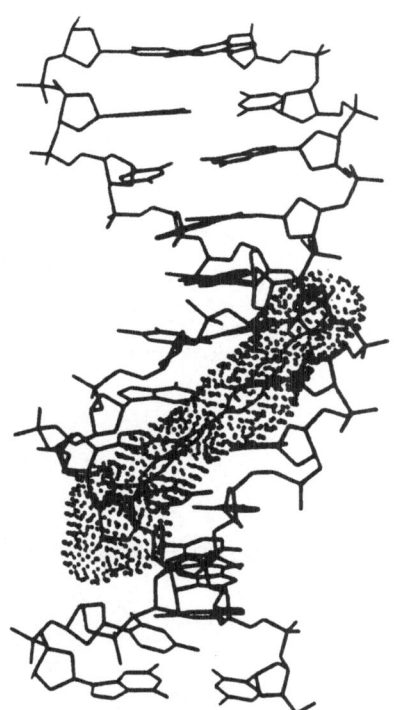

Figure 7. Skeletal drawing of the B-DNA double helix of sequence:
C-G-C-G-A-A-T-T-C-G-C-G, with a van der Waals dotted surface
enclosing the drug molecule Hoechst 33258. The drug fits within
the narrow A-A-T-T region of the minor groove, with its phenol
at the upper right, and its piperazine ring extending down at
lower left into the inferently wider C-G-C-G portion of the groove.

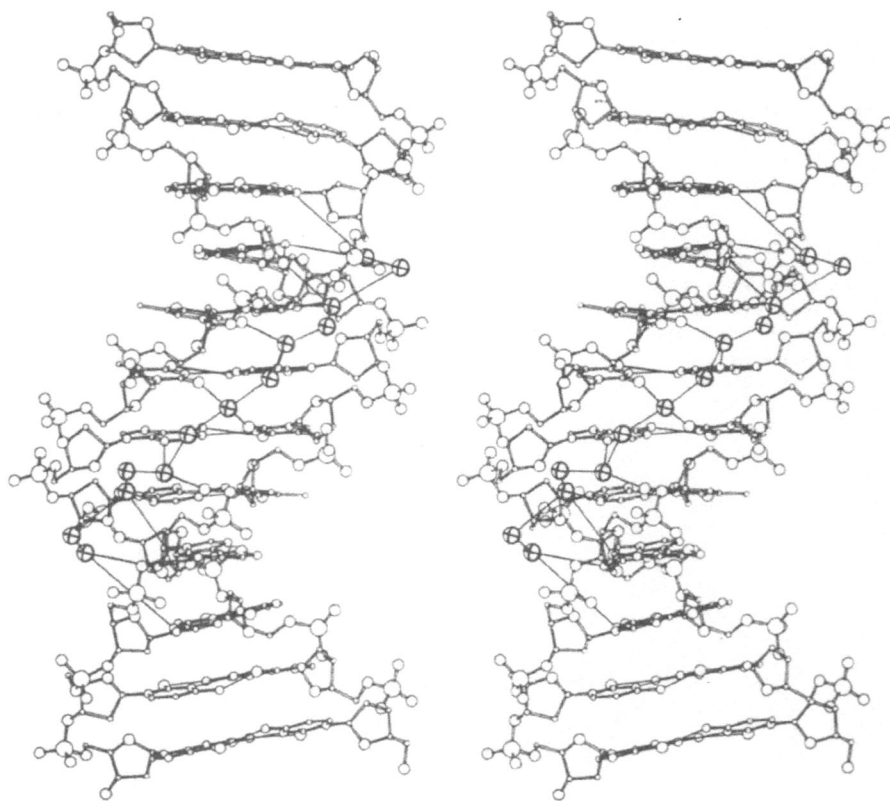

Figure 8. The minor groove spine of hydration in B-DNA. Above: Location of water molecules (crossed spheres) down the narrow A.T central region of the minor groove as observed in C-G-C-G-A-A-T-T-C-G-C-G. Below: Schematic showing how water molecules (W) bridge adenine N3 and thymine O2 atoms. Strand 1 of the helix is labeled C1 to G12 from top to bottom, and strand 2 is labeled C13 to G24 from bottom to top.

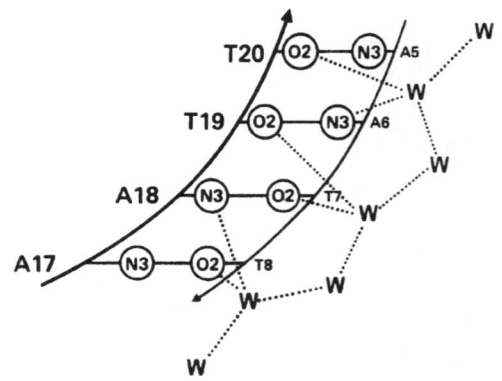

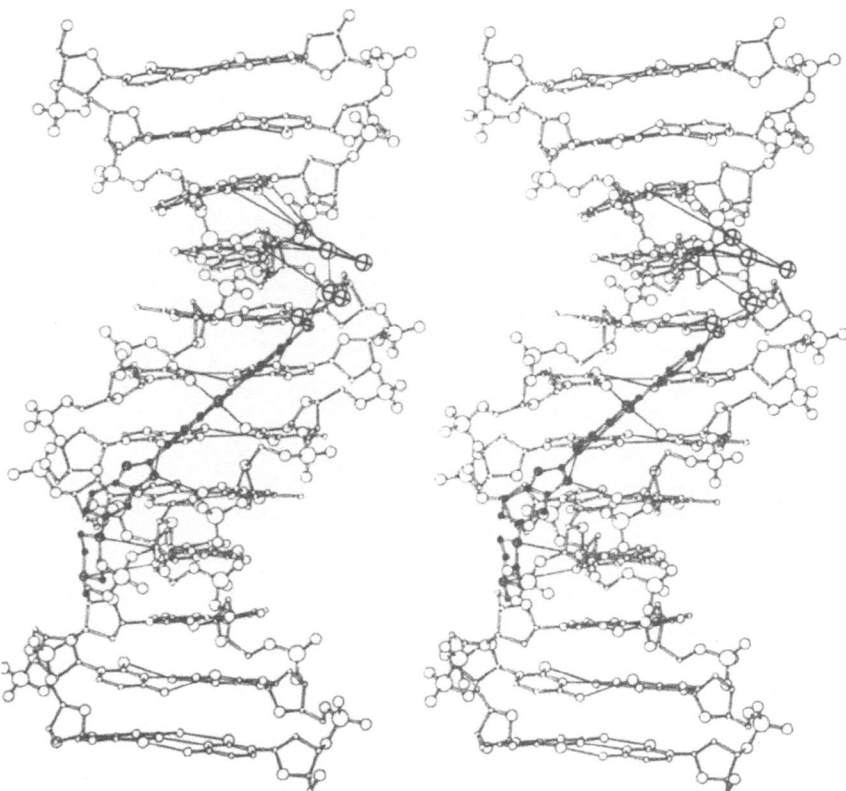

Figure 9. Hoechst 33258 displaces part of the spine of hydration and binds in its place down the A-T-T-C region of the
minor groove. It makes the same hydrogen-bonded bridges between
adenine N3 and thymine O2 as did the spine of hydration. Rem-
nants of the spine (crossed spheres) can be seen at upper right
in the stereo drawing above.

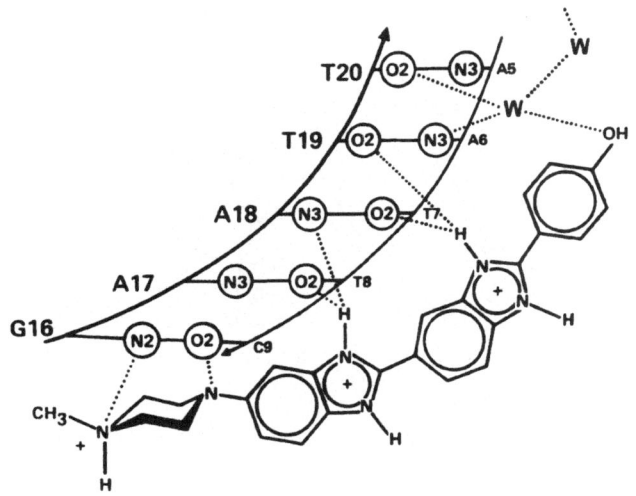

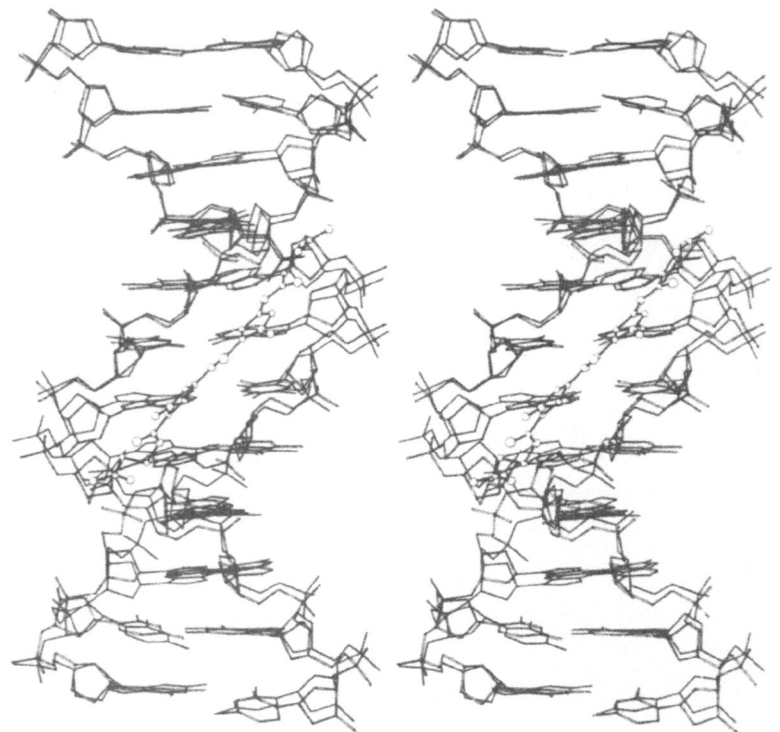

Figure 10. Netropsin binds squarely within the A-A-T-T center
of the dodecamer: C-G-C-G-A-A-T-T-C-G-C-G, almost entirely
displacing the spine of hydration, but again making the same
hydrogen-bonded bridges between N3 and O2. The cationic ends
also assist binding, as is true for Hoechst as well. The stereo
drawing above shows the DNA structure without netropsin (open
bonds) and with it (filled bonds).

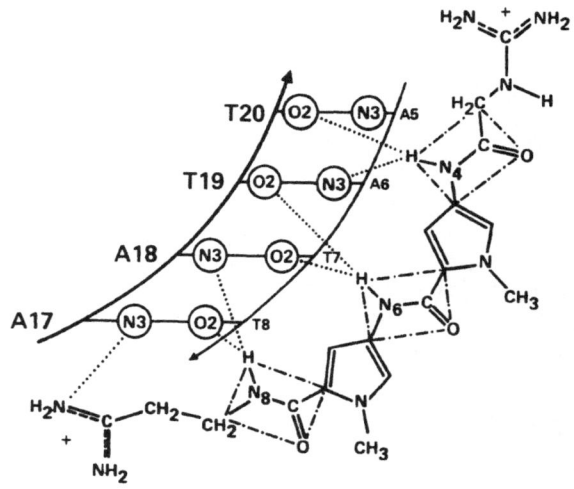

The spine of hydration provides an explanation for both observations. The spine (Figure 8) is observed in the crystal structure of C-G-C-G-A-A-T-T-C-G-C-G to be well developed in the central A-A-T-T region, but to break up gradually, the farther one goes into the C-G-C-G ends.[20, 21] This seems to be the consequence of steric hindrance between water molecules of the spine and the guanine N2 amine groups. In support of this, and of the idea that the spine stabilizes the B helix relative to conversion of the A form, are the findings of Leslie et al.[23] that fibers drawn from polymers containing only G.C base pairs can always be converted from the B to the A form by drying, whereas polymers containing only A.T pairs resist the conversion down to the point where excessive drying leads to complete disorder. Moreover, I.C base pairs, which are identical to G.C pairs except for deletion of the N2 amine, behave identically to A.T base pairs, resisting the B-to-A helix transition upon drying. Hence the very same groups that disrupt the spine of hydration also destabilize the B helix and facilitate conversion to the A form.

Underscoring the importance of the minor groove in stabilizing the B helix is the observation of Bartenev et al[24] that certain cations can displace the water molecules of the spine of hydration under high salt conditions, and that this spine of cations stabilizes the B form even more strongly than did the water molecules. Furthermore, these observations are limited only to polymers made up of A.T base pairs, not G.C. Hence it seems abundantly clear that, although the economy of hydration of the phosphate backbone does contribute to relative stabilization of the B vs. A and Z helices in a non-sequence-dependent manner, the spine of hydration must also be invoked to account for the important sequence-dependent hydration behavior of DNA.

Binding of Hoechst 33258 and Netropsin

Drug molecules such as netropsin and Hoechst 33258 can be regarded as more extreme, covalently-bonded analogues of the spine of hydration or the Bartenev cation string. Netropsin sits squarely in the middle of the A-A-T-T region of C-G-C-G-A-A-T-T-C-G-C-G (Figure 10), leaving very little of the spine left, but Hoechst slips down one base pair to bind to A-T-T-C, with a residue of the spine remaining intact at the upper end of the groove (Figure 9). The hydrogen bond patterns are represented schematically at the bottom of Figures 8-10. Bridging between adenine N3 and/or thymine O2 atoms, which in the spine of hydration was accomplished by two hydrogen bonds from a single water molecule, is accomplished in the two drug molecules by bifurcated or three-center hydrogen bonds donated by N-H. As both Jeffrey [25] and Kennard[26] have pointed out, such three-center hydrogen bonds are common in biological molecules when a donor is encountered in a donor-poor and acceptor-rich environment, which is the case on the floor of the minor groove in an A.T-rich region.

Hydrogen bonds are not the only source of stabilization of the DNA/drug complex. Electrostatic attraction between the cationic drug and anionic DNA also contributes. Lavery and Pullman[27] have shown that the potential minima over the surface of a B-DNA double helix lie in the bottom of the major and minor grooves, ideal for burying a linear polycation. A third source of binding stability is the hydrophobic interactions consequent upon burying the nonpolar rings of the drug molecule deep within the cleft of the minor helix (Figures 11-14). This will include an entropic factor arising from the displacement of water molecules from the surface of the drug and from within the minor groove as the two molecules come together. Breslauer and coworkers[28] have measured the thermodynamics of binding of netropsin to various DNA polymers and oligomers with the following results:

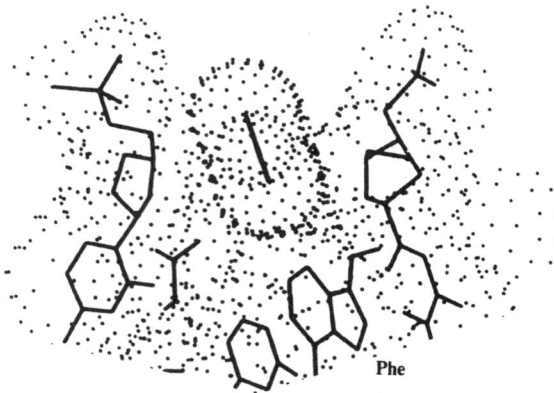

Figure 11. Cross section through the minor groove and the phenol ring of the DNA/Hoechst complex. The opening of the minor groove, at top, is flanked by two phosphates, with deoxyribose rings below, and with obliquely sectioned base pairs at bottom. The phenol ring is not coplanar with the groove walls because resonance delocalization keeps it coplanar instead with the first benzimidazole.

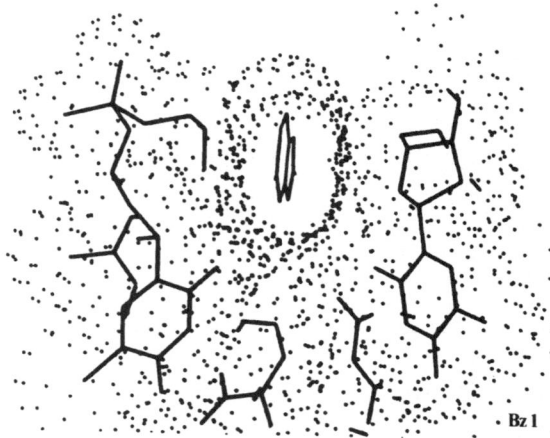

Figure 12. Cross section through the minor groove and the first benzimidazole ring of the DNA/Hoechst complex. The benzimidazole ring is fitted snugly into the narrow A-A-T-T minor groove, parallel to the groove walls.

56

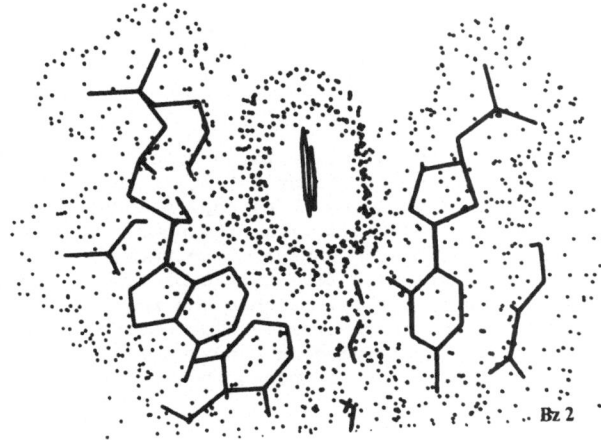

Figure 13. Cross section through the minor groove and the second
benzimidazole ring of the DNA/Hoechst complex. The view in all
four of these van der Waals packing drawings is from the back of
the Hoechst molecule (the piperazine end) toward the front.

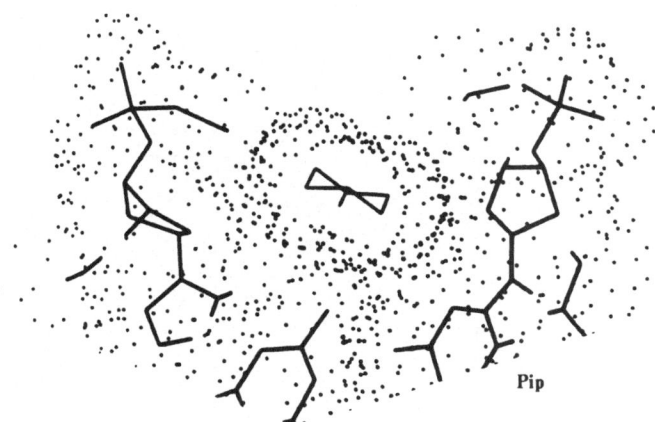

Figure 14. Cross section through the minor groove and the piperazine ring of the DNA/Hoechst complex. Steric hindrance about
the bond to Bz2 wants to keep the best mean planes through Bz2
and Pip roughly at right angles, but this would locate Pip crosswise across the minor groove. A compromise is reached, as shown
here. Note how much wider the groove is in this G.C region than
in the A.T regions in Figures 11-13.

	$\Delta G°$ (kcal/mol)	$\Delta H°$ (kcal/mol)	$-T\Delta S°$ (kcal/mol)	$\Delta S°$ (cal/deg mol)
poly(dA)·poly(dT)	-12.1	- 2.2	-10.0	+33.6
(G-C-G-A-A-T-T-C-G-C)$_2$	-11.5	- 9.3	- 2.2	+ 7.4
poly(dAdT)·poly(dAdT)	-12.7	-11.2	- 1.5	+ 5.1

In each case there is an entropic contribution to netropsin binding of at
least 5 cal/deg mol. Moreover, from the standpoint of thermodynamics of
binding, the G-A-A-T-T-C binding site looks to netropsin more like an alternating A-T polymer than a poly(dA)·poly(dT) homopolymer. This may be
a clue that the homopolymer itself has an unusual structure, perhaps one
that is converted into a more classical B form by the binding of netropsin.
A "super-hydrated" form is one possibility, because the release of even
more water molecules would provide the abnormally high entropy drive, and
the energy required to break all these hydrogen bonds would make the enthalpy of binding much less negative, as is indeed observed with the homopolymer.

Figures 12 and 13 show just how tightly the two benzimidazole rings
fit within the minor groove. It was observed in the structure of B-helical
C-G-C-G-A-A-T-T-brC-G-C-G[29] that the minor groove was inherently narrower
in AT regions than in GC, a feature that may be attributable in part to the
greater propeller twist in AT regions. This difference in groove width is
diagrammed in Figure 15. As a consequence, AT regions are intrinsically

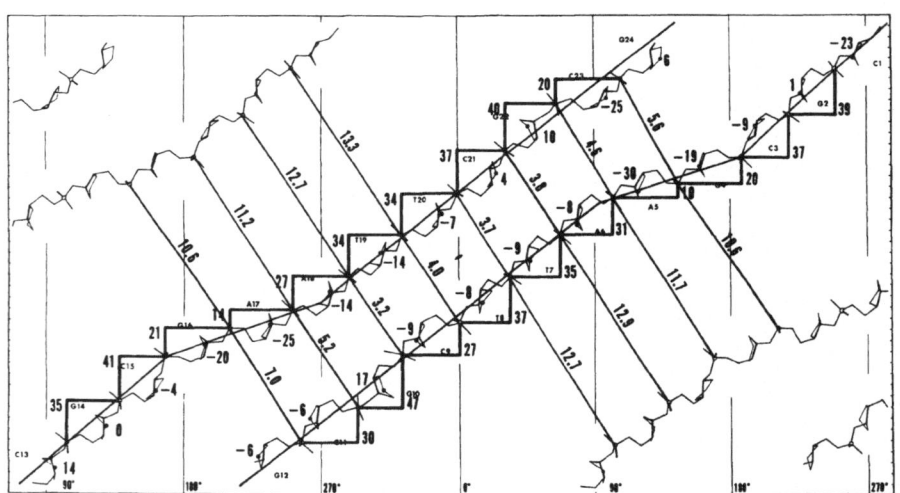

Figure 15. Unrolled-helix diagram of the backbone chains of
the B-DNA helix C-G-C-G-A-A-T-T-brC-G-C-G.[29] The two chains
run from upper right to lower left, with the minor groove cross-
ing through the center of the diagram. Diagonal lines from
upper left to lower right connect phosphate atoms across the
minor and major grooves. Numerical labels on these lines give
P-P distances, less 5.8 Å to represent two phosphate group radii.
These net distances hence measure the width of the opening available to an outside molecule. For explanation of other features
of this diagram, see reference 29.

better binding sites for flat organic ring molecules because they allow a snug fit, and are particularly inhospitable to bulky structures such as alpha helices of proteins.

Each benzimidazole ring tries to sit parallel to the walls of the minor groove in its own region, and as a consequence, the two rings are twisted by about 36° about the bond that connects them. For the same reason, the normal vectors to the two pyrrole rings of netropsin in its DNA complex make an angle of 33° relative to one another. But the phenol and piperazine rings at the two ends of Hoechst 33258 are not coplanar with the walls of the groove. Phenol (Figure 11) is coplanar instead with the first benzimidazole ring, Bz1. This probably means that phenol and Bz1 comprise one large delocalized or conjugated electronic system, and that the extra stabilization produced because of this is sufficient to compensate for the strain energy of keeping the phenol nonparallel to the walls.

At the other end of the molecule, steric hindrance about the bond connecting Bz2 to piperazine would make the most stable conformation in isolation one in which the best planes through the piperazine and Bz2 were at right angles to one another. But this would require that Pip lie transverse to the minor groove in the DNA complex, and not even the wider GC region where Pip binds could accommodate that situation. Hence a compromise is reached, in which the piperazine ring is twisted 30° away from a crosswise orientation relative to the groove walls, as seen in Figure 14. This crosswise arrangement is barely possible in the wide GC region of the groove; it would be totally out of the question in a narrow AT region. Hence, by its very shape, piperazine in this context is a GC-reading element.

Reading of Sequence Information by the Bound Drug

So far we have said nothing about the way in which a groove-binding drug such as netropsin or Hoechst recognizes a particular base sequence, other than to mention the NMR study that showed most groove-binding drugs to be AT-specific. It turns out that these drugs in the minor groove do not read the hydrogen-bonding pattern as repressors do in the major groove. Rather, they simply sense the presence or absence of the N2 amine group of guanine. This group is virtually the only factor that differentiates a G.C base pair from an A.T in the minor groove (Figure 4). A careful study of the stereo drawings in Figures 9 and 10 reveals that the parts of the drug molecules nearest the C2 atoms of adenines are not the hydrogen-bonding N-H of pyrrole or imidazole. Close contact with adenine C2-H is provided by two -CH_2- methylenes and two pyrrole rings in netropsin, and by the six-membered rings of Phe, Bz1 and Bz2 in Hoechst 33258. Figure 16 shows how these close contacts prohibit the substitution of guanine for adenine. There is no room for an amine on the C2 position of the base, and introducing one at that point would push the drug away from the bottom of the groove. The hydrogen bonds shown in Figures 9 and 10 are necessary to help hold the drug down, and to position it in the correct reading frame. But the actual reading is done by touch rather than by hydrogen bonding.

Netropsin is admirably engineered to require a binding site that consists of four adjacent A.T base pairs, without regard to individual base pair reversals. With Hoechst, in contrast, the last base pair must be G.C rather than A.T, because of the bulk of the piperazine ring. (Compare groove widths in Figure 14 vs. Figures 11-13.) Added stability is provided by interactions between the two piperazine nitrogens and base edge atoms on the floor of the groove, including the guanine N2 amine.

Hence the pyrroles and methylenes of netropsin and the phenol and benzimidazoles of Hoechst are AT-reading structural elements, while piper-

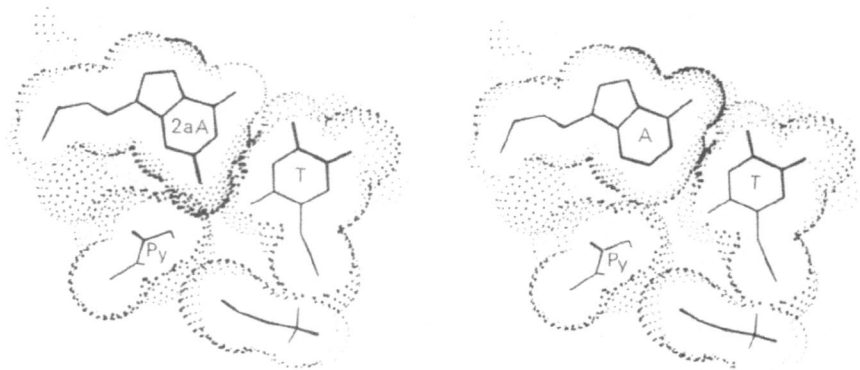

Figure 16. Left: Van der Waals packing diagram of one A.T base pair and the adjacent first pyrrole ring, showing the touching contact between pyrrole C-H and adenine C2-H. Right: Addition of a guanine-like N2 amine group, to form 2-aminoadenine, would create a steric clash with pyrrole that is symbolized here by overlapping van der Waals surfaces. If guanine is present instead of adenine, then the pyrrole cannot bind at this locus.

azine is a weak GC-reading element. Is it possible to design drug analogues with more discriminating GC-readers, so that an organic polymer could be synthesized having any desired base sequence as its required binding site? One line of thought is outlined in Figure 17. If one pyrrole were to be replaced by an imidazole or furan, then room would be generated to fit the guanine amine, which would also be capable of forming a new hydrogen bond with the ring N or O atom. This is the "lexitropsin", or sequence-reading netropsin analogue, approach that was proposed simultaneously by our group and that of J. William Lown of Alberta. But would such a substitution site in a longer netropsin analogue actually <u>require</u> a GC base pair, or only permit it? To say that netropsin requires an A.T base pair at a given position is really to say that it forbids a G.C. pair--the exclusion arising from steric clash. But in what sense would the imidazole site in the lexitropsin in Figure 17 <u>exclude</u> an A.T. base pair? Binding studies and crystal structure analyses will have to be carried out between synthetic lexitropsins and short lengths of specially tailored DNA, but

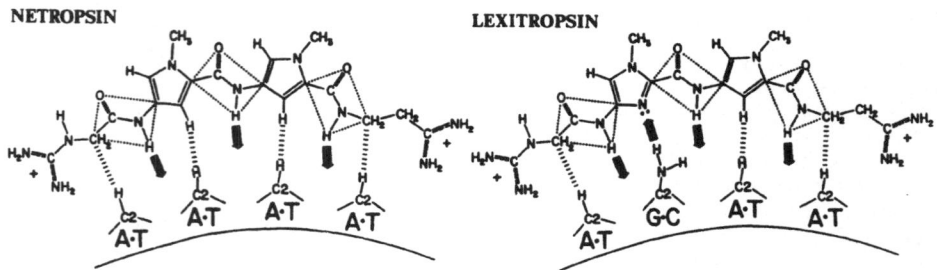

Figure 17. Left: Schematic of base sequence recognition and bonding in netropsin. Right: Possible alteration of netropsin to allow a G.C. base pair at one position. Black arrows are hydrogen bonds, and dashed lines mark close nonbonded packing contacts.

it may be that the kind of replacement of pyrrole in Figure 17 simply creates a nondiscriminating site rather than a G.C-specific site.

The piperazine of Hoechst suggests another possibility for G.C-sensing: a bulky group that can fit snugly in a G.C region of the minor groove, but is too large for an A.T region. But there may be a good explanation for the fact that nearly all groove-binding drugs are AT-specific: it is very hard to fashion a GC-requiring site in a drug molecule.

Summary

The two channels for intrinsic information readout from the DNA double helix are the major and minor grooves. The major groove has twice the information content of the minor, in the sense that hydrogen-bonding probes from repressors and other control proteins can discriminate among four possibilities: A.T, T.A, G.C, and C.G. In the minor groove, near-identity of hydrogen bonding groups upon base pair reversal means that A.T pairs can be differentiated from G.C pairs, but that the orientation of the pair cannot be sensed.

Control proteins use almost exclusively the major groove for base sequence reading, inserting one or more alpha helices with amino acid side chains that can sense the pattern of hydrogen bond donors and acceptors on the floor of the major groove. Drug molecules, in contrast, use the minor groove, even though it contains less information. Flat aromatic rings can slip into the narrow minor groove, whereas the alpha helices of proteins cannot. Drug molecules sense the presence of A.T base pairs by the absence of an N2 amine on the bottom of the minor groove. It is easy to design an A.T-requiring locus on a drug molecule, but more difficult to design a site that requires a G.C base pair. Indeed, virtually all groove-binding drugs that exhibit sequence specificity do so by demanding A.T base pairs.

There is an interesting and probably valid evolutionary reason why drugs such as netropsin utilize the minor groove of DNA rather than the major. Streptomyces netropsis and similar microorganisms synthesize and excrete toxic compounds as a means of dealing with competitors for living space and resources. The fact that H. sapiens has also found them to be useful pharmacologically is interesting but irrelevant to their original intended application. The competitors, themselves, use the major groove of their DNA for control purposes, with restriction enzymes, repressors, polymerases, and related proteins binding to the major groove. It seems only rational (if such terminology can be applied to S. netropsis) that killer molecules should be synthesized that attack the victim's DNA through channels that are not competed for by the victim's own proteins. The minor groove is an unused "back door" giving access to the DNA of the organism with which the secretor of the antibiotic is in competition.

Acknowledgements

This research was performed with the financial assistance of American Cancer Society Grant No. NP-504, and NIH Grant GM-31299. We would like to thank David Goodsell for advice and assistance in preparing many of the van der Waals packing surface diagrams, which were photographed from a computer graphic screen on slide film, and then printed in black-white reversal by using the slide as a negative.

REFERENCES

1. Dickerson, R.E., Sci. Amer. (December), 94-111 (1983).
2. Seeman, N.C., Rosenberg, J.M., and Rich, A., Proc. Natl. Acad. Sci. USA 73, 804-808 (1976).

3. Dickerson, R.E., _J. Mol. Biol._ 166, 419-441 (1983).
4. Kopka, M.L., Yoon, C., Goodsell, D., Pjura, P. and Dickerson, R.E., _Proc. Natl. Acad. Sci. USA_ 82, 1376-1380 (1985).
5. Ohlendorf, D.H., Anderson, W.F., Fisher, R.G., Takeda, Y. and Matthews, B.W., _Nature_ 298, 718-723 (1982).
6. McKay, D.M., Weber, I.T. and Steitz, T.A., _J. Biol. Chem._ 257, 9518-9524 (1982).
7. Pabo, C.O. and Lewis, M., _Nature_ 298, 443-447 (1982).
8. Schevitz, R.W., Otwinowski, Z., Joachimiak, A., Lawson, C.L. and Sigler, P.B., _Nature_ 317, 782-786 (1985).
9. Anderson, J.E., Ptashne, M. and Harrison, S.C., _Nature_ 316, 596-601 (1985).
10. Frederick, C.A., Grable, J., Melia, M., Samudzi, C., Jen-Jacobson, L., Wang, B.C., Greene, P., Boyer, H.W. and Rosenberg, J.M., _Nature_ 309, 327-333 (1984).
11. McClarin, J.A., Frederick, C.A., Grable, J., Samudzi, C.T., Wang, B.-C., Greene, P., Boyer, H.W. and Rosenberg, J.M., _Proc. Fourth Conversation on Biomolecular Stereodynamics_, 4-8 June 1985, SUNY Albany, Adenine Press, Albany (1986).
12. Quigley, G.J., Wang, A.H.-J., Ughetto, G., van der Marel, G., van Boom, J.H. and Rich, A., _Proc. Natl. Acad. Sci. USA_ 77, 7204-7208 (1980).
13. Sobell, H.M. and Jain, S.C., _J. Mol. Biol._ 68, 21-34 (1972).
14. Wang, A.H.-J., Ughetto, G., Quigley, G.J., Hakoshima, T., van der Marel, G.A., van Boom, H.J. and Rich, A., _Science_ 225, 1115-1127 (1984).
15. Feigon, J., Denny, W.A., Leupin, W. and Kearns, D.R., _J. Medicinal Chem._ 27, 450-465 (1984).
16. Kopka, M.L., Yoon, C., Goodsell, D., Pjura, P. and Dickerson, R.E., _Proc. Natl. Acad. Sci. USA_ 82, 1376-1380 (1985).
17. Kopka, M.L., Yoon, C., Goodsell, D., Pjura, P. and Dickerson, R.E., _J. Mol. Biol._ 183, 553-563 (1985).
18. Kopka, M.L., Pjura, P.E., Goodsell, D.S. and Dickerson, R.E., _In:_ "Nucleic Acids and Molecular Biology", D. Lilley and F. Eckstein, Editors, Vol. 1, Springer-Verlag, Berlin (1986).
19. Pjura, P.E., Grzeskowiak, K. and Dickerson, R.E., submitted (1986).
20. Drew, H.R. and Dickerson, R.E., _J. Mol. Biol._ 151, 535-556 (1981).
21. Kopka, M.L., Fratini, A.V., Drew, H.R. and Dickerson, R.E., _J. Mol. Biol._ 163, 129-146 (1983).
22. Saenger, W., Hunter, W.N., and Kennard, O., _Nature_, submitted (1986).
23. Leslie, A.G.W., Arnott, S., Chandrasekaran, R. and Ratliff, R.L., _J. Mol. Biol._ 143, 49-72 (1980).
24. Bartenev, V.N., Golovamov, Eu.I., Kapitonova, K.A., Mokulskii, M.A., Volkova, L.I. and Skuratovskii, Y.Ya., _J. Mol. Biol._ 169, 217-234 (1983).
25. Jeffrey, G.A. and Maluszynska, H., _Int. J. Biol. Macromol._ 4, 173-185 (1982).
26. Taylor R., Kennard, O., and Versichel, W., _J. Am. Chem. Soc._ 106, 244-248 (1984).
27. Lavery, R. and Pullman, B., _Int. J. Quant. Chem._ 20, 259-272 (1981).
28. Marky, L.A., Curry, J., Breslauer, K.J., _in_ "Molecular Basis of Cancer, Part B: Macromolecular Recognition, Chemotherapy, and Immunology", 155-173 (1985).
29. Fratini, A.V., Kopka, M.L., Drew, H.R., and Dickerson, R.E., _J. Biol. Chem._ 257, 14686-14707 (1982).

UNUSUAL DNA STRUCTURES AND GENE REGULATION

Robert D. Wells

University of Alabama at Birmingham
Department of Biochemistry
Schools of Medicine and Dentistry
Birmingham, Alabama 35294

ABSTRACT

Left-handed Z-DNA, cruciforms, purine•pyrimidine (pur•pyr) sequences, and bent DNA have been studied with supercoiled recombinant plasmids, restriction fragments, and DNA polymers. Different types of sequences of cellular and viral origin as well as synthetic polymers and plasmid inserts have been investigated.

Left-handed Z-DNA is stabilized by negative supercoiling for a variety of suitable sequences. Strictly alternating purine-pyrimidine sequences are neither necessary nor sufficient for Z-DNA formation. The junctions between right-handed B and left-handed Z helices contain few, if any, nonpaired bases. Consecutive AT pairs can adopt left-handed Z conformations.

DNA cruciforms are stabilized by negative supercoiling at inverted repeat sequences. The influence of salts, temperature, and stem length indicates a sensitive interrelationship between these factors for cruciform stability.

Segments of recombinant plasmids containing pur•pyr sequences adopt unusual helical conformations under the influence of negative supercoiling. This feature is recognized and specifically cleaved by S1 nuclease. The major late promoter of adenovirus exhibits this property at the TATA box. A variety of other studies led to the conclusion that the B-helix deformation was not due to a previously recognized DNA conformation. Furthermore, the direct repeat sequences at the cleavage-packaging signal of herpes simplex virus type 1 adopts a novel DNA structure under the influence of negative supercoiling. The properties of this direct repeat have been previously unrecognized with any other system. This DNA (anisomorphic DNA) has different conformations in the complementary strands, which give rise to an aberrant DNA structure in the center of the direct repeat sequences.

Bent DNA occurs in fragments of kinetoplast DNA from <u>Crithidia</u> as well as other trypanosomes. Most fragments contain numerous runs of 4 to 6 A's which are spaced with a periodicity of 1 per helical repeat.

The chemical and biological properties of these unusual DNA conformations are described.

INTRODUCTION

DNA is a polymorphic molecule and can adopt a variety of conformations depending mainly on base sequence and environmental conditions. Recent interest in this polymorphism has been prompted to a large part by the discovery of Z-DNA, a structure possessing a left-handed helix sense as opposed to the canonical right-handed state (B-DNA) (Fig. 1). Several findings imply a biological function for Z-DNA or B-to-Z transitions (reviewed in (7,8)). Other types of unusual structures shown in Fig. 1 are cruciforms which occur at inverted repeat sequences, slipped structures which have been postulated to occur at direct repeats, bent DNA, thermolabile regions, and long range structural and thermodynamic effects (telestability). Furthermore, a new DNA conformation named anisomorphic DNA has been discovered in the Herpes Simplex Virus genome (discussed below).

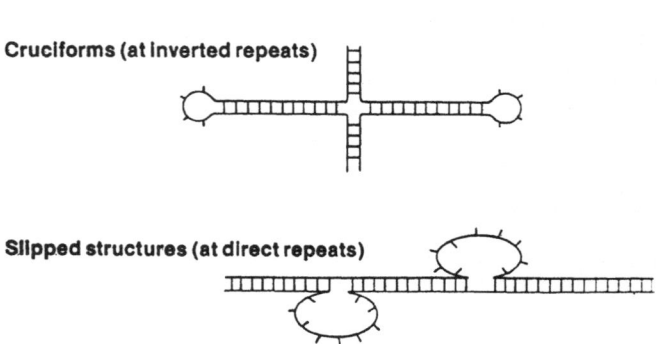

Left handed Z-DNA (therefore B-Z junctions)

Cruciforms (at inverted repeats)

Slipped structures (at direct repeats)

Fig. 1A.   Unusual Secondary Structures of DNA.
Schematic representations of non-B DNA structures
reported in the literature (See Text).

The structures of a segment of DNA in a right-handed B or in a left-handed Z helix are shown in Fig. 2. The structural features have been reviewed in detail previously (2-8). However, it might be pointed out that virtually all features of the two structures differ, including the sugar conformations, the helical sense, the orientation of some of the glycosidic bonds, the phosphodiester bond torsion angles, etc.

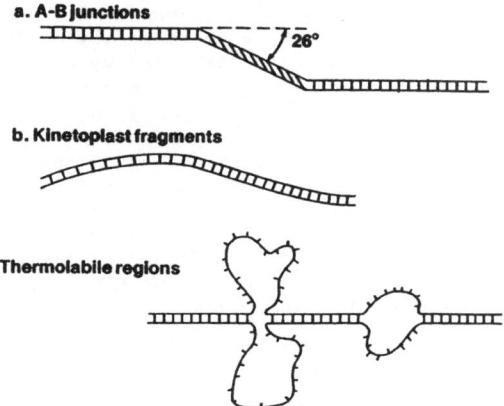

Bent DNA

a. A-B junctions

26°

b. Kinetoplast fragments

Thermolabile regions

**Long range structural
and thermodynamic effects
(telestability)**

Fig. 1B. Unusual Secondary Structures of DNA.
Schematic representations of non-B DNA structures
reported in the literature (See Text).

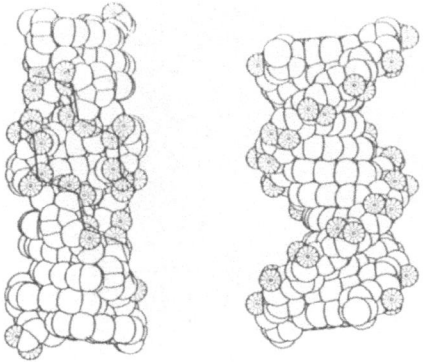

Fig. 2. Computer drawn models of left-handed Z-DNA (left
side) and right-handed B-DNA (right side). We are
grateful to Dr. Stephen C. Harvey (this department)
for these figures. Reprinted with permission from
(70).

A possible model for the junction between a right-handed B and a left-handed Z helix is shown in Fig. 3. Several investigators (reviewed in 2-8) have calculated that the transition can be accomplished within several base pairs (less than 5) and may be effected without unstacking or unpairing of bases.

Left-handed DNA can exist at physiological salt conditions under the influence of supercoiling (9,10 and reviewed in 4,7,8). The discovery that negative supercoiling induces the formation of left-handed Z-DNA was pivotal to the studies described in this review, namely the capacity of different enzymes to recognize and utilize left-handed Z-DNA. The vast majority of studies described herein require the use of negative supercoiling to generate the left-handed helices. Furthermore, although the original studies on Z-DNA used alternating (dC-dG) sequences, it has become increasingly clear that perfectly alternating purine-pyrimidine stretches are not a requirement for B-Z transitions (11,12,27-29). Sequences with the potential to adopt left-handed structures are widespread in nature (13,14) and Z-DNA has been detected in chromosomes from a variety of organisms (reviewed in (4,7)). Also, antibodies to Z-DNA have been found in sera from patients with autoimmune diseases (18) and from autoimmune MRL/1 mice (19). There are a number of ways to detect left-handed DNA, including immunological and chemical methods (1-8).

This section will review the known interactions of Z-DNA and other conformations with enzymes and enzymatic probes which are available as determinants of structure. A general overview of nuclease sensitivity is significant at this point since these types of probes have been extremely useful for the detection of cruciforms, left-handed Z-DNA, oligopurine·oligopyrimidine regions, and slipped structures.

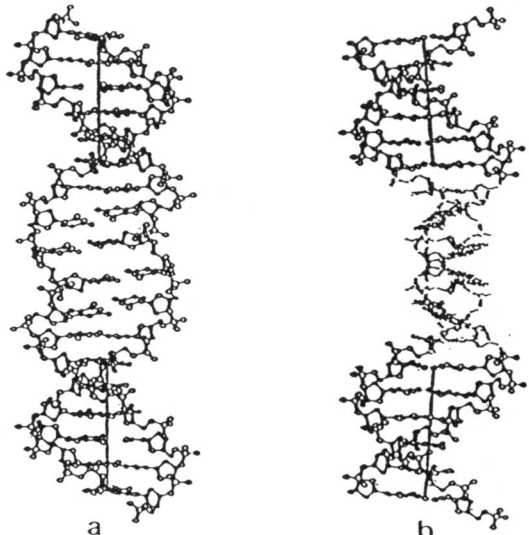

a   b

Fig. 3. Models of structural interface between right-handed B and left-handed Z helices. Panels A and B are two mutually perpendicular projections of a minimal patch of Z-DNA embedded in a B-DNA helix. The two central basepairs represent the Z patch. The basepairs immediately above and below are in a transition zone. All other basepairs are regularly helical B segments. This model was published previously; we appreciate the help of Dr. Struther Arnott in providing this figure. Republished with permission from (70).

NUCLEASES AS STRUCTURAL PROBES FOR Z-DNA

Sl Nuclease from Aspergillus oryzae

Sl nuclease recognizes and cleaves structural perturbations at the junctions between contiguous left- and right-handed DNA. This has been demonstrated for junctions involving left-handed $(dC-dG)_n$ blocks (9,13), $(dC-dA)_n$ $(dT-dG)$ blocks (14) and mixed sequences. Sl nuclease is therefore widely used to probe Z-DNA stretches embedded within right-handed helical regions. The cleavage sites for various junctions map within 3 to 20 base pairs outside of the first alternating residue. More detailed mapping (15, M.J. McLean, personal communication) shows Sl susceptibility over a wider range, with some cleavage sites extending into the left-handed region. Comparison of the extent of cleavage at several junctions of $(dC-dG)_n$ or $(dC-dA)_n$ with various non-alternating regions demonstrated different susceptibilities. In the case of $(dT-dG)_n$ $(dC-dA)_n$ blocks that flank the rat somatostatin gene, the entire left-handed segment is sensitive to attack by Sl nuclease (22). The B-Z junctions appear to be located within the alternating purine-pyrimidine stretches and the Sl cleavage pattern is strongly dependent on the nature of the adjacent non-Z-DNA sequences. Recent studies on the thermodynamics of B-Z transitions (11,12,23) also indicate conformational heterogeneity of junctions. The flexibility of junctions is therefore sequence- and supercoil dependent (13,24).

The mechanism of the Sl reaction with the B-Z interface is not known. The enzyme was originally isolated as a single-strand specific endonuclease. This observation, together with the finding that $\Delta G_{junction}$ increases with increasing NaCl concentrations (25), and the theoretical unwinding of the adjacent right-handed helix by about 0.4 turns/junction (23,26), would be consistent with some single-stranded character of the B-Z junctions. However, chemical studies (24) using bromoacetaldehyde (BAA) have demonstrated that the B-Z interface contains few, if any unpaired bases. Furthermore, the Sl nuclease hypersensitivity of other types of double-stranded sequences (27,29) indicates that single-strandedness is not necessary for cleavage. Thus, the precise structure recognized by Sl nuclease remains to be elucidated.

Other Nucleases

The B-Z interface is recognized by a number of nucleases besides Sl. Use of these enzymes revealed conformational differences between various junctions (14,30). Mung bean nuclease catalyzes cleavage of junctions adjacent to left-handed $(dC-dG)_n$ regions but does not cleave within junctions flanking $(dC-dA)_n$ $(dT-dG)_n$ regions (14,30).

Although Sl nuclease is a highly specific enzyme, its major disadvantage as a probe for left-handed DNA lies in the fact that the pH optimum for endonucleolytic cleavage is around 4.6. The BAL31 nuclease has been used at neutral pH values to detect the conformational aberrations within B-Z junctions (30). This enzyme is active at high NaCl concentrations (several molar) and can therefore be used to investigate the salt-induced B-Z transition. However, the associated potent exonuclease activity limits its use for fine mapping of junction regions at the base pair level. However, the BAL31 nuclease is an excellent probe if long sequences (several kbp) are explored, especially for salt-induced conformational changes.

Nuclease Pl from Penicillium citrinum recognizes a number of structural features at neutral pH (31). Although its pH optimum for double strand cleavage is acid, the enzyme will catalyze site-specific cleavage of junctions between right-and left-handed DNA sequences at neutral pH (J.A. Blaho, personal communication). This result is important since it demonstrates that B-Z junction (or Z-helices) are not favored by low pHs.

RESTRICTION ENDONUCLEASES AND METHYLASES

HhaI and BssHII

Because of the dramatically altered DNA conformation in Z-DNA compared to B-DNA, several groups have investigated the capacity of left-handed DNA to serve as a substrate for modification methylases and restriction endonucleases. Studies on DNA polymers containing alternating G and 5-methyl-C moieties (see (32,33)) and on recombinant plasmids (34) demonstrate the stabilizing effect of methyl groups on Z-DNA. In negatively supercoiled DNA, the HhaI modification methylase (MHhaI) will not methylate its target site d(GCGC) when it is in a left-handed conformation (35,36). Likewise, the HhaI restriction endonuclease does not catalyze cleavage at left-handed d(GCGC) sites. The BssHII restriction endonuclease recognition sequence is d(GCGCGC). It is not cleaved by the restriction enzyme when it is part of a left-handed structure (35). In the case of MHhaI, a low amount of methylation of (dC-dG)$_n$ tracts is observed in conditions under which these are completely left-handed (36). This may be due to the B-Z dynamic equilibrium.

In summary, these results show that some features of the Z-helix (the helical sense or the sugar conformation or the glycosidic bond orientations, etc.) are not compatible with cleavage, as expected. However, this observation may be used as a probe for left-handed helices in other genomes (12,36).

Other Restriction Endonucleases

Structural perturbations within B-Z junction regions have also been demonstrated using restriction endonucleases. When a BamHI site (GGATCC) neighbors a (dC-dG)$_n$ tract, cleavage by the BamHI restriction enzyme was inhibited by the formation of left-handed DNA in the flanking sequences (13). In a related study (37), the BamHI recognition sequence was moved various distances from a left-handed segment. Cleavage by BamHI was inhibited at a distance of 4 or less base pairs from a left-handed (dG-dC)$_n$ block, but no inhibition was observed when the site was 8 bp away. This result supports the notion of a maximal length of 8 bp for the junction. However, prior Raman spectroscopic studies [38] revealed that the B-structure was perturbed at a distance of several turns of helix (30-40 bp) away from the Z-helix.

Similar results were found with EcoRI when an EcoRI site (GAATTC) replaced the BamHI site in the work described above (M. Caserta, W.T. Hsieh, and R.D. Wells, unpublished studies).

ROLE OF SEQUENCE IN STABILITY OF LEFT-HANDED Z-DNA

Left-handed Z-DNA has been studied in recombinant plasmids, restriction fragments, and DNA polymers and oligomers (1-15). The types of sequences proven to adopt left-handed Z conformations under appropriate conditions (Fig. 4) are generally of the strictly alternating purine-pyrimidine type and include C-G, 5-methyl-C-G, TG·CA and mixtures of C-G and T-G (1-15). Also, we demonstrated (13) that a BamHI site (GGATTC) within a track of 58 base pairs of CG adopted a non-B DNA structure which may be fully left-handed and stabilized by negative supercoiling. Further, the IVS2 sequences of human fetal globin genes (15) contain GTTTG and GACTG sequences which were proven to adopt left-handed Z structures by supercoil induced topoisomer relaxation studies (on two dimensional gels), S1 nuclease mapping of B-Z junctions, and Z-DNA antibody binding studies. The general rationale of the supercoil relaxation studies which have been used extensively in this laboratory as well as others is shown in Fig. 5. In summary, even early studies with various types of DNA sequences demonstrated that an alternating

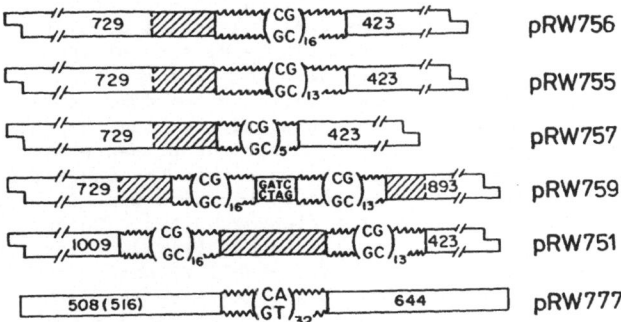

Fig. 4. Plasmids used in some of the studies reviewed herein. Reprinted with permission from (71).

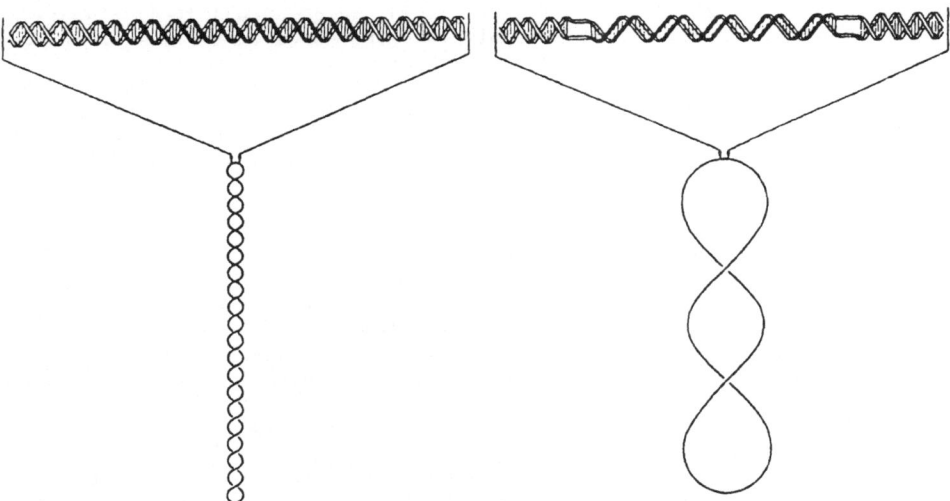

Fig. 5. Cartoon showing the relationship of B to Z transition in a small segment of a recombinant plasmid to the supercoiled state of the DNA. Republished with permission from (71).

purine-pyrimidine sequence is not <u>necessary</u> for a Z-helix. Since the alternating A-T duplex DNA has not <u>been demonstrated</u> to adopt a left-handed Z-helix, an alternating purine-pyrimidine is not sufficient.

This laboratory has conducted further investigations (12) to attempt to identify further features of the sequence requirements for left-handed Z-DNA. The capacity of six sequences with different numbers and orientations of A•T pairs flanked by alternating C-G pairs to adopt left-handed structures was evaluated in recombinant plasmids (12). A series of synthetic oligodeoxynucleotides were cloned into the <u>Bam</u>HI site of pRW790, a small plasmid ($\approx$2 kilobases) prepared especially <u>for</u> conformational studies of this type (Fig. 6). Supercoil relaxation studies by two-dimensional gel electrophoresis on topoisomers of each plasmid revealed the energetics and structures of the left-handed helices. Also, the presence of supercoil-induced altered DNA conformations within the inserts of topoisomer populations of the plasmids was detected by reaction with S1 nuclease followed by restriction mapping of the cleavage sites. We conclude that consecutive T-A base pairs, whether alternating (TATA) or continguous (TTTT), can adopt a left-handed conformation (presumably Z) when flanked by reasonably short runs of alternating $(C-G)_n (n=3-5)$. Thus, these results substantially broaden the range of DNA sequences that can adopt left-handed Z conformations.

Recent investigations with G and C containing inserts as well as one molecule with consecutive A-T pairs confirm our notions (39,40).

The thermodynamic parameters (entropy and enthalpy changes) which contribute to the B to Z transition were determined for three recombinant plasmids containing $(dC-dG)_{16}$ tracts and for a plasmid containing a pair of $(dT-dG)_{20}$ regions (41). The absolute values for the enthalpy and entropy changes for the formation of a B-Z junction in the dC-dG and dT-dG plasmids suggest that the properties and possibly the structures of the junctions are different [41]. Calculations using the enthalpy and entropy changes determined in this study reveal that the B to Z transition in plasmids containing (dC-dG) blocks are more temperature dependent than the transitions in plasmids with (dT-dG) blocks. Surprisingly, at temperatures above 60°C, calculations indicate that the B to Z transitions in (dT-dG) plasmids should be energetically favored over that transition in (dC-dG) plasmids (41).

OLIGOPURINE•OLIGOPYRIMIDINE SEQUENCES

A variety of biochemical, spectroscopic, and physical studies with 14 different DNA polymers (reviewed in (42-44)) demonstrated that DNAs containing oligopurine•oligopyrimidine sequences had unusual properties compared to the sequence isomeric DNAs which had both purines and pyrimidines in both complementary strands. Furthermore, in some cases the DNAs had altered conformations. Also it was demonstrated (44) that the pur•pyr polymer T-C•G-A was very unusual since it existed in two metastable states as determined by density gradient centrifugation and that one of these states could be converted into the other form. It was postulated (44) that the DNA existed in two conformational states.

Nine recent studies are reviewed in Table I which demonstrate that a variety of naturally occurring pur•pyr sequences are unusually sensitive to specific cleavage by S1 nuclease. The majority of these sites are dependent upon negative supercoiling. In some cases, (45-47,49) the structure responsible for this cleavage as been postulated to be a slipped structure (Fig 1.). However definitive evidence for this structure has not yet been provided. Furthermore, other studies reviewed in Table I (51-53) clearly indicate that the structure is a previously unrecognized unusual conformation which is neither melted (a bubble) nor a slipped structure (Fig. 1).

70

Table 1. S1 Nuclease Sensitive Pur·Pyr Sites Mapped to Base Pair Level

Reference	Gene	Sequence	Negative Supercoil Dependent Cleavage	Site of Cleavage	Postulated Role
Hentschel	Sea Urchin Histone Gene	$(GT)_{10}(GA)_{22}$ and $(GA)_{16}$	No	Trimodal	Unequal Crossover or Gene Conversion
Travers et al. (45)	5'-flanking Region of Drosophila Heat Shock Genes	Pur·Pyr	Yes	Bimodal	Uncertain
de Crombrugghe et al. (46)	5'-flanking Region of Chicken and Mouse $\alpha 2(I)$ Collagen Genes	Pur·Pyr	Yes	Bimodal	Transcription Regulation
Shen (47)	Human $\alpha 2-\alpha 1$ Globin Intergenic Region	Pur·Pyr	Yes	Distributed Across Pur·Pyr Region	DNA Rearrangements in Evolution
Worcel et al. (48)	Drosophila Histone Gene Repeat Unit	$(GA)_{10}$	Yes	Uncertain	Site for Single Strand Binding Protein
Schlesinger et al. (49)	Chicken Repetitive DNA	$(AGAGG)_{32}$	Yes	Distributed Across Pur·Pyr	Recombination
Elgin et al. (50)	Drosophila Heat Shock Locus 67B1	Pur·Pyr (71 out of 81)	Yes	Uncertain	Chromatin Structure

(continued)

Table 1. (Continued)

Reference	Gene	Sequence	Negative Supercoil Dependent Cleavage	Site of Cleavage	Postulated Role
Nickol and Feisenfeld (51)	5'-end of Chicken Adult β-Globin	G_{16}	Yes	Distributed in G_{16} Tract	Regulation and Organization of Chromatin
Efstratiadis et al. (52)	Adult Chicken β-Globin	Three Pur·Pyr sequences including G_{18}	Yes	Distributed Across Pur·Pyr Regions	Transcription Regulation
Dahlberg et al. (53)	Human U1 Small Nuclear RNA Genes	$(AG)_{15-25}$ (1.8 Kb downstream from coding region)	No	Distributed Across Pur·Pyr Regions	Gene Homogenization
		Pur·Pyr (0.3 Kb upstream from coding region)	Yes	Distributed Across Pur·Pyr Regions	Transcription Regulation

pRW 790   A B C D E F G H

pRW 1001   A B C D $\left(\frac{CG}{GC}\right)_9$ D E F G H

pRW 1002   A B C D $\left(\frac{CG}{GC}\right)_4$ D E F G H

pRW 1003   A B C D $\left(\frac{CG}{GC}\right)_5$ $\frac{TT}{AA}$ $\left(\frac{CG}{GC}\right)_5$ D E F G H

pRW 1004   A B C D $\left(\frac{CG}{GC}\right)_5$ $\frac{TTTT}{AAAA}$ $\left(\frac{CG}{GC}\right)_5$ D E F G H

pRW 1007   A B C D $\left(\frac{CG}{GC}\right)_5$ $\frac{TA}{AT}$ $\left(\frac{CG}{GC}\right)_5$ D E F G H

pRW 1008   A B C D $\left(\frac{CG}{GC}\right)_5$ $\frac{TATA}{ATAT}$ $\left(\frac{CG}{GC}\right)_5$ D E F G H

pRW 1009   A B C D $\left(\frac{CG}{GC}\right)_5$ $\frac{TATATA}{ATATAT}$ $\left(\frac{CG}{GC}\right)_4$ D E F G H

pRW 1101   A B C D $\left(\frac{CG}{GC}\right)_4$ $\frac{TAC}{ATG}$ $\left(\frac{CG}{GC}\right)_3$ D E F G H

Fig. 6. List of plasmids showing the multiple cloning site and positions and sequences of inserts. A, EcoRI; B, Sst I; C, Sma I; D, BamHI; E, Xba I; F, HincII; G, Pst I; H, HindIII. Republished with permission from (12).

Work in this laboratory on related problems has followed two avenues including studies on a series of direct repeat sequences which occur in the site of segment inversion of the HSV-1 genome and studies with the sequence surrounding the major late promoter of adenovirus.

HERPES SIMPLEX VIRUS TYPE 1 JOINT REGION: ANISOMORPHIC DNA

HSV-1 is a member of a large group of enveloped DNA viruses (herpesviruses) including cytomegalovirus, Epstein-Barr virus and HSV-2 (54,55). The genome of HSV-1 (Fig. 7) is a linear DNA duplex of approximately 155,000 bp (56) which replicates in the nucleus of the infected cell. It can be divided into a long and short segment of unique sequences (U_L and U_s), both flanked by inverted repeats (b and c, respectively) and ending in tandem reiterations of variable length (the a sequences) (58). The overall G+C-content of the DNA is 67% (55), while the joint region (the bac-junction) has close to 80% G+C (60,61).

The presence of large, internal, inverted duplications of terminal sequences correlates with high-frequency inversion of unique sequence elements enclosed by these duplications and thus with the presence of approximately equimolar concentrations of four isomeric forms of linear virion DNA which differ in the relative orientation of unique sequence components. There is now considerable evidence that segment inversion around the "joint" is mediated by a site-specific recombination system operating between the mutually inverted copies of a recognition/cleavage sequence (62). This sequence contains a 17 to 21 bp direct repeat (63) which is cleaved asymmetrically to give genome termini and multiple direct reiterations of other (G+C)-rich sequences (Fig. 8).

It has recently been demonstrated that both the signal for packaging and cleavage of the concatemeric replicative form of the viral DNA as well as the signal for segment inversions reside within the a sequence. More precise

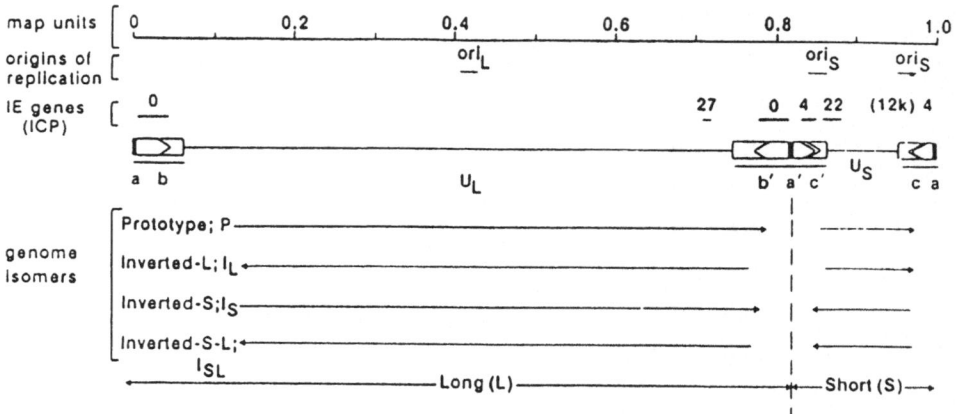

Fig. 7. Major structural features of the genome of HSV-1. The genome
(approx. 150 kbp) is composed of unique sequences (solid line) and
inverted repeats (rectangular boxes) and has a small (270 to 540 bp,
depending on the virus strain) direct terminal redundancy (the a
sequence). This sequence is also represented internally in an
inverted orientation at the junction between inverted copies of
subterminal sequences. This junction ("joint") divides the molecule
into a long (L, 125 kbp) and short (S, 25 kbp) segment each composed
of unique components (U$_L$ and U$_S$) enclosed by inverted
duplications of subterminal sequence (b and b'; c and c').
Populations of isomers obtained by inversions of the L and S
components relative to each other. The nomenclature for these isomers
is relative to an arbitrarily chosen prototype (P). The physical map
positions of the origins of replication (Ori$_L$, Ori$_S$) and the α or
immediate early (IE) proteins on the P isomer are indicated.

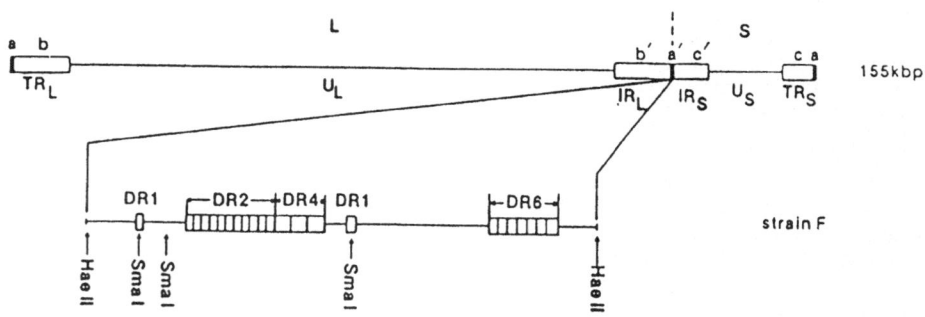

Fig. 8. Structure of the joint region of HSV-1 Strain F. The upper line of
the figure shows the prototype map of HSV-1. The lower line is a
schematic representation of the a sequence. The direct repeats
designated as DR1, DR2, DR4, and DR6 (see text) are shown as open
boxes. SmaI and HaeII cleavage sites are indicated below the line.
TR$_L$ and TR$_S$: long and short terminal repeats. IR$_L$ and IR$_S$:
long and short internal repeats. DR2 contains 19 reiterations of the
12-mer CGCTCCTCCCCC and DR4 contains 3 copies of the 37-mer
CGCTCCTCCCCGCTCCCGCGGCCCCGCCCCCAACGCC.

analysis of this region located the signal for segment inversion in or near a set of directly repeated sequence components within the a region. This set of tandem repeats, designated DR2, is a highly conserved region which, depending on the strain examined, consists of 12 or 11 nucleotides which are repeated 8 to 23 times (60,61). This repeat is very rich in G+C (92%) and shows a high bias for purines on one of the strands (92%); a closely related and neighboring repeat (DR4) shows similar properties.

We have recently determined (64) that the 12 bp tandem direct repeat sequences (DR2) at the joint region (a sequence) of Herpes Simplex Virus Type 1 (Strain F) adopts a new type of DNA conformation under the influence of negative supercoiling. The novel conformation is dependent on the number of the DR2 repeats; the 19mer (228 bp total) and the 14 mer (168 bp) readily form the alternate structure whereas pentamer, trimer, and dimer repeats show somewhat different properties (Fig. 9). S1 and P1 nuclease studies reveal that the new conformation has a major structural aberration at its center and conformational periodicities which are identical on the complementary strands. Also, the effect of salt and pH, the location of reaction with bromo- and chloroacetaldehyde, the type of sequence (direct repeat) involved, and the nature and extent of supercoil induced relaxations, demonstrate that this structure differs from previously recognized conformations including left-handed Z helices, cruciforms, bent DNA, and slipped structures. We propose the existence of a novel conformation, anisomorphic DNA, with different structures on the complementary strands which elicits a structural aberration at the physical center of the tandem sequences. Since the oligopurine oligopyrimidine sequence may be inherently inflexible, this supercoil-induced structural change and the physical stress on these inserts in recombinant plasmids tends to deform (crack) the DR2 sequences at their centers. Possible roles for anisomorphic DNA in the functions of this segment of intense biological activity are proposed (64).

pRW1201   ...CGTTTT (CGCTCCTCCCCC)$_{19}$CGCTCC...

pRW1202   ...CGTTTT (CGCTCCTCCCCC)$_{14}$CGCTCC...

pRW1203   ...CGTTTT (CGCTCCTCCCCC)$_5$ CGCTCC...

pRW1214   ...GATCCC (CGCTCCTCCCCC)$_3$ CGCTCCTCGCCC CGGATC...

pRW1212   ...GATCCC CGCTCCTCCCCC CGCTCCTCCCTC CGATC...

pRW1250   ...GATCCC CGCTCCTCCCCC CGCC CGCTCCTCCCCC CGATC...

Fig. 9. Copy number variants of the HSV-1 DR2 plasmid inserts used in the studies described in Reference 64. The insert in pRW1202 is in the opposite orientation compared to other plasmids. Republished with permission from (64).

75

Figs. 10 and 11 schematically show some of the features of anisomorphic DNA. The two complementary strands of the double helix have different conformations which are induced by negative supercoiling (Fig. 11). The nature of this asymmetry is unclear as yet, but it is possible that the angles between consecutive bases are different on the two strands due to different stacking energies. This will lead to the helix axis describing a curve in space, thereby establishing a periodicity of bases positioned on the outside of the helix. This hypothesis is consistent with the S1 nicking data. The very strong central nick on the pyrimidine-rich strand may be explained by a small structural aberration in the structure which is propagated from both ends towards the center of the insert. Such a gradual

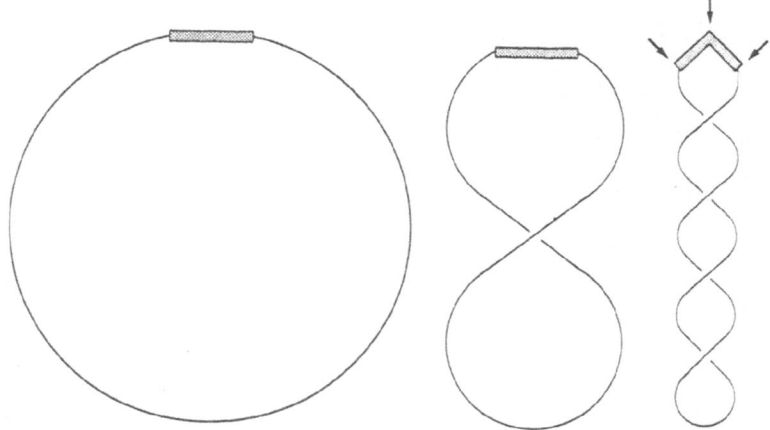

Fig. 10. Schematic representation of supercoil induced anisomorphic DNA. Left side; relaxed plasmid with insert (open box) in right-handed B structure. Center; moderately supercoiled plasmid, below the structural transition. Right side; plasmid at supercoil densities above the transition which induces the conformational aberrations (three arrows). Reprinted with permission from (64).

structural amplification could for example be due to a different rise/residue on the two strands. The purine-rich strand would remain stacked but the pyrimidine-rich strand would be unstacked as a consequence of the unequal structure and thus give rise to the observed pattern of nuclease sensitivity. Also, a sufficient length of DR2 sequences (as for example for pRW1201 and 1202) would be required to provide a sufficient amplification of the structural aberration in order to elicit the pronounced nicking on the pyrimidine-rich strand. Long range interactions in DNA were reviewed in Wells et al. (1980).

In addition to the asymmetry of the two helix strands, we propose that the DR2 repeats are relatively stiff and inflexible compared to the rest of the plasmid. This resistance to bending will manifest itself upon supercoiling (Figure 10, center) and lead to an unequal distribution of supercoil density along the plasmid. As a consequence, torsional stress will be exerted on the ends of the DR2 repeats, leading to a distorted vector-insert interface region (Figure 10, right side). In addition, above a certain length of insert, this bilateral stress will tend to physically generate a structural discontinuity at the center as schematically indicated in Fig. 10, right side. This physical effect may act in concert with the

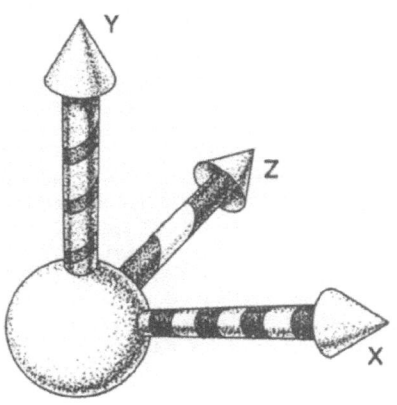

Anisomorphic:
    exhibiting different forms in different parts.

Anisomorphic DNA

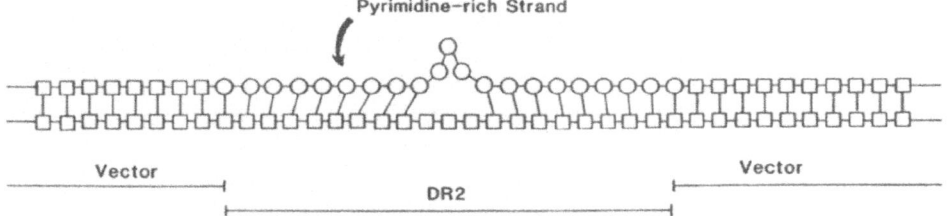

Fig. 11. Possible structural features of anisomorphic DNA. We hypothesize that negative supercoiling causes a structural transition in the DR2 sequences of HSV-1. The non-identical conformations on the two complementary strands of the DR2 sequences embody features (such as unequal rise per residue or angle of rotation between consecutive basepairs) that cause an unstacking of the oligopyrimidine-rich strand. The oligopurine-rich strand remains stacked. Due to the influence of flanking vector sequences, the structural aberration will be most pronounced at the center of the DR2 region.

unstacking of the pyrimidine-rich strand in the center of the DR2 repeats described above to generate the observed S1 nuclease nicking patterns (64).

MAJOR LATE PROMOTER OF ADENOVIRUS DNA

Adenovirus is a well characterized tumor virus which contains a linear duplex DNA genome of 36 kb (Fig. 12). The entire sequence of adenovirus serotype 2 DNA has recently been established and is deposited in GenBank.

This laboratory has recently surveyed the adenovirus genome for the presence of unusual conformations (65). More than 80% (approximately 29 kilobase pairs) of the adenovirus serotype 2 genome was surveyed for the presence of unusual DNA conformations. Seven recombinant DNAs containing the largest HindIII fragments of Ad2 DNA were analyzed for the presence of negative supercoil-dependent S1 nuclease-sensitive sites. Four plasmids each contained a specific site of S1 nuclease sensitivity whereas the other three showed no reaction. Further investigation was focused on a plasmid containing one of the positively reacting fragments (fragment C) which contained the major late promoter at coordinate 16.4 on the genome; three serotypes (Ad2, Ad7, Ad12) were studied. Fine mapping studies revealed the S1-sensitive sites to be a small region (~6 base pairs) located at the TATA box of the major late promoter in all three cases (Fig. 13). Other determinations (supercoil relaxation, T7 gene 3 product sensitivity, bromoacetaldehyde reactivity, anomalous gel mobility, the influence of negative superhelical density on nuclease sensitivity) led to the conclusion that the B-helix deformation was not due to a previously recognized DNA conformation (left-handed Z-DNA, cruciforms, bent DNA), but may be accounted for by the homopurine·homopyrimidine nature of this region (65).

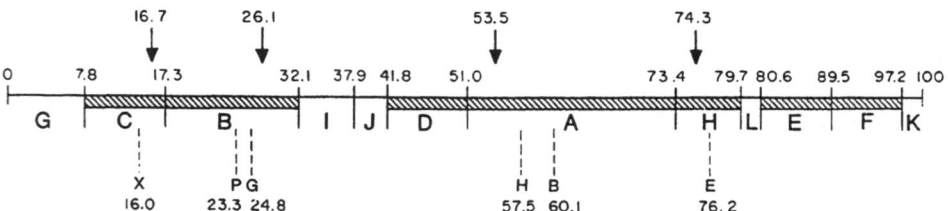

Fig. 12. Map of the adenovirus serotype 2 genome showing HindIII fragments which were cloned into pBR322. Regions designated by cross-hatching are those HindIII fragments which were surveyed herein (80% of the genome). The bold arrows at 16.7, 26.1, 53.5, and 74.3 map units designate the principal sites of S1 nuclease cleavage. The broken lines show the locations of some of the restriction sites used for mapping studies (X=XhoII; P=PstI; G=BglII; H=HpaII; B=BamHI; E=EcoRI). Map is not shown to scale. Republished with permission from (65).

```
          A AG GGG GGC TATA A A AGGG G GTGGGGGCGC
Ad 2      T TC C CCC C GA TAT T T TCCCC CACCCCCGCG
```

```
          CCG G GGG GG TAT A A A AGGGG GCGG A CCTCT
Ad 7      GGC C C CCC CATA T T T TCCC CCGCC T GGAGA
```

```
          GTGG TG G GC TATA AA A AGGGGCGGG T CCTT
Ad 12     CACC AC C CGATA TTT T TCCCCGCCCAGGAA
```

Fig. 13. DNA sequences surrounding the S1 nuclease-sensitive sites in the major late promoters of adenovirus serotypes 2,7, and 12. Locations of the S1 nuclease-sensitive sites in pRW784, pRW786, and pRW787 (containing the major late promoters of Ad2, Ad7, and Ad12, respectively) was determined as described in the legend to Figure 5. The horizontal bars represent the positions of S1 nuclease cleavage. Reprinted with permission from (65).

Yu and Manley (66) have recently conducted S1 nuclease sensitivity studies on plasmids containing the major late promoter of adenovirus as well as on a series of deletion mutations. These investigators hypothesize that a slipped structure accounts for their experimental observations. However, slipped structures have only been characterized by S1 nuclease sensitivity and not by chemical probes, physical determinations, or spectroscopic measurements. It will be interesting in the future to probe these and other deletion mutants with a variety of tools in order to determine the type and properties of the physical structures which are present.

BENT DNA: A CRITHIDIA KINETOPLAST FRAGMENT

Bent DNA was discovered in 1979 (67). A variety of recent studies have been conducted, particularly with restriction fragments from kinetoplast DNA. Kinetoplast DNA is the mitochonrial DNA of trypanosomes and related parasitic protozoa. This unusual DNA consists of a network of thousands of interlocked circles. A small number of these circles are maxicircles which resemble, in structure and genetic function, the mitochondrial DNA of other eukaryotes. The function of minicircles is still unknown.

A region of bent DNA has been discovered in the kinetoplast DNA minicircles of Leishmania tarentolae (68). The unusually slow electrophoretic mobility of some of the minicircle restriction fragments on polyacrylamide gels is a convenient way of detecting this unusual property. The result has been interpreted as due to the natural curvature of the helix which hinders the snaking of the molecule through the gel pores. The effect is most easily observed in gels which contain a higher percent of acrylamide (8 or 10%). Other physical determinations including electric dichroism

relaxation have also supported a bent DNA structure. The physical basis at
the base pair level for explaining this phenomenon is a matter of intense
scrutiny in a variety of laboratories.

Kinetoplast DNA minicircles from <u>Crithidia fasciculata</u> contain a single
major region of bent helix. Restriction fragments containing this bent helix
have electrophoretic behavior on polyacrylamide gels which is much more
anomalous than that of previously studied bent fragments. Therefore, the <u>C.
fasciculata</u> fragments probably have a more extreme curvature. Sequencing
part of a cloned minicircle revealed an usual structure for the bent region.
In a sequence of 200 bases, the bent region contains 18 runs of 4 to 6 A's
with 16 of these runs in the same strand (Fig. 14). In some parts of this
sequence the A runs are regularly spaced with a periodicity of about 10 bp.

```
        10          20          30          40          50
  GCGGAAAATG TCAGAAGTCC ATTTCTGTCA AACCCCCCAA AAATCCAGAA

        60          70          80          90         100
  TACGACCGAA CGGAATACAC ATAAACACAC CCAGAAGCGA AANAGCAACC

       110         120         130         140         150
  AGGAAAGCGG TGAAAACACC CCCACCAAAC CCAAGGCAGG CCCAAAGTAC

       160         170         180         190         200
  CAAACCAGCG CAAATCACCT CTGTCCAGCA CAAACCCCGT CCAAACCAGC

       210         220         230         240         250
  ACTCAGACCC AGGAAAACCC CTCCCGGAGG CCCCGAAATC GGGCTAGAAC
                        StuI
       260         270         280         290         300
  CCCGCCAAAC CCCCTGCCAG GAGGCCTAAA ATTCCAACCG AAAATCGCGA

       310         320         330         340         350
  GGTTACTTTT TTGGAGCCCG AAAACCACCC AAAATCAAGG AAAAATGGCC

       360         370         380         390         400
  AAAAAATGCC AAAAAATAGC GAAAATACCC CGAAAATTGG CAAAAATTAA

       410         420         430         440         450
  CAAAAAATAG CGAATTTCCC TGAATTTTAG GCGAAAAAAC CCCCGAAAAT
                                  AccI
       460         470         480         490         500
  GGCCAAAAAC GCACTGAAAA TCAAAATCTG AACGTCTACG CCTCTGCTTT

       510         520         530         540         550
  AGAGTCTGTC TACCACCCGG GTCGGTTAAA TATGTCGGCC GTATTAAAGC

       560         570         580         590         600
  TCACATGACA AAATCATGCG ATGAGGTTAG GAAAGGGAAA GACAATGTAC

       610         620         630         640         650
  CTTTCCGCGA TCACCATCCT ATATCTGTTG CTTGACCCCC TCTGTCTCCA
```

  GGCAAGCTT

Fig. 14. Sequence of bent DNA fragment from pPK201. This sequence begins at
the first G derived from the <u>SstII</u> site used to linearize the
minicircle prior to cloning; it ends with the 3' terminal base of
the <u>HindIII</u> site at position 659 (69). Rightward pointing arrows
indicate sequences of 4-6 A's. Leftward pointing arrows indicate
similar sequences in the complementary strand. The StuI and AccI
cleavage sites used to produce the fragment cloned in pPK201/CAT are
indicated. Note that there is a second <u>Acc</u> site at nucleotide 509.
The ambiguous nucleotide at position 93 is probably C or G.
Reprinted with permission from (69).

This spacing is nearly in phase with the twist of the DNA helix. This same sequence arrangement has been observed in other bent fragments, but the number of A runs is much greater in this C. fasciculata sequence. It is likely that there are small bends associated with each A run which, because of their periodic spacing, add up to produce substantial curvature in this molecule.

In addition to having highly anomalous electrophoretic behavior, the fragment has unusual circular dichroism spectra. Its spectrum in the absence of ethanol is that of B DNA, but ethanol in the concentration range of 51 to 71% (w/w) induces changes to forms which are different from those of any well characterized DNA structure. The C. fasciculata bent helix is neither cleaved by S1 nuclease nor modified by bromoacetaldehyde under conditions in which other unusual DNA structures (such as cruciforms or B-Z junctions) are susceptible to attack by these reagents. Finally, a two-dimensional polyacrylamide gel analysis of a family of topoisomers of a plasmid containing the bent helix revealed no supercoil-induced relaxation.

ACKNOWLEDGEMENTS

This work was supported by grants from the National Institutes of Health (GM 30822) and the National Science Foundation (08644). A number of talented and dedicated co-workers from the author's laboratory have contributed very substantially to the success of the projects which are described in this review. The contributions from these individuals have been designated in the original research contributions. Without the diligent efforts of these highly skilled individuals, little progress would have been made. Most of the figures have been republished with permission from prior original research papers and a portion of the text on nuclease sensitivities was described previously (70).

REFERENCES

1. R.D. Wells, T.C. Goodman, W. Hillen, G.T. Horn, R.D. Klein, J.E. Larson, U.R. Muller, S.K. Neuendorf, N. Panayotatos, and S.M. Stirdivant, Prog. Nucleic Acids Res. Mol. Biol. 24, 167-267 (1980).
2. S.B. Zimmerman, Ann. Rev. Biochem. 51, 395-427 (1982).
3. R.E. Dickerson, H.R. Drew, B.N. Connor, R.M. Wing, A.V. Fratini, and M.L. Kopka, Science 216, 475-478 (1982).
4. C.K. Singleton in: Cell Proliferation: Recent Advances (Leffert, H.L. and Boynton, A.L., eds., 1984), Academic Press, New York.
5. A.H.-J. Wang, G.J. Quigley, F.J. Kolpak, J.L. Crawford, J.H. van Boom, G. van der Marel, and A. Rich, Nature 282, 680-686 (1979).
6. H.R. Drew, T. Takano, K. Itakura, and R.E. Dickerson, Nature 286, 567-573 (1980).
7. A. Rich, A. Nordheim, and A.H.-J. Wang, Ann. Rev. Biochem. 53, 791-846 (1984).
8. R.D. Wells, B.F. Erlanger, H.B. Gray, L.H. Hanau, T.M. Jovin, M.W. Kilpatrick, J. Klysik, J.E. Larson, J.C. Martin, J.J. Miglietta, C.K. Singleton, S.M. Stirdivant, C.M. Veneziale, R.M. Wartell, C.F. Wei, W. Zacharias, and D. Zarling, Gene Expression, UCLA Symp. Mol. Cell. Biol. 8, 3-18 (1983).
9. C.K. Singleton, J. Klysik, S.M. Stirdivant, and R.D. Wells, Nature 299, 312-316 (1982).
10. L.J. Peck, A. Nordheim. A. Rich, and J.C. Wang, Proc. Natl. Acad. Sci. USA 79, 4560-4564 (1982).
11. M.J. Ellison, R.J. Kelleher, III, A.H.-J. Wang, J.F. Habener, and A. Rich, Proc. Natl. Acad. Sci. USA 82, 8320-8324 (1985).

12. M. McLean, J.A. Blaho, M.W. Kilpatrick, and R.D. Wells, Proc. Natl. A Acad. Sci. USA 83, 5884-5888 (1986).
13. C.K. Singleton, J. Klysik, and R.D. Wells, Proc. Natl. Acad. Sci. USA 80, 2447-2451 (1983).
14. C.K. Singleton, M.W. Kilpatrick, and R.D.Wells, J. Biol. Chem. 259, 1963-1967 (1984).
15. M.W. Kilpatrick, J. Klysik, C.K. Singleton, D. Zarling, T.M. Jovin, L.H. Hanau, B.F. Erlanger, and R.D. Wells, J. Biol. Chem. 259, 7268-7274 (1984).
16. H. Hamada, M.G. Petrino, T. Kakunaga, and A. Novel, Proc. Natl. Acad. Sci. USA 79, 6465 (1982).
17. R. Thomas, S. Beck, and F.M. Pohl, Proc. Natl. Acad. Sci. USA 80, 5550-5553 (1983).
18. E.M. Lafer, R.P.C. Valle, A. Moller, A. Nordheim, P.H. Schur, A. Rich, and B.D. Stollar, J. Clin. Invest. 71, 314-321 (1983).
19. H.R. Bergen, III, M.J. Losman, T.R. O'Connor, W. Zacharias, J.E. Larson, R.D. Wells, and W.J. Koopman, J. Immun., submitted 1986.
20. B.H. Johnston and A. Rich, Cell 42, 713-724 (1986).
21. G. Galazka, E. Palacek, R.D. Wells, and J. Klysik, J. Biol. Chem. in press, 1986.
22. T.E. Hayes and J.E. Dixon, J. Biol. Chem. 260, 8145-8156 (1985).
23. T.R. O'Connor, D.S. Kang, and R.D. Wells, J. Biol. Chem., in the press 1986.
24. D.S. Kang and R.D. Wells, J. Biol. Chem. 260, 7783-7790 (1985).
25. F. Azorin, A. Nordheim, and A. Rich, EMBO J. 2, 649-655 (1983).
26. L.J. Peck and J.C. Wang, Proc. Natl. Acad. Sci. USA 80, 6206-6210 (1983).
27. E. Schon, T. Evans, J. Welsh, and A. Efstratiadis, A, Cell 35, 837-848 (1983).
28. D.E. Pulleyblank, D.B. Haniford, and A.R. Morgan, Cell 42, 271-280 (1985).
29. C.R. Cantor and A. Efstratiadis, Nucleic Acids Res. 21, 8059-8072 (1984).
30. M.W. Kilpatrick, C.-F. Wei, H.B. Gray, and R.D. Wells, Nucleic Acids Res. 11, 3811-3818 (1983).
31. D.B. Haniford and D.E. Pulleyblank, Nucleic Acids Res. 13. 4343-4363 (1985).
32. M. Behe and G. Felsenfeld, Proc. Natl. Acad. Sci. USA 78, 1619-1623 (1981).
33. J. Feigon, A.H.-J. Wang, G.A. van der Marel, J.H. van Boom, and A. Rich, Nucleic Acids Res. 12, 1243-1263 (1984).
34. J. Klysik, S.M. Stirdivant, C.K. Singleton, W. Zacharias, and R.D. Wells, J. Mol. Biol. 168, 51-71 (1983).
35. L. Vardimon and A. Rich, Proc. Natl. Acad. Sci. USA 81, 3268-3272 (1984).
36. W. Zacharias, J.E. Larson, M.W. Kilpatrick and R.D. Wells, Nucleic Acids Res. 12, 7677-7692 (1984).
37. F. Azorin, R. Hahn, and A. Rich, Proc. Natl. Acad. Sci. USA 81, 3268-3272 (1984).
38. R.M. Wartell, J. Klysik, W. Hillen, and R.D. Wells, Proc. Natl. Acad. Sci. USA 79, 2549-2553 (1982).
39. J. Feigon, A.H.-J. Wang, G.A. van der Marel, J.H. van Boom, and A. Rich, Science 230, 82-84 (1985).
40. M.J. Ellison, J. Feigon, R.J. Kelleher III, A. H.-J. Wang, J.F. Habener, and A. Rich, Biochem 25, 3648-3655 (1986).
41. T.R. O'Connor, D.S. Kang, and R.D. Wells, J. Biol. Chem., in the press (1986).
42. R.D. Wells, T.C. Goodman, W. Hillen, G.T. Horn, R. D. Klein, J. E. Larson, U.R. Muller, S.K. Neuendorf, N. Panayotatos, N., and S.M. Stirdivant, Prog. Nucleic Acids Res. Mol. Biol. 24, 167-267 (1980).

43. R.D. Wells, R.W. Blakesley, J.F. Burd, H.W. Chan, J.B. Dodgson, S.C. Hardies, G.T. Horn, K.F. Jensen, J. Larson, I.F. Nes, E. Selsing, and R.M. Wartell. Critical Reviews in Biochemistry 4, 305-340 (1977).

44. R.D. Wells, J.E. Larson, R.C. Grant, B.E. Shortle, and Cantor, C.R., J. Mol. Biol. 54, 465-497.

45. H.A.F. Mace, H.R.B. Pelham, and A.A. Travers, Nature 304, 555-557 (1983).

46. C. McKeon, A. Schmidt, and B. deCrombrugghe, J. Biol. Chem. 259, 6636-6640 (1984).

47. C-K.J. Shen, Nuc. Acids Res. 11, 7899-7911 (1983).

48. G.C. Glikin, G. Garguils, L. Rena-Descalzi, and A. Worcel, Nature 303, 770-774, (1983).

49. D. Dybvig, D.D. Clark, G. Aliperti, and M.J. Schlesinger, Nucl. Acids Res. 11, 8495-8505.

50. S.B. Selleck, S.C.R. Elgin, and I.L. Cartwright, J. Mol. Biol. 178, 17-33 (1984).

51. J.M. Nickol and G. Felsenfeld, Cell 35, 467-477, (1983).

52. E. Schon, T. Evans, J. Welsh, and A. Efstratiadis, Cell 35, 838-848 (1983).

53. H. Htun, E. Lund, and J.E. Dahlberg, Proc. Natl. Acad. Sci, USA 81, 7288-7292 (1984).

54. A. J. Nahmias. in: Viruses, Evolution and Cancer (E. Kurstak and K. Maramorosch, eds.) 605-634, Academic Press, New York (1980).

55. P.G. Spear and B. Roizman in: DNA Tumor Viruses (J. Tooze, ed.) 615-746, Cold Spring Harbor (1980).

56. S. Wadsworth, R.J. Jacob, and B. Roizman, J. Virol. 15, 1487-1497 (1975).

57. P. Sheldrick and N. Berthelot, Cold Spring Harbor Symp. Quant. Biol. 39, 667-678 (1975).

58. M.J. Wagner and W.C. Summers, J. Virol. 27, 374-387 (1978).

59. B.J. Graham, H. Ludwig, D.L. Bronson, M. Benyesh-Melnick, and N. Biswal, Biochim, Biophys. Acta 259, 13-23 (1972).

60. A.J. Davison and N.M. Wilkie, J. Gen. Virol 55, 315-331 (1981).

61. E.S. Mocarski and B. Roizman, Proc. Natl. Acad. Sci. USA 78, 7047-7051 (1981).

62. R.W. Honess, J. Gen. Virol. 65, 2077-2107 (1984).

63. E.S. Mocarski and B. Roizman, Cell 31, 89-97 (1982).

64. F. Wohlrab, M.J. McLean, and R.D. Wells, submitted (1986).

65. M.W. Kilpatrick, A. Torri, D.S. Kang, J.A. Engler, and R.D. Wells, J. Biol. Chem. 261, 11350-11354 (1986).

66. Y.-T. Yu and J.L. Manley, Cell 45, 743-751 (1986).

67. E. Selsing, R.D. Wells, C.J. Alden, and S. Arnott, J. Biol. Chem., 254, 5555-5561 (1979).

68. J.C. Marini, S.D. Levene, D.M. Crothers, and P.T. Englund, Proc. Natl. Acad. Sci. USA 79, 7664-7668 (1982).

69. P.A. Kitchin, V.A. Klein, K.A. Ryan, K.L. Gann, C.A. Raush, D.S. Kang, R.D. Wells, and P.T. Englund, J. Biol. Chem. 261, 11302-11309 (1986).

70. F. Wohlrab and R.D. Wells, Gene Amplification and Analysis, Vol. 5, in the press (1986).

71. T.R. O'Connor, M.W. Kilpatrick, J. Klysik, J.E. Larson, J.C. Martin, C.K. Singleton, S.M. Stirdivant, W. Zacharias, and R.D. Wells, J. of Biomole. Str. and Dyn., 1, 999-1009 (1983).

SPECIFICITY AND DYNAMICS

OF PROTEIN-NUCLEIC ACID INTERACTIONS

Dietmar Porschke

Max-Planck-Institut
für biophysikalische Chemie
3400 Göttingen, FRG

ABSTRACT

First, a survey is given on model experiments, which provide information on the contribution of individual amino acid side chains to interactions of amino acids with nucleic acids. The most sensitive approach proved to be an analysis of helix-coil transitions in the presence of various ligands. Measurements for different RNA and DNA helices in the presence of amino acid amides demonstrate the existence of an affinity scale for the interaction of amino acid residues with nucleic acids, which is mainly determined by the hydrophobicity. The data indicate a direct pathway for the evolution of melting proteins and can be used to construct a simple model for the evolution of the genetic code. Recent experiments also demonstrate a selective interaction of amino acid residues with tRNA^{Phe}.

In the second part, the dynamics of nucleic acid ligand interactions is discussed. As shown by electric field jump experiments, simple oligonucleotide-oligopeptide complexes are formed at a diffusion controlled rate, whereas "insertion" of aromatic residues between stacked bases is already a relatively slow reaction. An example is also given for the use of electro-optical procedures for analysis of structures in solution: rotation time constants demonstrate that binding of cyclic AMP receptor to specific DNA fragments leads to strong bending of the double helix around the protein. The importance of ligand mobility along nucleic acid chains is demonstrated for the example of a melting protein. Finally, the dynamics of a repressor-operator recognition is compared for the lac- and the tet-system; some mechanisms for the motion of ligands along polynucleotides are discussed.

INTRODUCTION

The genetic information is processed at all essential levels by protein nucleic acid interactions. Most of these interactions are surprisingly specific and accurate. Of course, the specificity and accuracy is due to molecular contacts between amino acids and nucleotides. It is generally accepted that these contacts are formed by hydrogen bonding-, hydrophobic-, stacking- and electrostatic-interactions. However, the contributions of these molecular contacts to specificity remain to be elucidated. For a quantitative description, knowledge is also required on the contributions of individual amino acids and nucleotides to these interactions. Furthermore, it would be important to know about the lowest level of molecular organisation, where substantial specificity can be observed. Due to the key function of protein nucleic acid interactions, these questions are relevant as well to the problem of self-organisation at an early stage of evolution. Usually, these questions are discussed from a merely static point of view. Since life is by definition far from equilibrium, the dynamics is at least as important.

One may learn about protein nucleic acid interactions both by analysis of model systems and of natural systems. Indeed, both approaches including investigations of model systems are necessary. This may be illustrated by an example of crystallized protein nucleic acid complex, which has been studied by X-ray analysis up to high resolution: even when all the contacts are visible, the X-ray analysis does not yet reveal the contribution of individual contacts to the interaction energy. Furthermore, X-ray structures do not provide the required information on the dynamics.

MODEL EXPERIMENTS

Contributions of individual amino acid residues to protein nucleic acid interactions

Ever since the biological functions of nucleic acids and proteins have been established, it has been attempted to provide experimental or theoretical evidence for a molecular code describing the interactions between nucleic acids and proteins. However, the experiments demonstrated that the interactions between individual components are rather weak[1-4] and in fact can hardly be demonstrated. Thus, the interactions at this elementary level can only be traced by particularly sensitive methods. The most sensitive approach for detection of ligand binding to nucleic acids appears to be the characterization of melting temperatures of double helices[5,6]. The outstanding sensitivity of this procedure is mainly due to the high cooperativity of the double helix-coil transition and its rather strong dependence on ligand binding. Another advantage of the melting procedure is the fact that the binding of any ligand without spectroscopic label can be characterized. The main problem of most techniques is their dependence upon such labels, which is, for example, the obvious reason for a strong overrepresentation of ligands with aromatic residues in the scientific literature. The melting approach has rarely been exploited at its full capacity, probably because the melting transition is influenced by a rather large number of parameters. However, this problem may be settled by measurements over a broad range of concentrations.

For an evaluation of individual contributions to protein nucleic acid interactions, the crucial step is of course the selection of an appropriate model system. Many model systems used in the past have an excessive contribution of electrostatic interactions, which may essentially cover the interactions to be studied. Electrostatic interactions may be useful to increase the interaction energies to a detectable level, but should not be dominant. In this respect amino acid amides provide a reasonable compromise. Actually, the positive charge at the amino group is identical to that of the natural activated precursors for loading of tRNA adaptors. The change in the molecular character of the amino acid residues by the amide function remains relatively small. Another advantage of the amino acid amides is their stability and easy accessibility.

When simple ligands bearing a positive charge are added to double helices, the stability of the helix increases. This is a standard observation, which is explained by the particularly high density of negative charges at the double helix and is described quantitatively by polyelectrolyte theory[7-9]. Thus, addition of amino acid amides with their positive charge should also increase the melting temperature. However, the amides of the aromatic amino acids clearly induce a decrease of the t_m-values of double helices formed from polyribonucleotides[10] (Fig. 1). Obviously these ligands do not show 'ion atmosphere binding' but 'site binding'. The experimental data can be described quantitatively by a simple site binding model. As directly indicated by the t_m-decrease, the affinity of these ligands to the single strands is higher than that to the double strands. Measurements with the different aromatic amino acids[10] show an increase of the affinity in the series Phe<Tyr<Trp.

Some special properties may have been expected for ligands with aromatic residues, but probably not for other amino acid amides with simple aliphatic side chains. However, a small but clearly detectable decrease of the melting temperature is also induced by the amides of leucin, isoleucin, methionine and even valine[11]. The expected ligand induced increase of the melting temperature was observed for the amides of e.g. glycine, alanine or serine. In general, the experimental data indicate the existence of an affinity scale for the interaction of amino acid residues with nucleic acids[11,12]. Hydrophilic amino acids preferentially bind to double helical structures and thus increase the melting temperature, whereas hydrophobic amino acids are associated more tightly with single strands, which induces a decrease of melting temperatures (Fig. 2).

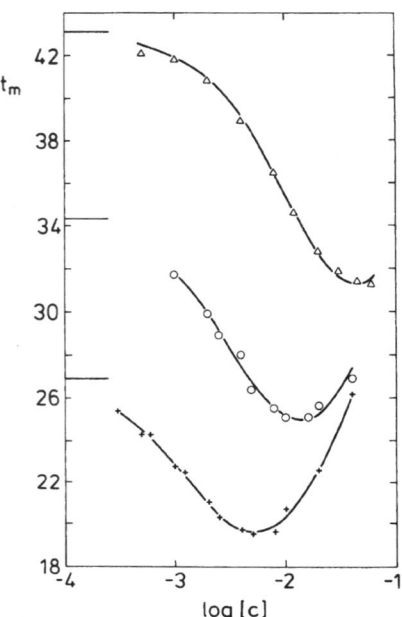

Fig. 1. t_m-values of poly(I)·poly(C) as a function of the logarithm of the Phe-amide concentration in 1mM NaCl, 1mM Na-cacodylate pH7, 0.2mM EDTA (buffer N; +), 2 times buffer N (o) and 9mM NaCl + buffer N (Δ). The solid lines show a simultaneous fit of all the data by a simple site model with consideration of a ionic strength dependence expected for unit charge compensation (for details cf. ref. 10; in the original figure legend Phe-amide was changed by error to Tyr-amide). The bars give the t_m-values in the absence of the ligand in the 3 different buffers. Binding parameters are compiled in Table I (from ref. 10).

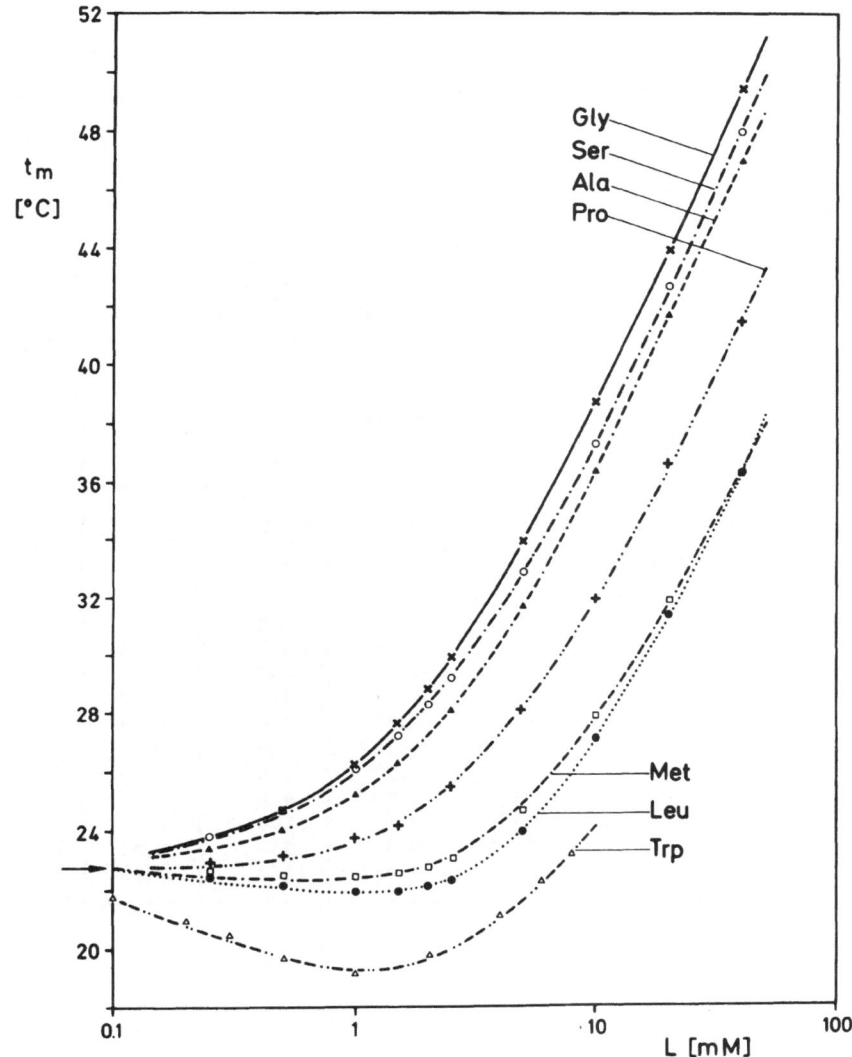

Fig. 2. t_m-values for poly(A)·poly(U) as a function of the logarithm of the concentration of various amides. Each line represents a least squares fit according to a site binding model. The arrow indicates the melting temperature in the absence of amide ligand (data from ref. 10 and 11).

This general result can be explained qualitatively by the fact that the surface of the double helix mainly exhibits hydrophilic residues like phosphates and sugars, which attract mainly hydrophilic ligands. The hydrophobic parts of the bases are much more exposed in the single stranded state, which leads to preferential association of hydrophobic ligands.

Table I. Parameters for the binding of amino acid amides to polynucleotides from melting data according to a site model (K_h (K_c) binding constant to the helix (coil form); n_c number of nucleotide residues per binding site in the coil form; 1 mM NaCl, 1 mM Na-cacodylate pH 7, 0.2 mM EDTA; data from ref. 10 and 11).

	polyA*polyU			polyI*polyC		
	K_h/K_c	$K_h[M^{-1}]$	n_c	K_h/K_c	$K_h[M^{-1}]$	n_c
Ala	1.30	150	1.14	1.20	210	1.15
Gly	1.15	350	1.17	1.15	480	1.16
Ser	1.07	380	1.17	1.16	410	1.16
Asn	1.00	260	1.17	1.00	360	1.17
Pro	0.84	510	1.19	0.95	480	1.18
Val	0.80	420	1.20	0.85	490	1.19
Met	0.80	460	1.18	0.76	380	1.21
Ile	0.76	440	1.20	0.78	470	1.20
Leu	0.75	430	1.20	0.75	440	1.19
Phe	0.70	480	1.15	0.59	290	1.19
Tyr	0.71	750	1.18	0.71	730	1.13
Trp	0.75	1480	1.11	0.71	1060	0.95

The binding parameters (cf. Table I) obtained from the interpretation of t_m-values by a simple site model also show some differences in the affinity of individual amino acid side chains to individual base pairs or nucleotide residues[11]. Unfortunately these differences are at the limit of accuracy even for the sensitive melting approach and thus should be regarded with caution.

Biological implications

The strong decrease of the melting temperature induced by aromatic amino acid amides demonstrates the existence of an individual character of these residues in their interaction with nucleic acids. It should be expected that these properties are useful for some biological function. Indeed, a well known class of proteins, which have been called melting proteins or single strand binding proteins (abbreviation SSB-proteins), apparently make extensive use of this property. The main function of SSB-proteins is dissociation of stable base-paired nucleic acid structures by selective binding to single strands[13,14]. In all cases, which have been analyzed so far, the residues of aromatic amino acids provide an important contribution to the binding of single strands. The major mode of binding appears to be insertion of the aromatic side chains between the nucleic acid bases[13]. Since the same function is already exhibited by simple derivatives of aromatic amino acids, there is a direct and facile evolutionary pathway from these amino acids (or small peptides containing these amino acids) to SSB-proteins.

The results of our model experiments can also be used to construct a simple mechanism for the evolution of the genetic code. The assignment of the codon triplets to the various amino acids does not follow any strict rules, but is at least partly ordered[15-18]: hydrophilic amino acids are coded preferentially by G/C containing codons, whereas hydrophobic amino acids are mainly coded by A/U codons. This partial order may be explained by our data according to the following argument: Let's assume the existence of two types of primitive adaptors, one of them mainly formed from A/U residues and the other one mainly from G/C residues. Because of the high stability of G/C pairs, the G/C rich adaptors are more base paired and contain more double helical structure than the A/U rich adaptors. Since double helices preferentially attract hydrophilic residues and single strands preferentially hydrophobic residues, the G/C adaptors are loaded mainly with hydrophilic amino acids and the A/U adaptors by hydrophobic amino acids. Thus, the simple mechanism predicts selective loading of adaptors, which is in general consistent with the genetic code table.

Selective interactions of amino acid residues with tRNA

If there is a simple stereo-chemical relation between nucleic acids and amino acids, indications for such a relation may be detectable in tRNA adaptors. Although these adaptors are now loaded with their cognate amino acids by complex enzyme systems, tRNA's may have some reminiscence of an early, simple selection mechanism. This possibility may be tested by experiments on the binding of simple amino acid derivatives to tRNA's. The experiments proved to be quite simple in the case of tRNA[Phe], where binding could be analyzed using the fluorescence of the Wye base - a natural label located at the anticodon. Addition of Phe-, Tyr- and

Trp-amide to tRNA^{Phe} induces a clearly detectable quenching of the Wye base fluorescence[19] (cf. Fig. 3), which is not observed for the other amino acid amides. A quantitative evaluation of the concentration dependence provided binding constants (Phe-amide 100 M^{-1}, Tyr-amide 110 M^{-1}, Trp-amide 300 M^{-1}; at 0.13 M salt and 7.2^{o}C) which are higher by a

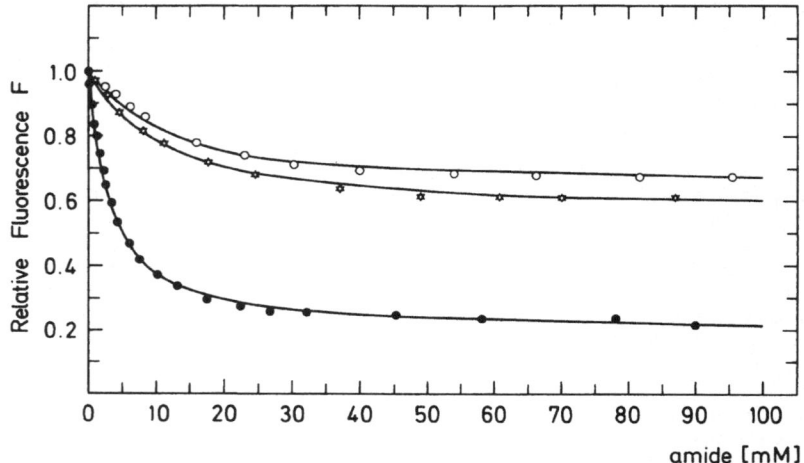

Fig. 3. Relative fluorescence intensity of tRNA^{Phe} as a function of the amide concentration: Phe-amide (o), Tyr-amide (*) and Trp-amide (●). The lines show a least squares fit according to a one step binding model (50mM NaClO$_4$, 80mM Tris-cacodylate pH 6.5; 7.2^{o}C; from ref. 19)

factor of 6 than that observed independently for simple single stranded polynucleotides and higher by a factor of 10 than for simple double stranded polynucleotides. Thus, tRNA^{Phe} selectively binds aromatic amino acid residues. Although selective interactions remain to be established in other cases, the results obtained for tRNA^{Phe} strongly supports the possibility of a stereo-chemical basis for the evolution of the genetic code.

Parameters obtained for oligonucleotide-oligopeptide complexes by the electric field jump method

The dynamics of simple model complexes between oligonucleotides and oligopeptides has been studied by the electric field jump method: electric field pulses of amplitudes up to 75 kV/cm are used to dissociate the complexes[3,20]; amplitudes and relaxation times of the reaction may be used to evaluate both thermodynamic and kinetic parameters for the complexation reaction. The results may be briefly summarised as follows:

Equilibrium parameters: In most cases the binding specificity observed for oligonucleotide-oligopeptide complexes is rather low. An example of a relatively high specificity has been observed[3,21] for the binding of a arginine trimer Arg_3 to an oligoinosinic acid $I(pI)_5$. The affinity of Arg_3 to this oligomer is higher by a factor of 4 than that found for Lys_3, which may form the same number of electrostatic contacts. A comparison of binding parameters obtained for various complexes indicates that the specificity observed in this case is due to hydrogen bonding of the guanidino group to the 06 and N7 acceptor sites of inosine. Corresponding hydrogen bonds can be formed with guanine. The experimental factor of 4 gives an indication for the degree of specificity accessible by hydrogen bonding at the level of simple oligomers.

Dynamics: The rate constants for formation of most oligomer complexes[3] studied by electric field jumps are around 10^{10} $M^{-1}s^{-1}$. These high rate constants indicate a diffusion controlled reaction: virtually all collisions between the reaction partners lead to complex formation. An exception to this rule is found for peptides containing aromatic amino acids[22,23]: in some cases a second slow intramolecular reaction step has been detected. Apparently the intramolecular step with forward rate constants around 10^4 s^{-1} reflects insertion of aromatic amino acids between bases of the oligonucleotide. Thus, a separate slow reaction step may be expected for the formation of protein nucleic acid complexes with insertion or intercalation of aromatic residues.

The structure of protein nucleic acid complexes from measurements of rotation time constants

Electric field pulses do not only induce reactions but also orientation effects. The orientation effects are particularly strong for large anisotropic molecules like DNA double helices. The degree of orientation and the orientation time constants can be characterized by absorbance measurements with polarized light ('linear dichroism')[24]. In the case of DNA helices, the orientation time constants provide direct and very accurate information on the length of the molecule, since the rotation time constant of rodlike molecules increases essentially with the third power of their length[25]. Thus, ligand induced changes of the DNA structure can be detected at a very high sensitivity.

This approach has been used recently to analyze the complex of the cyclic AMP receptor with its specific DNA recognition sites[26]. One of these complexes formed with a 80 bp fragment may be used here as an example: the rotation time τ_r of the free DNA is 0.49 µs; addition of an equivalent protein induces an increase to 0.52 µs, which is due to formation of the protein DNA complex; addition of the activator cyclic AMP induces a strong reduction to 0.13 µs. These results show that formation of the specific complex induces strong bending of the DNA around the protein. The distortion of the DNA structure by specific proteins seems to be important for the regulation of gene activity[27].

SOME SPECIAL EXAMPLES FOR THE DYNAMICS OF PROTEIN – NUCLEIC ACID INTERACTIONS

Saturation of a polymer lattice by a 'melting' protein

Usually protein ligands cover more than a single nucleotide residue. As shown by various authors, this type of binding leads to a relatively complex equilibrium distribution with exclusion of binding sites[28]. The kinetics of binding observed in these cases is particularly complex[29], as illustrated in Fig. 4. The first phase of ligand binding to the empty polymer lattice can still be described by standard second order kinetics. The ligands are attached to the polymer at random, until free sites with a sufficiently large number of nucleotide residues are not available anymore. Further binding of ligands is only possible via rearrangement of ligands along the polymer. The ligands may be rearranged by dissociation and reassociation: this pathway does not require any special properties of the ligand; however, it is a very slow reaction, since dissociation of ligands is necessary. A more elegant and also much faster rearrangement requires a special quality of the ligand: mobility along the lattice in the complexed state.

Experimental data on the binding dynamics of a protein, which covers 4 nucleotide residues, have been collected by stopped flow measurements for 'gene 5 protein' – a representative of single strand binding proteins. The results[30] depend very much upon the degree of binding approached in the experiment. At low degrees of binding

(<10%) the binding curves can be described by standard
second order kinetics, which provides the rate constants of
binding. When the degree of binding is increased to about
40%, the kinetic curves are much more complex: fitting by
exponentials requires at least 3 separate time constants.
This result may lead to an erroneous interpretation by 3

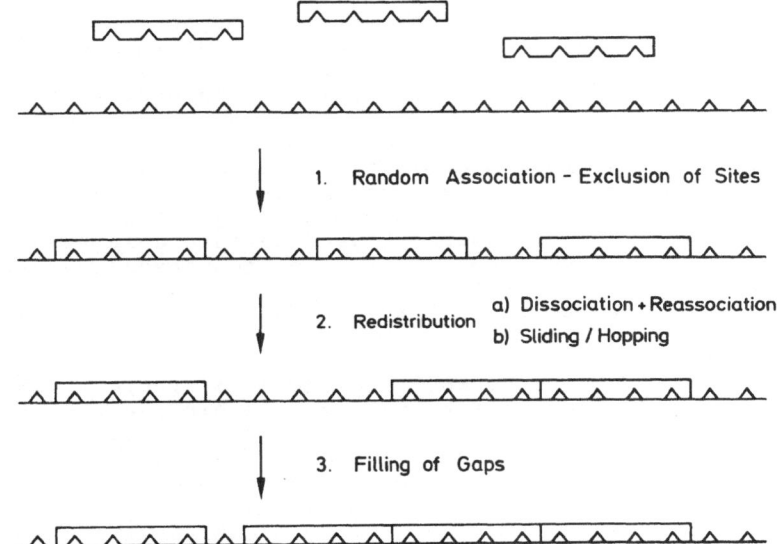

Fig. 4. Dynamics of excluded site binding for a ligand that
covers 4 residues.

independent reactions. Simulations of the binding dynamics
according to Fig. 4 by Monte Carlo procedures[29] shows that
the broadened reaction progress curve is described within
the limits of accuracy by excluded site phenomena. Since
the experimental data could be simulated without any
assumption using a direct binding mechanism and
consideration of excluded site phenomena, this result also
demonstrates that rearrangements of gene 5 protein complexes
are not accelerated by motion along the lattice at the
millisecond time scale of the stopped flow experiment.

However, mobility of gene 5 protein complexes is observed at a slower time scale. When an excess of gene 5 protein is added to the polymer poly(dT), the lattice snould be covered by ligands completely. However, this reaction is not at all simple due to excluded side effects, which are dominant at high degrees of binding. In the case of the gene 5 protein the kinetics at high Θ is mainly determined by cooperatively bound ligands, which dissociate only at a very low rate. This is reflected in Monte Carlo simulations of the binding dynamics by an extremely slow progress of the reaction at high Θ: the reaction simulated without ligand mobility is not yet complete even after 1000 s (cf. Fig. 5a). In contrast to these simulations, the slowest reaction component detected for binding of gene 5 protein to poly (dT) up to high Θ is 6 s (cf. Fig. 5b). Thus, the stopped flow experiments demonstrate that gene 5 protein complexes are mobile at the s-time scale.

The absence of gene 5 mobility at the ms time scale is not surprising in view of the strong evidence that this protein interacts with polynucleotides by insertion of aromatic amino acid side chains between nucleic acid bases. Interactions via insertion clearly do not favor any fast sliding reaction. The mobility observed at the s-time scale is apparently due to one of the other mechanisms discussed below.

The simulations performed for gene 5 protein as a representative of non-specific nucleic acid binding proteins demonstrate that in general any protein ligand, which has to cover a polynucleotide at high degrees of binding, must be mobile for a dynamics compatible with the biological time scale.

Recognition of an operator by a repressor

Until recently the lac-repressor has been the only example of a specific transcription control protein, which has been studied with respect to the dynamics of operator recognition. The results obtained originally by Riggs et al.[31] and later extended by Barkley[32] and by Winter et al.[33] demonstrate that lac repressor finds its operator target at a higher rate than expected for a simple diffusion controlled reaction. This phenomenon is explained by sliding of the lac repressor along unspecific DNA, which leads to an increase of the effective operator target size[34-36].

Since the experimental data for the lac system had to be collected by the filter binding technique, one should like to analyze another system by a more direct technique. The tet repressor regulating the tetracycline resistance[37] in the E.coli transposon Tn10 proved to be an appropriate system owing to its tryptophan fluorescence, which is quenched upon binding to DNA[38]. The rate constants obtained for the binding of tet repressor to its operator by stopped

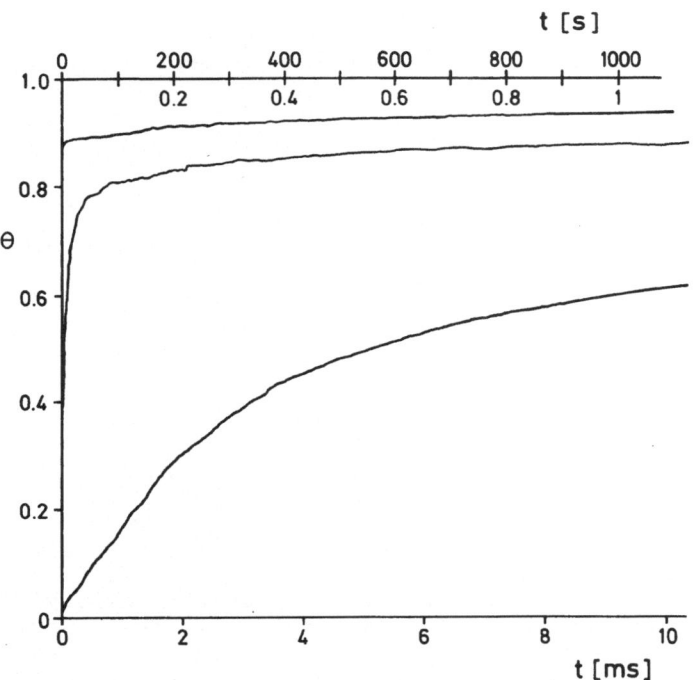

Fig. 5a. Binding kinetics of a ligand (2µM, covers 4 sites) to a polymer lattice (4µM monomer units, chain length 100 nucleotides) simulated by a Monte Carlo procedure according to Epstein[29]. Rate constants for isolated sites: association $k_a^i = 2.7 \cdot 10^7$ $M^{-1}s^1$, dissociation $k_d^i = 4.2$ s^{-1} and for cooperative sites: association $k_a^c = 2.7 \cdot 10^7$ $M^{-1}s^{-1}$, dissociation $k_d^c = 0.0084$ s^{-1} (cf. ref. 30). The degree of binding Θ is given in 3 different time scales.

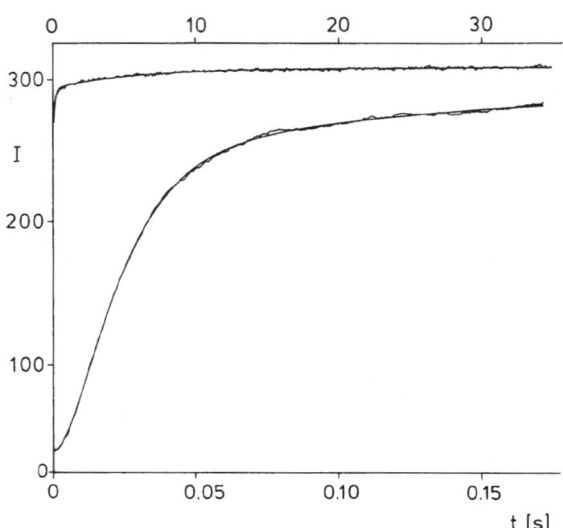

Fig. 5b. Change of the fluorescence intensity (arbitrary units) as a function of time observed in a stopped flow apparatus after mixing of gene 5 protein and poly(dT) to concentrations of 2μM and 4μM respectively. The signal is convoluted with a detector rise time of 10ms. The upper and lower time scales are valid for the upper and lower curve, respectively. The line without noise represents a least squares fit with 3 exponentials with time constants 15.2ms, 120ms and 5.8 s and amplitudes 75.8%, 19.1% and 5.1% respectively (0.1 M NaCl, 5mM Na-cacodylate pH 6.7, 2mM EDTA; 20°C; cf. ref. 30)

flow measurements[38] do not indicate any contribution from a sliding mechanism. Some enhancement of the observed association rates above the usual diffusion controlled limit is due to hopping phenomena, which are expected for any ligand with the capacity of unspecific binding.

A detailed analysis of the data found for lac and tet repressor indicate that the differences are probably not due to the different experimental techniques but more due to different modes of binding. The unspecific binding affinity of lac repressor is much higher than that found for tet repressor, which is due to a higher number of ion contacts in the unspecific complex of the lac repressor than that of the tet repressor. Another obvious difference exists with respect to the molecular weight and the number of subunits. A comparison of some important parameters for the two repressors is given in Table II.

Table II. Comparison of lac and tet repressor. Data are from ref. 32, 33, 38 and other references cited therein.

		LAC	TET	
BINDING CONSTANTS AT	OPERATOR SITE	$7*10^{12}$	$2*10^{11}$	M^{-1}
AT 0.16 M SALT	UNSPECIFIC SITE	$2*10^{4}$	$3*10^{2}$	M^{-1}
NUMBER	OPERATOR SITE	8-9	7-8	
OF ION	UNSPECIFIC SITE	11-12	3-4	
CONTACTS	PRE-EQUILIBRIUM	5-6	3-4	
	SLIDING	YES	NO	
	TARGET SIZE	300-1000	$\leqslant 40$	BP
(INTRAMOLECULAR) OPERATOR RECOGNITION RATE		?	$6*10^{4}$	s^{-1}
	MOLECULAR WEIGHT	150,000	47,000	DALTON
	SUBUNIT NUMBER	4	2	

Mechanisms of motion along a polynucleotide

The clear differences observed for the mobility of protein ligands along DNA indicate that the modes of binding and also of the binding dynamics are different. Thus, it should be useful to compare various binding modes, which are associated with different mechanisms of mobility.

The free energy of unspecific binding is known to result mainly from electrostatic contacts between lysine or arginine residues of the protein and phosphate residues of the nucleic acid. Since the positive charges of lysine or arginine residues are attached to the peptide backbone via methylen chains, the charge centers may be quite mobile and may form contacts at either one of two adjacent phosphates. When all the contact sites have this type of mobility, the whole protein ligand may move by subsequent shifts of all individual contact sites. Because the majority of contacts remain closed at any time, the activation energy is relatively low and thus the 'crawling' mechanism is expected to be quite efficient.

If the positive charge centers of the protein are not mobile but as fixed in their position as the phosphate residues of DNA, the protein may move by another mechanism. The charges of the protein at the contact site to the DNA may be arranged to form an interaction potential with the DNA, which is independent of the protein position along the DNA. In this case the protein moves without any activation barrier and the rate of motion is only limited by hydrodynamic friction[39]. The equipotential surface for the interaction with DNA, which is required for this 'sliding' mechanism , can be constructed more and more smoothly with an increasing number of charge centers.

Proteins with more than a single binding site for DNA may move by still another mechanism. If the binding sites are flexible, contacts of individual sites may be dissociated and formed again independent of each other. As shown in Fig. 6c, dissociation of one binding unit at one side and its reassociation at the other side leads to motion with activation barriers, which are least clearly lower than required for motion by complete dissociation and reassociation. For a protein like lac repressor with 4 binding sites alternative dissociation and formation of contacts may result in a 'rolling' mechanism (cf. ref. 40).

Finally another mechanism should be mentioned, which requires more than a single binding site. In this case the contacts are not dissociated and formed at adjacent residues of the DNA helix, but at different segments of a long DNA molecule as shown in Fig. 6e. This mechanism should be particularly efficient, when the DNA is partly covered by other ligands. These other ligands would seriously inhibit 'crawling', 'sliding' or 'rolling', but not interfere much with 'segment jumping' (cf. ref. 36). Thus, segment jumping is a very attractive mechanism for fast recognition of control elements in chromatin, for example.

a

b

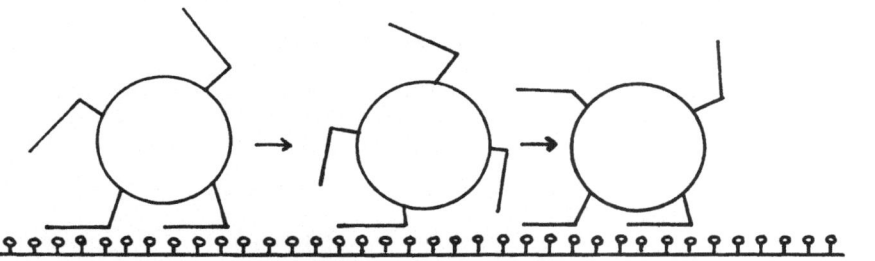

c

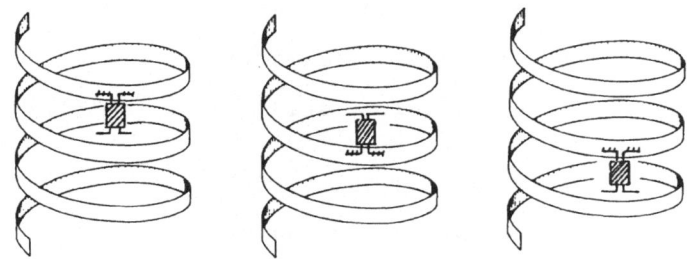

d

e

Fig. 6. Schematic representation of motion mechanisms (cf. text): a)"crawling"; b)"sliding"; c)"twirling"; d)"rolling"; e)"segment jumping".

This compilation illustrates that there is a large number of mechanisms for motion of protein ligands along DNA. However, it is also obvious that a high mobility requires special qualities of the ligand. Since there are clearly other demands on a gene regulating protein than high mobility along DNA, it is certainly not surprising to find proteins like tet repressor without high mobility. Probably the tet repressor represents an example, where evolutionary pressure lead to the construction of a relatively small protein - an economy version without special accessories.

ACKNOWLEDGEMENT

The results described in this short review have been obtained together with several colleagues and coworkers: W. Bujalowski, W. Hillen, M. Jung, C. Kleinschmidt, H. Rauh and M. Takahashi. The author is also greatly indebted to J. Ronnenberg, who contributed to many projects by his expert technical assistance.

REFERENCES

1. C. Helene and J.C. Maurizot, CRC Crit. Rev. Biochem. 10:213 (1981)

2. J.C. Lacey and D.W. Mullins Jr., Orig. Life 13:3 (1983)

3. D. Porschke, Eur. J. Biochem 86:291 (1978)

4. E.J. Gabbay, in Bioorganic "Chemistry", Vol. 3, Macro- and Multimolecular Systems, Ed. van Tamelen, Academic Press, New York, p. 33

5. Y.S. Lazurkin, M.D. Frank-Kamenetskii and E.N. Trifonov, Biopolymers 9:1253 (1970)

6. J.D. McGhee, Biopolymers 15:1345 (1976)

7. L. Kotin, J. Mol. Biol. 7:309 (1963)

8. C. Schildkraut and S. Lifson, Biopolymers 3:195 (1965)

9. G.S. Manning, Q. Rev. Biophys. 11:179 (1978)

10. D. Porschke and M. Jung, Nucl. Acids Res. 10:6163 (1982)

11. D. Porschke, J. Mol. Evol. 21:192 (1985)

12. D. Porschke and J. Ronnenberg, Biopolymers 22:2549 (1983)

13. S.C. Kowalczykowski, D.G. Bear and P.H. von Hippel, " The Enzymes", Vol. XIV, p. 373 (1981)

14. J.W. Chase and K.R. Williams, Ann. Rev. Biochem. 55:103 (1986)

15. M.V. Volkenstein, Biochim. Biophys. Acta 119:421 (1966)

16. C. Woese,"The genetic code", Harper and Row, New York (1967)

17. T.H. Jukes,"The amino acid code", in: Neuberger, A., Ed., Comprehensive biochemistry (Elsevier, Amsterdam) Vol. 24:235 (1977)

18. R.V. Wolfenden, P.M. Cullis and C.C.F. Southgate, Science 206:575 (1979)

19. W. Bujalowski and D. Porschke, Nucl. Acids Res. 12:7549 (1984)

20. D. Porschke, Ann. Rev. Phys. Chem. 36:159 (1985)

21. D. Porschke, Biophys. Chem. 10:1 (1979)

22. D. Porschke, Nucl. Acids Res. 8:1591 (1980)

23. D. Porschke and J. Ronnenberg, Biophys. Chem. 13:283 (1981)

24. E. Fredericq and C. Houssier, "Electric Dichroism and Electric Birefringence", Clarendon, Oxford (1973)

25. S. Broersma, J. Chem. Phys. 32:1626, 1632 (1960)

26. D. Porschke, W. Hillen and M. Takahashi, EMBO J. 3:2873 (1984)

27. M. Ptashne, Nature 322:697 (1986)

28. J.D. McGhee and P.H. von Hippel, J. Mol. Biol. 86:469 (1974)

29. I.R. Epstein, Biopolymers 18:2037 (1979)

30. D. Porschke and H. Rauh, Biochemistry 22:4737 (1983)

31. A.D. Riggs, S. Bourgeois and M. Cohn, J. Mol. Biol. 53:401 (1970)

32. M.D. Barkley, Biochemistry 20:3833 (1981)

33. R.B. Winter, O.G. Berg and P.H. von Hippel, Biochemistry 20:6961 (1981)

34. G. Adam and M. Delbrück, in "Structural chemistry and molecular biology", Eds. A. Rich and N. Davidson, Freeman, p. 198 (1968)

35. P.H. Richter and M. Eigen, Biophys. Chem. 2:255 (1974)

36. O.G. Berg, R.B. Winter and P.H. von Hippel, Biochemistry 20:6929 (1981)

37. W. Hillen, K. Schollmeier and C. Gatz, J. Mol. Biol. 172:185 (1984)

38. C. Kleinschmidt, K. Tovar, W. Hillen and D. Porschke, J. Mol. Biol. submitted

39. J.M. Schurr, Biophys. Chem. 9:413 (1979)

40. R. Roemer, U. Schomburg, G. Kraus and G. Maass, Biochemistry 23:6132 (1984)

DNA AND ITS COUNTERIONS: METAL IONS, AMINO ACIDS, HISTONES

AND PROTAMINES

Juan A. Subirana

Escuela T.S.de Ingenieros Industriales
Diagonal 647
08028 Barcelona, Spain

INTRODUCTION

The conformation of DNA is influenced both by its sequence
and by the counterions and other small molecules (in particular water) which interact with it. In this article we will review the effect on DNA of counterions, including basic proteins.
We will not deal with sequence effects and thus we will not
consider specific protein-DNA interactions. We will also not
consider the influence of ions on the denaturation of DNA in
solution.

METAL IONS

The various conformations of DNA found in the presence of
metal ions are summarized in Fig. 1. In the early fiber diffraction studies carried out in the laboratory of Prof. Wilkins,
the alkali metal salts of DNA were throughly studied. It was
then established that:
- at high humidities DNA usually remains in the B form for all
  these counterions (Langridge et al, 1960). In aqueous solution in fact only the B form is observed with any of these
  counterions at any ionic strength.
- upon moderate dehydration DNA changes its conformation to the
  C form in the case of the lithium salt (Marvin et al, 1961;
  under some well controlled conditions, this form has also
  been observed in the sodium salt by Rhodes et al, 1982) and
  to the A form in the case of the sodium salt (Fuller et al,
  1965). More recently it has been shown that the A to B transition depends on the amount of excess salt in the fibers
  used for X-ray diffraction, so that the A form is preserved
  even at high humidities, so long as the amount of excess
  NaCl present in the fibers is minimal (Cooper and Hamilton,
  1966; Fornells et al, 1983).These results indicate that the
  A form might be stable under conditions of biological inte-
  rest.However, nobody has succeeded in observing this form
  of DNA in aqueous solution, only in the presence of ethanol
  it is possible to observe it in solution (Ivanov et al,1983).
  From these studies, later confirmed by a large body of ex-

Low Humidity form of DNA	Counterion	
	Metal	Amino
A	Na	Not found
B	Unstable	Arg Several charges
C and related	Li, (Na)	Poly One charge
Destabilized	Transition	Hydrophobic
D, Z	Peculiar Sequences	

Fig.1. Forms of DNA found upon dehydration of different DNA salts. Under high humidity conditions the only regular conformation found in DNA is the B form. Upon dehydration it may loose this conformation, as indicated in the figure.The changes found depend on the counterions present. "Amino" refers to amino acid and oligopeptide counterions (Campos et al, 1986). In the figure only the most common situations are shown. The text and the references quoted should be consulted for details.

perimental results, it was established that the normal conformation of DNA in a hydrated environment is the B form. Upon dehydration, and depending on the counterion, the A and C forms are found. In the case of some regular base sequences it is also possible to find other dehydrated structures, such as the D and Z conformations, as reviewed by Leslie et al (1980).

In the presence of transition metals the situation is complicated by the fact that they interact with the nitrogen atoms of the bases and the B conformation is destabilized (Eichhorn and Shin, 1968). No X-ray data are available on the resulting conformations.

BASIC AMINO ACIDS AND OLIGOPEPTIDES

We have recently reviewed our work with these substances (Campos et al, 1986), so that here we will only provide a summary of the results we have obtained, which are schematically presented in Fig. 1.We have monitored the changes in conformation of DNA by measuring its change in pitch as a function of water content in fibers.A set of representative results is shown in Figs. 2 and 3. Our main conclusions can be summarized as follows:
(1) At high levels of relative humidity, the B form of DNA is always found, with about ten base-pairs per helical turn. We have never found a higher pitch as it has been reported in solution (Strauss et al, 1981, and references therein). I suggested an explanation for this discrepancy (Subirana, 1982) as due to the fact that in an X-ray ex-

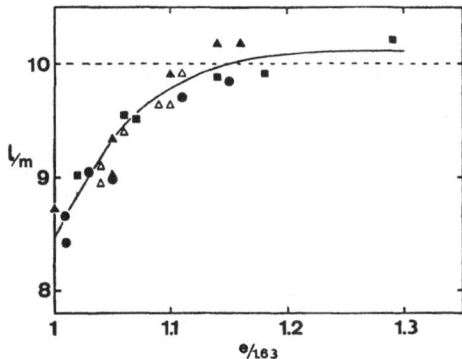

Fig.2. Change in pitch of DNA as a function of water concentration for different counterions which alter the conformation of DNA (Portugal and Subirana, 1985). The abscissa measures indirectly the water concentration, as the equatorial spacing determined by fiber diffraction divided by 1.63 nm, which is the lowest equatorial spacing found at 0% relative humidity. The pitch of DNA is given in the ordinate as the number of base pairs per turn. It is measured on fiber diagrams as the ratio of the layer line spacing (l) divided by the meridional spacing (m). The following complexes are presented: Histidine-amide (●); alaninol (△); serinol (▲) and "Tris" (■)

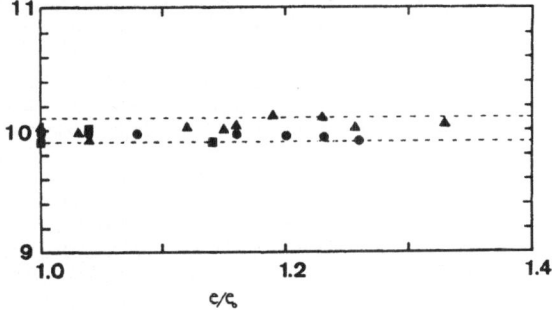

Fig.3. Stability of the B form of DNA when associated with either arginine (▲), lysine (■) histidine (●). Data given as in Fig. 2.

periment we measure the pitch of molecules which are maintained in a straight position, whereas in solution the DNA molecules are bent as a result of Brownian motion.

(2) The A form of DNA is never observed upon dehydration, as long as the DNA phosphate charges are fully neutralized by the peptide ions.

(3) Upon   dehydration, the B form of DNA may be either stabilized, altered or destabilized. Arginine oligopeptides, as well as other oligopeptides containing several charges, tend to stabilize the B form. On the other hand, oligopeptides with a single charge tend to alter the conformation of DNA by reducing the number of base pairs per helical turn. It is striking that practically never the pitch of DNA is found to increase. In some cases the altered conformation corresponds to the C form of DNA. Surprisingly, large basic proteins (histones, protamines) and synthetic basic polypeptides which contain many positive charges, also show this reduction of pitch of DNA upon dehydration. Finally, when DNA is neutralized by some peptides with a hydrophobic nature, its conformational order is lost upon dehydration, since no layer-lines are detected.

PROTAMINES

As reviewed by Subirana (1983), protamines may show a large compositional variety, although they always have a high content of basic amino acids (45 to 80%). In vertebrates, clusters of basic amino acids are usually found, although in some species regions with a different sequence organization are common. Recent work in the sequence of some mollusc protamines (Sautiere, private communication) has shown extreme cases in the sequential organization of these protamines. Thus regions containing clusters of 10 arginines are found in squid protamine, whereas in the major component of Mytilus edulis most of the basic amino acids are either isolated or organized in clusters of two amino acids. Part of this molecule has a long string of alternating basic-non basic residues, which had only been previously found in the N-terminalend of galline.

In spite of these changes in composition and sequential organization, the nucleoprotamine complexes appear to have always a simple structure, as a set of parallel DNA molecules packed in a simple hexagonal system (Suau and Subirana, 1977). From a comparison with the structure of DNA-arginine and DNA polyarginine complexes, we have concluded that protamines are associated with DNA on the wide groove (Fita et al, 1983). We have also shown that it is possible to neutralize all the phosphate charges with the basic residues present and at the same time accomodate the neutral residues on the wide groove without steric hindrances. However the exact conformation of the protamine cannot be determined from the X-ray patterns, since they lack enough detail. It appears that an undistorted α-helical structure can not be accomodated among the DNA molecules because its diameter is too large for the available space. On the other hand, the models that we have considered thus far (Fita et al, 1983; Puigjaner et al, 1986), although they are stereochemically correct, do not take into account the hydrogen bonding capability of the peptide units.

In mammalian spermatozoa, the presence of cystine residues results in a reduced crystallinity of the nucleoprotamine complexes (Loir et al, 1985), but the DNA molecules are still organized as bundles of parallel molecules.

BASIC POLYPEPTIDES

As shown in Fig. 1, basic polypeptides and proteins alter the conformation of DNA upon dehydration as observed with many small counterions. But we may now ask what will happen to the shape of a basic polypeptide chain with a well defined conformation (α, β, globular) when it interacts with DNA. We have recently studied in detail this question (Azorín et al, 1985).

When we analyze the interaction of DNA with sequential oligo- and polypeptides which form β-sheets, the DNA usually forms layers which are stacked between the β-sheets. There can be one or two polypeptide sheets between the DNA layers, depending on the charge density as it is schematically indicated in Fig. 4. However, in the case of poly (Ala-Lys) we found that the β-sheets were not stable in presence of DNA, so that this polypeptide apparently formed α helices which are hexagonally packed together with the DNA molecules, as it is schematically shown in Fig.4.

When we study polypeptides with an α-helical conformation, such a conformation may again be preserved or modified upon interaction with the DNA. Two types of ordered arrays of α helices and DNA molecules that we have observed are shown in Figs. 4 c and d.

In conclusion it appears that basic polypeptides with a defined conformation (α or β) usually maintain it upon interaction with DNA, but it may be modified depending on the amino acid sequence.

In the more complex case of histone H1, the whole protein and the N-terminal half show a complex behaviour which we will not discuss here. On the other hand, the highly charged C-terminal region behaves like a protamine, but surprisingly, at high humidity it develops a cross-β structure, indicating that some regions of the protein (probably with a small charge density) aggregate in this conformation. These results have been reported in detail by Azorín et al (1985).

CHROMATIN

In the last twelve years, the association of histones and DNA in nucleosomes has been studied in detail. The crystallographic structure of the nucleosome core has been reported by Richmond et al (1984), who have confirmed that the DNA is wrapped around a histone core containing a complex of eight histones. Whereas in the case of protamines these proteins follow the grooves of DNA without distorting it, here the opposite is found: the DNA bends in order to follow the highly charged surface of the histone core. We are thus confronted with another case in which the DNA adapts itself to the protein conformation, a situation which is reminiscent of those described in the previous section and shown in Fig. 4.

In inactive chromatin under physiological conditions the nucleosomes are aggregated and give rise to chromatin fibers with a diameter of about 30 nm. The nucleosomes are stacked on their edges along these fibers (Subirana and Martinez,1976). Finch and Klug (1976) have suggested a simple helical model, but unfortunately it has not been stablished with certainty,

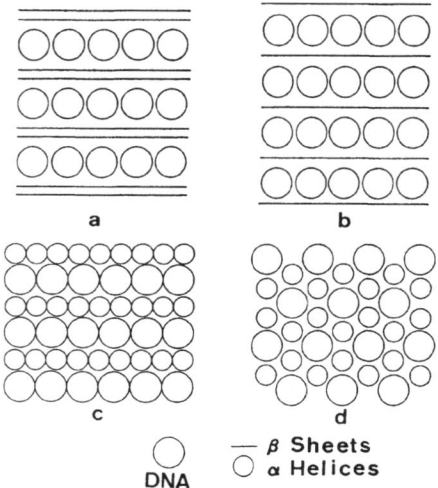

Fig.4. Some of the packing schemes found in complexes of DNA with basic polypeptides (Azorín et al, 1985). In (a) pairs of β-sheets are present between DNA layers. The double sheets are stabilized by hydrophobic interactions. This organization has been found with N-acetyl-Lys-Ala-Tyr-Ala-Lys-ethylamide. In (b) single β-sheets are stacked between layers of DNA molecules. Found in poly (Leu-Lys) and in poly (Val-Lys). In (c) and (d) are shown different arrangements of α-helices and DNA molecules found in polypeptides which contain lysine and alanine in different sequential order.

since it has not yet been possible to experimentally determine the organization of the linker DNA which binds one nucleosome core to the next. Under these conditions, although all workers agree that the nucleosome cores are tightly packed inside the chromatin fiber, quite different topological models have been proposed. Several of these models take into account that extended chromatin fibers show the nucleosome cores organized in a zig-zag fashion, as demonstrated for example by Thoma et al (1979). It is thus logical to suggest that the zig-zag may be preserved in the chromatin fibers, either folded in a helical fashion (Azorín et al, 1980; Worcel et al, 1981; Woodcock et al, 1984; Williams et al, 1986; Sen et al, 1986) or in an accordion fashion (Subirana et al, 1985). Other workers have suggested (Bordas et al, 1986) that the chromatin fibers form a loose helix at low ionic strength which collapses on itself at physiological ionic strength. A decision on the adequacy of these various models awaits a clear cut determination of the spatial arrangement of spacer DNA in chromatin fibers.

REFERENCES

Azorín,F., Martinez,A.B., Subirana,J.A., 1980, Int.J.Biol Macromol, 2:81

Azorín,F.,Vives, J.,Campos,J.L.,Jordan,A., Lloveras,J.,Puigjaner, L.,Subirana,J.A.,Mayer,R.,and Brack,A.,1985, J.Mol.Biol, 185: 371

Bordas,J.,Perez-Grau,L.,Koch,M.H.J.,Vega,M.C.,and Nave,C.,1986 Eur.Biophys J ,13:175

Campos,J.L.,Portugal,J.,Subirana,J.A.,and Mayer,R.,1986,J.Mol. Biol., 187:441

Cooper, P.J.,and Hamilton,L.D., 1966, J.Mol.Biol, 16:562

Eichhorn,G.L., and Shin,Y.A., 1968,J.Amer.chem.Soc.,90:7323

Finch,J.T., and Klug,A., 1976, Proc Natl Acad Sci USA, 73:1897

Fita,I., Campos,J.L., Puigjaner L.C., and Subirana J.A.,1983, J.Mol.Biol., 167: 157

Fornells,M., Campos,J.L., Puigjaner,L.C., and Subirana,J.A., 1983, J.Mol. Biol., 166: 249

Fuller,W., Wilkins,M.H.F.,Wilson,H.R., and Hamilton, L.D.,1965, J.Mol.Biol, 12: 60

Ivanov,V.I., Minchenkova,L.E., Minyat,E.E., and Schyolkina, 1983, Cold Spring Harbor Symposia on Quantitative Biology,47:243

Landgridge,R., Wilson,H.R., Hooper,C.W.,Wilkins,M.H.F., and Hamilton,L.D., 1960, J.Mol.Biol, 2: 19

Leslie,A.G.W., Arnott, S., Chandrasekaran,R., and Ratliff,R.L., 1980, J.Mol.Biol, 143: 49

Loir,M.,Bouvier,D., Fornells,M., Lanneau,M., and J.A. Subirana, 1985, Chromosoma, 92: 304.

Marvin, D.A.,Spencer, M., Wilkins,M.H.F.,and Hamilton, L.D., 1961, J.Mol.Biol., 3: 547

Portugal J., and Subirana J.A., 1985, The EMBO Journal, vol.4 n.9:2403

Puigjaner, L.C.,Fita,I., Arnott,S.,Chandrasekaran,R.,and Subirana,J.A., 1986, J.Biomol.Str.Dyn., 3: 1067

Richmond, T.J., Fintch, J.T., Rushton,B., Rhodes,D., and Klug, A., 1984, Nature 311: 532

Rhodes,N.J., Mahendrasingam,A., Pigram,W.J., Fuller,W.,Brahms, J., Vergne,J., and Warren, R.A.J., 1982, Nature 296:267

Sen,D., Mitra,S., and Crothers, D.M., 1986, Biochemistry, 25: 3441

Strauss, F., Gaillard,C., and Prunell,A., 1981, Eur.J.Biochem, 118:215

Suau,P., and Subirana, J.A., 1977,J.Mol.Biol , 117:909

Subirana, J.A., 1982, "Structural Molecular Biology Methods and Applications", Davies,D.B.,Saenger,W., and Danyluk,S.S., eds.Plenum Press, New York & London, 455

Subirana, J.A.,1983, "The Sperm Cell", ed. J.André, Martinus,N., BV, The Hague, 197

Subirana, J.A., and Martinez,A.B., 1976, Nucl.Ac.Res., 3: 3025

Subirana, J.A., MUñoz-Guerra,S., Aymamí,J., Radermacher,M., and Frank,J., 1985, Chromosoma, 91:377

Thoma,F., Koller,Th., Klug,A., 1979, J.Cell,Biol 83: 403

Williams, S.P.,Athey,B.D., Muglia,L.J., Scott Schappe,R.,Gough,
    A.H., and Langmore,J.P., 1986, Biophys.J., 49: 233
Worcel,A., Strogatz,S., Riley, D., 1981,Proc Natl Acad Sci USA,
    78: 1461
Woodcock,C.L.F.,Frado,L.L., and Rattner, J.B., 1984, J.Cell Biol,
    99:42

DYNAMIC ASPECTS OF ANTIBIOTIC-DNA INTERACTION

Michael Waring

University of Cambridge   Department of Pharmacology
Medical School, Hills Road, Cambridge CB2 2QD   England

INTRODUCTION

The DNA molecule is a target for a variety of chemotherapeutic agents, including numerous antibiotics (Figure 1). Among them are antibacterial, antiviral and antiprotozoal compounds, but probably most important as regards the frontiers of chemotherapy are the anticancer drugs. A quarter of a century has elapsed since actinomycin was first identified as a DNA-binder whose antitumour activity could be attributed to its capacity to distort the structure and function of DNA (reviewed by Reich and Goldberg, 1964; Sobell, 1973; Gale et al., 1981). During that time increasingly sophisticated experiments have been performed to elucidate the molecular mechanisms of antibiotic-DNA interaction and to attempt to relate those interactions to biological activity (Gale et al., 1981; Waring, 1981). Here we shall be mainly concerned with a group of cyclic depsipeptide antibiotics called the quinoxalines, of which echinomycin (Figure 2) is the best-known member (Waring, 1979; Waring and Fox, 1983).

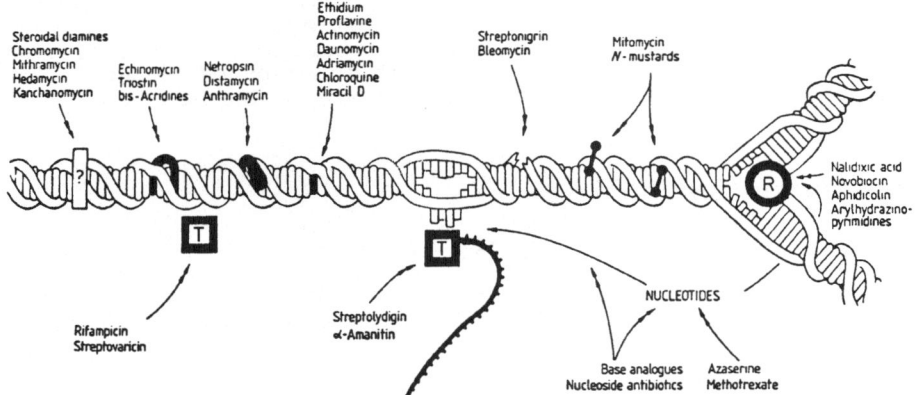

Figure 1. Diagrammatic representation of a molecule of DNA engaged in transcription by RNA polymerase (T) and undergoing duplication by the replicating enzyme complex (R), subject to interference by various antibiotics and drugs whose principal sites of action are indicated by double-headed arrows. From Gale et al. (1981).

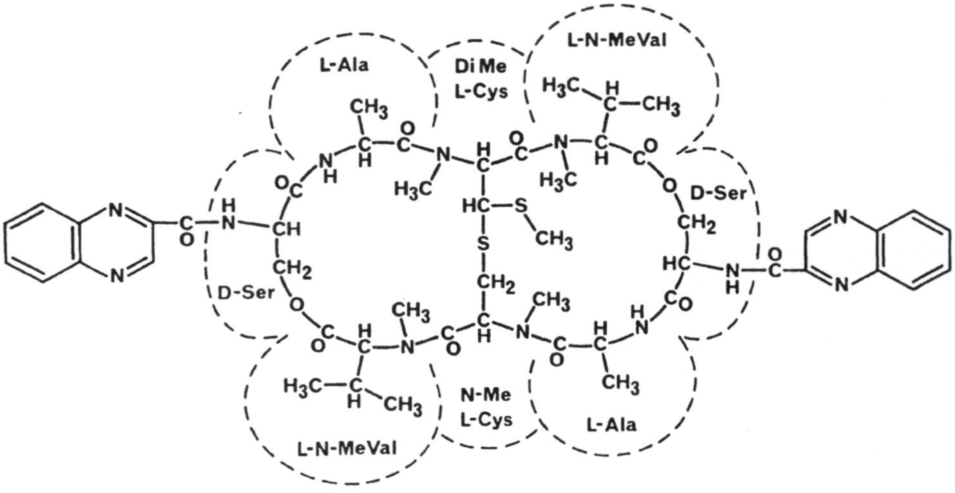

Figure 2. Structural formula of echinomycin (quinomycin A).

Early work established that quinoxaline antibiotics are powerful antitumour drugs (reviewed by Katagiri et al., 1975) and it was not long before their potency was attributed to interaction with DNA resulting in selective inhibition of DNA-directed RNA synthesis (Ward et al., 1965; Sato et al., 1967; Waring and Makoff, 1974). The binding of echinomycin to DNA was subsequently investigated in detail, culminating in the discovery that it binds to DNA by the hitherto unknown mechanism of bifunctional (bis) intercalation (Waring and Wakelin, 1974) and that it seemed to be selective for GC-rich sequences (Wakelin and Waring, 1976). The true nature of that selectivity (actually better called preference) only became fully apparent with the advent of "footprinting" methods for detecting antibiotic binding sites on natural DNA fragments of known nucleotide sequence, employing either DNAase I or the synthetic reagent methidium-propyl-EDTA-Fe(II) as a probe for accessibility of internucleotide bonds to cleavage (Low et al., 1984; Van Dyke and Dervan, 1984).

SEQUENCE-SELECTIVE BINDING TO DNA

The results of a classical footprinting experiment with echinomycin are illustrated in Figure 3. Six or seven antibiotic binding sites can be clearly identified on the 160 base-pair DNA molecule employed here (visualised as gaps in the ladder of fragments produced by DNAase I

114

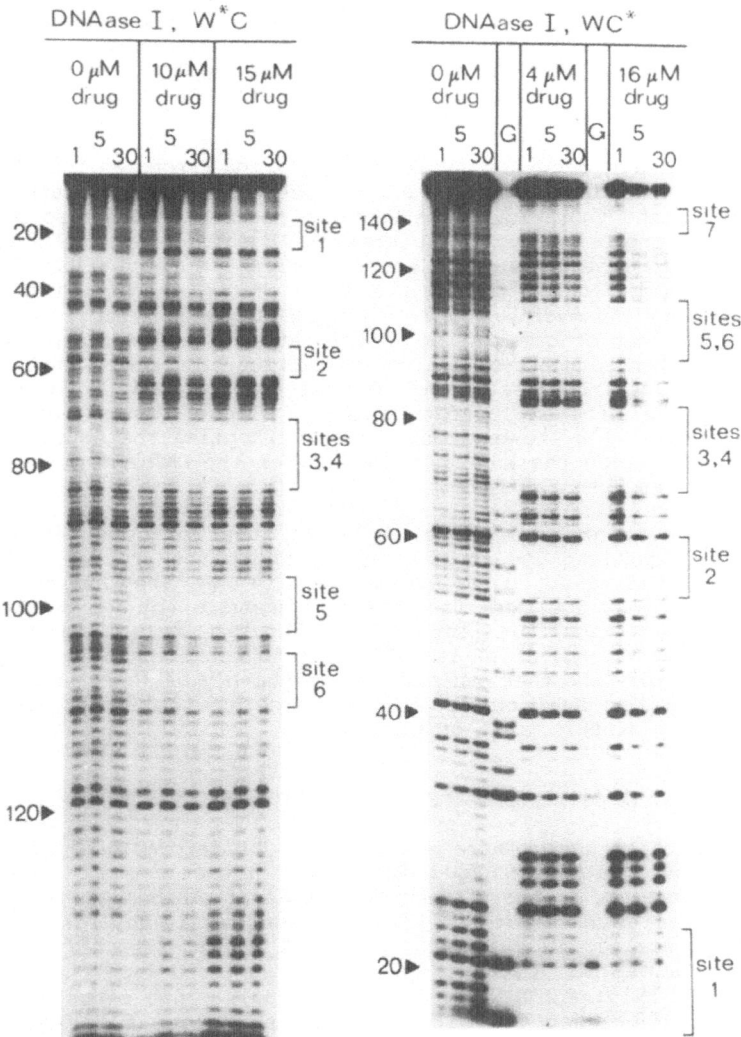

Figure 3. DNAase I footprinting of echinomycin bound to a 160-base-pair DNA fragment from E.coli containing the tyrT promoter sequence. Symbols W*C and WC* indicate which of the two strands (Watson or Crick) bears a radioactive 3'-end label. Time in minutes (1, 5, 30) after the addition of enzyme is shown at the top of each gel lane. Tracks labelled "G" are dimethyl sulphate-piperidine markers specific for guanine. Numbers on the left refer to the numbering scheme shown in Figure 4, while sites of protection from DNAase I digestion are identified on the right. For details see Low et al. (1984).

digestion) and each is centred around one or more CpG steps in the nucleotide sequence (see the differential cleavage plot represented in Figure 4). With the possible exception of the region around position 35, all CpG sequences are strongly protected from nuclease attack by the binding of the antibiotic. Also in evidence is the curious phenomenon of enhanced cutting at some sites when echinomycin is present; these regions of enhancement always occur at sequences flanking strong antibiotic binding sites, but not all flanking sequences are affected. Enhanced cutting is not an artefact of the footprinting methodology because it can also be seen in experiments using other cutting probes: DNAase II (Low et al., 1984) and even MPE-Fe(II) (Van Dyke and Dervan, 1984). Moreover, it occurs at sequences flanking the binding sites of other antibiotics such as actinomycin and distamycin, whose nucleotide sequence selectivity is quite different from that of quinoxaline antibiotics (Fox and Waring, 1984a). We conclude that when sequence-selective antibiotics bind to DNA it is a common phenomenon for local changes in the structure of the helix to be propagated into regions flanking the binding site(s). Those changes can result in enhanced susceptibility to nuclease attack, prompting the notion that they might also influence the interaction between the DNA and other types of protein molecules (see later).

It is important to realise, however, that although footprinting analysis has identified CpG steps as constituting the critical determinant of preferred binding sites for echinomycin in DNA, they are not the only possible sequences. The mere fact that echinomycin (in common with other quinoxaline antibiotics) will bind to polydG.polydC as well as to poly(dA-dT) (Wakelin and Waring, 1976) is sufficient to establish that the CpG step cannot be mandatory and that other sites, presumably of lower affinity, must exist. In a later section we will examine the consequences of this situation from a kinetic point of view.

THE MOLECULAR BASIS OF SEQUENCE RECOGNITION

Until quite recently there was a dearth of information pertaining to the precise conformation of any quinoxaline antibiotic (made worse by the existence of an error in the structural formula of echinomycin as originally published – reviewed in Waring, 1979). Consequently any ideas about how echinomycin might recognise the CpG step in a bis-intercalated complex were of necessity somewhat speculative, if not inspired (see the reviews of Waring, 1977, 1979; and Waring and Fox, 1983). That situation changed radically in 1984 when the first crystal structure for a quinoxaline antibiotic was published (Sheldrick et al., 1984), sharing many features in common with a structure previously reported for the closely related synthetic depsipeptide TANDEM (Viswamitra et al., 1981; Hossain et al., 1982). It was soon followed by crystal structures for intermolecular complexes formed between quinoxaline antibiotics and self-complementary oligonucleotides (Wang et al., 1984; Ughetto et al., 1985; Quigley et al., 1986). In all these structures the earlier predictions of indirect work based on solution studies were nicely borne out, i.e. that the antibiotic molecules should adopt a staple-like conformation having the intercalative quinoxaline rings attached to the fairly rigid peptide moiety in such a fashion that they could neatly sandwich two DNA base-pairs between them (Figure 5). It was a reasonable guess that the sandwiched base-pairs would take the form of a CpG step, as turned out to be the case in the antibiotic-oligonucleotide complexes, and a fairly detailed picture of intermolecular contacts which govern the sequence-selectivity of binding can now be drawn up. Steric complementarity in the form of van der Waals contacts is important, reinforced by specific hydrogen-bonding interactions between the carbonyl oxygen and NH groups of alanine residues in the antibiotic structure (see Figure 5) and the 2-NH$_2$ and N(3) functionalities of guanine

116

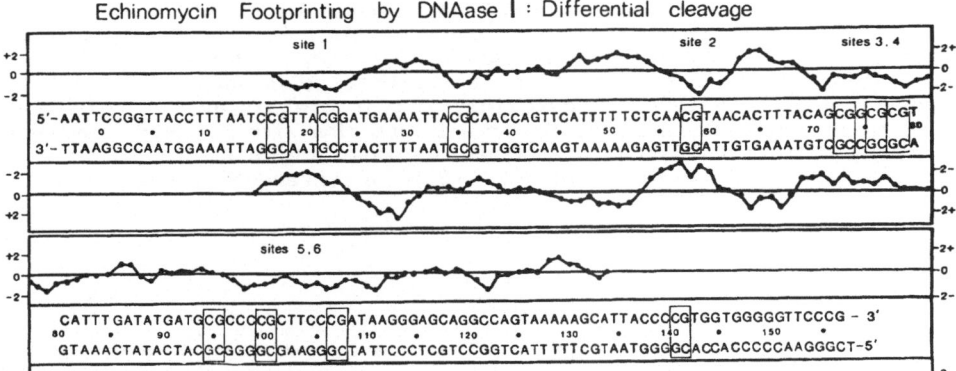

Figure 4. Echinomycin-induced differences in the susceptibility of tyrT DNA to DNAase I digestion. The upper Watson strand reads 5' to 3' left-to-right, while the lower Crick strand reads 5' to 3' right-to-left. Vertical scales on both sides are in units of $\ln(f_a) - \ln(f_c)$, where f_a is the fractional cleavage at any bond in the presence of 15μM antibiotic and f_c is the fractional cleavage of the same bond in the control, given closely similar extents of overall digestion (approx. 30% of the starting material in both cases). Positive values indicate enhancement, negative values blockage. Sites of protection are labelled as in Figure 3. From Low et al. (1984).

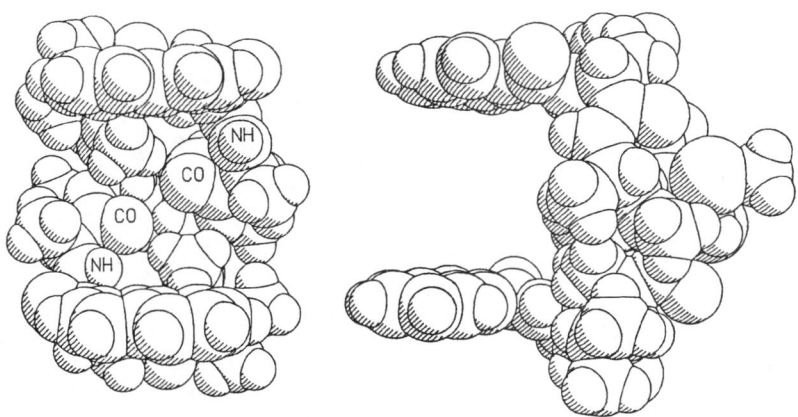

Figure 5. Crystal structure of a bis-quinoline analogue of echinomycin prepared by directed biosynthesis (Gauvreau and Waring, 1984). In the illustration on the left the molecule is viewed down its quasi-dyad axis with the aromatic chromophores projecting out towards the viewer; this reveals the face of the peptide ring presented towards the DNA base-pairs, with the NH and CO groups of the two L-alanine residues lying in a diagonal array from lower left to upper right. In the illustration on the right the molecule is viewed from the "side", i.e. having been turned through 90° about a vertical axis, so as to show its staple-like arrangement which is ideally suited for bis-intercalation. Small rotations have been applied to the points of attachment of the aromatic rings so as to bring their planes exactly parallel. Unpublished work of G.M. Sheldrick, P.G. Jones and M.J. Waring.

117

attained: for practical purposes it is not feasible to measure a pattern at all with less than 5-10 seconds exposure to antibiotic at low temperature. Under such conditions some time-dependent changes with echinomycin are detectable but their significance is equivocal (Fox and Waring, 1986). However, the technique is certainly applicable to actinomycin and nogalamycin, two other antibiotics whose interaction with DNA bears the same hallmarks of putative shuffling as does the interaction with echinomycin (Fox and Waring, 1984b,d; Fox et al., 1985). A representative experiment

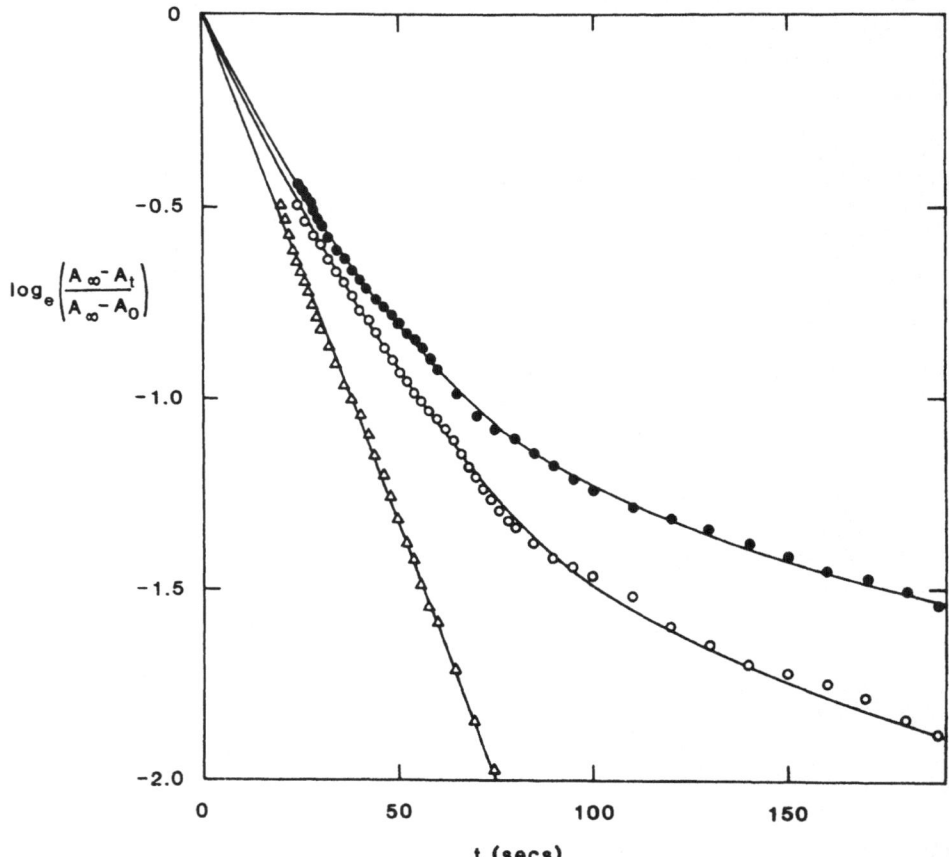

t (secs)

Figure 6. Detergent-induced dissociation of echinomycin from calf thymus DNA. The complexes were left to equilibrate for various periods of time before addition of sodium dodecyl sulphate. The ordinate represents the natural logarithm of the fractional absorbance change. (Δ) The complex was left for 30s before dissociating. The fitted line is a single exponential with a time constant of 37.7s. (O) The complex was left for 5 mins before dissociating. The fitted curve is described by the sum of two exponentials with parameters τ_1=37s (80%), τ_2=597s (20%). (●) The complex was left to equilibrate for several hours before dissociating. The fitted curve represents the sum of two exponentials with parameters τ_1=31s (66%), τ_2=757s (34%). From Fox and Waring (1985).

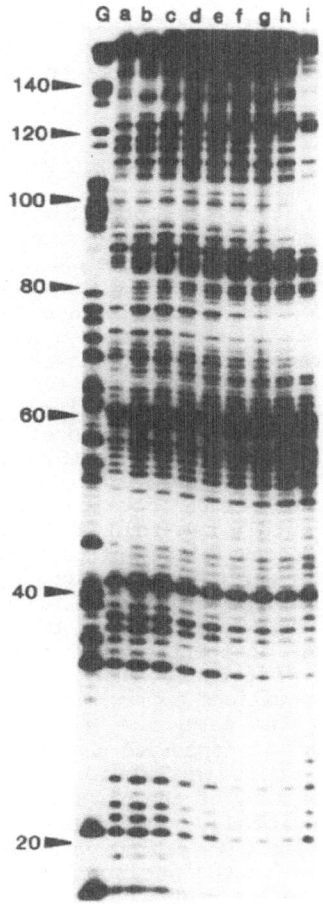

Figure 7. DNAase I digestion pattern of tyrT DNA in the presence of 7.5μM actinomycin D at various times after mixing at 4°C. All samples were digested for 5 seconds with an enzyme concentration of 1.8 units/ml. Lane (a) represents a control performed in the absence of antibiotic. Actinomycin and DNA were left in contact for the following periods of time before digestion: (b) 5 seconds; (c) 10 seconds; (d) 30 seconds; (e) 1 minute; (f) 2 minutes; (g) 5 minutes; (h) 10 minutes; (i) 30 minutes. The track labelled "G" represents a dimethylsulphate-piperidine marker specific for guanine. From Fox and Waring (1986).

with actinomycin is illustrated in Figure 7. Sites of protection are clearly visible around positions 36, 74 and 118 but they do not develop at the same rate. Likewise, enhanced cleavage occurs around positions 55 and 82 but takes several minutes to develop. Perhaps most interestingly there are occasional bands like that at position 103 which show a definite short-lived enhancement of cleavage as if the antibiotic were transiently bound to a neighbouring site but only 'passing through'. The technical difficulty of performing experiments of this type renders thorough-going analysis rather daunting, but at least the present results serve to reinforce the belief that shuffling can be directly observed in propitious circumstances (Fox and Waring, 1986).

EFFECTS OF ANTIBIOTIC BINDING TO NUCLEOSOME CORE PARTICLES

No-one would deny that crystallographic and solution studies have convincingly established the nature of the interaction between antibiotics and purified DNA, but in living cells the DNA is bound to a collection of histones and other chromosomal proteins. Can antibiotics bind equally well to DNA in situ, as it were, and what effects might they have on nucleosome structure? To address this question we have investigated the interaction between echinomycin and reconstituted nucleosome core particles containing the same 160 bp tyrT DNA molecule employed for the experiment depicted in Figure 3. Happily the same method of analysis is applicable - limited DNAase I digestion followed by electrophoretic separation of labelled fragments - enabling direct comparison with the effects of the antibiotic on naked DNA. Typical results are shown in Figure 8. Comparing the control core track ("0 μM") with the free DNA control track it is readily apparent that there is an approximately ten-fold periodicity in intensity of the bands, which reflects the accessibility of particular internucleotide bonds to nuclease cleavage. This periodic variation in cutting results from the specificity of nucleosome positioning, i.e. the "phasing" of the DNA sequence with respect to its winding around the protein core, as described by Drew and Travers (1985). In the presence of echinomycin the products of digestion are distinctly different. Many new bands appear (marked with asterisks in Figure 8) at positions approximately mid-way between those cut well in the control. At the same time there is a 50-70% reduction in intensity of the old bands. The observed changes cannot be explained on the basis of displacement of a certain fraction of the DNA from attachment to the histone octamer because many of the new bands (those at positions 61, 71 and 82, for example) are totally new in the core digest and virtually missing in free DNA complexed with echinomycin; others (such as 51, 103 and 114) are strongly enhanced in the nucleosome core but suppressed by the antibiotic in free DNA. We conclude that echinomycin must have caused a substantial fraction of the DNA to change its positioning on the surface of the protein by a translational shift of about five base-pairs, which corresponds to a change in rotational orientation of about half a helical turn (180°). To confirm this interpretation we measured corresponding data for the complementary ("Crick") strand and calculated a differential cleavage plot representing the antibiotic-induced changes in susceptibility of each bond to nuclease attack (Figure 9). Visual inspection of this plot suggests a regular fluctuation in relative accessibility of internucleotide bonds, modulated with a periodicity of about ten nucleotides, and staggered bases in the DNA. A curious feature of the antibiotic-oligonucleotide complexes is the occurrence of Hoogsteen pairing between the purine and pyrimidine bases on either side of the sandwiched CpG step (Quigley et al., 1986). Whether this unorthodox pairing arises as a necessary consequence of the formation of a bis-intercalated complex, and indeed whether it could occur in macromolecular DNA on binding echinomycin, is an open question - as is its possible relation to the propagated structural changes in the double helix mentioned above.

THE KINETICS OF ANTIBIOTIC-DNA INTERACTION

It is one thing to identify the nature of preferred binding sites for an antibiotic such as echinomycin, but quite another to explain how the ligand locates its preferred sites among a plethora of varied potential binding sites, mostly of lesser affinity, such as must occur in natural DNA molecules of enormous size and sequence complexity. For many years it has been known that attainment of equilibrium between actinomycin and DNA is a complicated process, requiring numerous exponential terms to describe the forward (binding) reaction and only a couple less to account for the dissociation behaviour of the antibiotic (Müller and Crothers, 1968; see also Fox and Waring, 1984b). By contrast, the reaction between echinomycin and DNA appears deceptively simple, at least as viewed in a stopped-flow spectrophotometer: a single exponential is observed and the entire optical change is essentially complete within a second or two - much faster than for actinomycin (Fox and Waring, 1984c). Does this mean that despite the presumed existence of multiple classes of binding sites, and the undoubtedly complicated nature of the bis-intercalation reaction, the antibiotic speedily and efficiently binds directly to its preferred (CpG-containing) sites without distraction or having to search for them? Apparently not, as an investigation of its dissociation behaviour reveals (Fox et al., 1981). In the first place, the echinomycin-DNA dissociation profile requires a minimum of three exponentials for its proper description, just like actinomycin, and the same is true for all quinoxaline antibiotics investigated (Fox and Waring, 1981). There is every reason to believe that these three exponentials correspond to dissociation of the antibiotic molecules from different classes of binding sites - crudely strong, medium and weak. Secondly, if echinomycin-DNA complexes are prepared and allowed to "mature" for different periods of time before dissociation is initiated, we see a steady time-dependent conversion of fast-dissociating antibiotic to a slowly-dissociating form (Figure 6), and the rate of that conversion is of the correct order for migration of bound antibiotic from weaker binding sites to stronger binding sites. Thirdly, no such time-dependence of the dissociation profile occurs with synthetic DNAs which contain only one (or at most two) types of binding sites (Fox and Waring, 1985). Although we cannot entirely exclude models based upon sequential conformational adjustments between antibiotic and DNA, these results strongly suggest a mechanism whereby antibiotic molecules initially associate relatively nonspecifically with many types of binding sites but subsequently "shuffle" along the DNA lattice to locate their preferred (most tightly binding) sites. The fact that the migration process is for practical purposes invisible with echinomycin, but spectroscopically detectable with actinomycin, is not in itself significant and may merely reflect the quite different optical properties of the two antibiotics (Fox and Waring, 1984b, 1985).

The ideal way to gain hard evidence of shuffling, i.e. migration of antibiotic molecules between different binding sites on DNA, would be to demonstrate time-dependent changes in footprinting patterns. Unfortunately, the exigencies of footprinting techniques are such that a finite time is required for the digestion reaction, which limits the resolution that can be

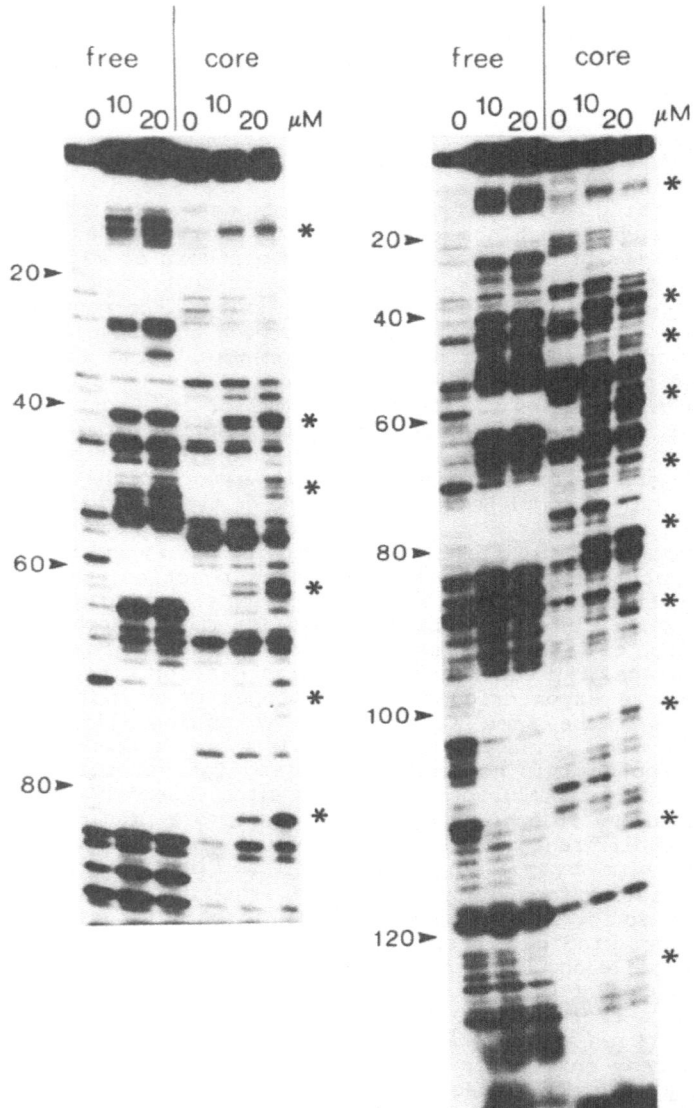

Figure 8. Effects of echinomycin on the DNAase I digestion pattern of free tyrT DNA or reconstituted nucleosome core particles. Two gels derived from the same set of digestion mixtures are shown: that on the left was run for a long time so as to improve resolution of the longer fragments (bands 10-90). In this experiment the transcribing "Watson" strand (upper sequence, 5' to 3' left-to-right in Figure 4) was labelled at its 3' end. Each set of three tracks represents a control (no antibiotic) together with samples containing 10 or 20μM echinomycin as shown at the top of each lane. Numbers at left refer to the numbering scheme shown in Figure 4. Asterisks indicate the new bands which appear in digests of core particles treated with echinomycin. From Low et al. (1986).

across the helix by about 2-3 bonds towards the 3'-end of each strand, as occurs with free DNA (cf. Figures 3 and 4). Subjecting the data in Figure 9 to Fourier analysis confirmed the existence of a strong, regular variation with maximum amplitude at a period of 10.64 bonds. No other periodic variations modulated within the range 5.0 to 19.0 bonds were detectable even with amplitudes half the maximum value.

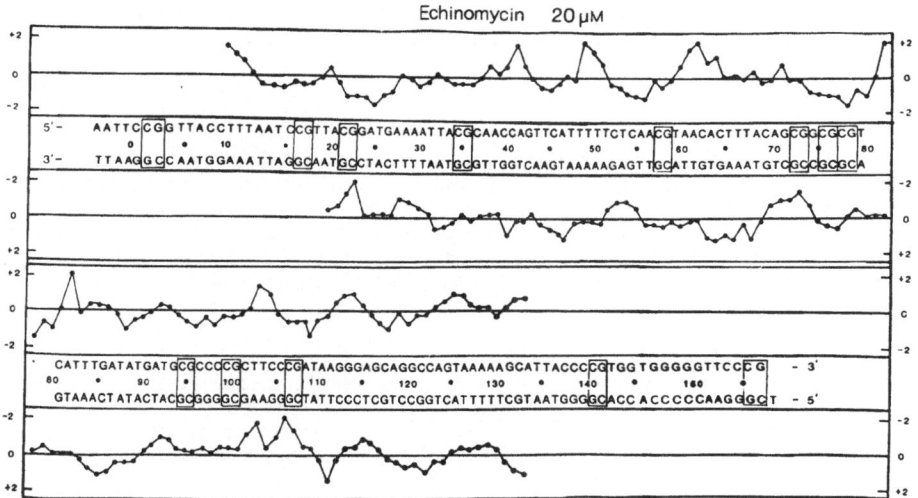

Figure 9. A plot of differential cleavage representing the extent to which echinomycin (20μM) affects the susceptibility of nucleosome core particles to attack by DNAase I. The method of plotting is as described in the legend to Figure 4. From Low et al. (1986).

Why should exposure to echinomycin cause DNA to rotate by about half a turn on the surface of the nucleosome, and how much antibiotic binding does it take? Our best estimate for the latter is about 2 - 2.5 antibiotic molecules on average per nucleosome core particle, based on the change in Fourier periodicity from unperturbed DNA (Low et al., 1986). Granted that each echinomycin molecule which bis-intercalates into DNA unwinds the helix by 48° (Gale et al., 1981) it is natural to suspect that the induced rotation of nucleosomal DNA in some way reflects the unwinding process. That does not seem to be the case. Very similar changes in rotational orientation of the tyrT DNA fragment have been observed with distamycin (Low et al., 1986), netropsin and berenil (Portugal and Waring, 1986) yet none of these substances affects the helical winding of DNA to any great extent. Conversely, actinomycin and nogalamycin, antibiotics which are well-known intercalators causing 26° and 18° helical unwinding respectively (Gale et al., 1981), fail to produce the characteristic pattern of changes in digestion products from nucleosomal DNA (Portugal and Waring, 1986). So far as echinomycin is concerned, the most likely explanation available at present is that it binds to certain sites containing the recognition sequence CpG that are exposed on the outside of the DNA superhelix in the antibiotic-free core particle, and induces them to turn inwards to face the histone octamer. Presumably this serves to optimise non-bonded contacts between the octapeptide ring of the antibiotic and the polynucleotide backbones. Further work with a variety of ligands and different preparations of nucleosome core particles, containing quite different DNA sequences, will be required to probe the generality (or otherwise) of this phenomenon and the nature of the forces which drive it.

ACKNOWLEDGMENTS

Original work reported here was supported by grants from the Cancer Research Campaign, the Royal Society and the Medical Research Council. Mark Freeman provided able technical assistance. The experiments, and their interpretation, involved the participation of Drs K.R. Fox, C.M.L. Low and J. Portugal; they were materially aided by gifts of samples and frequent discussion with Drs H.R. Drew, A. Klug, D. Rhodes and A.A. Travers.

REFERENCES

Drew, H.R. and Travers, A.A. (1985) DNA bending and its relation to nucleosome positioning. J. Mol. Biol. 186, 773-790.
Fox, K.R., Brassett, C. and Waring, M.J. (1985) Kinetics of dissociation of nogalamycin from DNA: comparison with other anthracycline antibiotics. Biochim. Biophys. Acta 840, 383-392.
Fox, K.R., Wakelin, L.P.G. and Waring, M.J. (1981) Kinetics of the interaction between echinomycin and DNA. Biochemistry 20, 5768-5779.
Fox, K.R. and Waring, M.J. (1981) Kinetics of dissociation of quinoxaline antibiotics from DNA. Biochim. Biophys. Acta 654, 279-286.
Fox, K.R. and Waring, M.J. (1984a) DNA structural variations produced by actinomycin and distamycin as revealed by DNAase I footprinting. Nucleic Acids Res. 12, 9271-9285.
Fox, K.R. and Waring, M.J. (1984b) Kinetic evidence for redistribution of actinomycin molecules between potential DNA binding sites. Eur. J. Biochem 145, 579-586.
Fox, K.R. and Waring, M.J. (1984c) Stopped-flow kinetic studies on the interaction between echinomycin and DNA. Biochemstry 23, 2627-2633.
Fox, K.R. and Waring, M.J. (1984d) Evidence of different binding sites for nogalamycin in DNA revealed by association kinetics. Biochim. Biophys. Acta 802, 162-168.
Fox, K.R. and Waring, M.J. (1985) Kinetic evidence that echinomycin migrates between potential DNA-binding sites. Nucleic Acids Res. 13, 595-603.
Fox, K.R. and Waring, M.J. (1986) Footprinting reveals that nogalamycin and actinomycin shuffle between DNA binding sites. Nucleic Acids Res. 14, 2001-2014.
Gale, E.F., Cundliffe, E., Reynolds, P.E., Richmond, M.H. and Waring, M.J. (1981). The Molecular Basis of Antibiotic Action, 2nd ed., Wiley, London.
Gauvreau, D. and Waring, M.J. (1984) Directed biosynthesis of novel derivatives of echinomycin by Streptomyces echinatus. Part I. Effect of

exogenous analogues of quinoxaline-2-carboxylic acid on the fermentation. Can. J. Microbiol. 30, 439-450.

Hossain, M.B., van der Helm, D., Olsen, R.K., Jones, P.G., Sheldrick, G.M., Egert, E., Kennard, O., Waring, M.J. and Viswamitra, M.A. (1982) Crystal and molecular structure of the quinoxaline antibiotic analogue TANDEM (des-N-tetramethyl triostin A). J. Amer. Chem. Soc. 104, 3401-3408.

Katagiri, K., Yoshida, T. and Sato, K. (1975) Quinoxaline antibiotics. In Antibiotics III. Mechanism of Action of Antimicrobial and Antitumour Agents (eds J.W. Corcoran and F.E. Hahn), 234-251. Springer-Verlag, Heidelberg.

Low, C.M.L., Drew, H.R. and Waring, M.J. (1984) Sequence-specific binding of echinomycin to DNA: evidence for conformational changes affecting flanking sequences. Nucleic Acids Res. 12, 4865-4879.

Low, C.M.L., Drew, H.R. and Waring, M.J. (1986) Echinomycin and distamycin induce rotation of nucleosome core DNA. Manuscript submitted for publication.

Müller, W. and Crothers, D.M. (1968) Studies of the binding of actinomycin and related compounds to DNA. J. Mol. Biol. 35, 251-290.

Portugal, J. and Waring, M.J. (1986) Antibiotics which can alter the rotational orientation of nucleosome core DNA. Manuscript submitted for publication.

Quigley, G.J., Ughetto, G., van der Marel, G.A., van Boom, J.H., Wang, A.H.J. and Rich, A. (1986). Non-Watson-Crick G·C and A·T base pairs in a DNA-antibiotic complex. Science 232, 1255-1258.

Reich, E. and Goldberg, I.H. (1964) Actinomycin and nucleic acid function. Prog. Nucleic Acid Res. Mol. Biol. 3, 184-234.

Sato, K., Shiratori, O. and Katagiri, K. (1967) The mode of action of quinoxaline antibiotics. Interaction of quinomycin A with DNA. J. Antibiotics, Ser. A., 20, 270-276.

Sheldrick, G.M., Guy, J.J., Kennard, O., Rivera, V. and Waring, M.J. (1984) Crystal and molecular structure of the DNA-binding antitumour antibiotic triostin A. J. Chem. Soc. Perkin II, 1601-1605.

Sobell, H.M. (1973) The stereochemistry of actinomycin binding to DNA and its implications in molecular biology. Prog. Nucleic Acid Res. Mol. Biol. 13, 153-190.

Ughetto, G., Wang, A.H.J., Quigley, G.J., van der Marel, G.A., van Boom, J.H. and Rich, A. (1985) A comparison of the structure of echinomycin and triostin A complexed to a DNA fragment. Nucleic Acids Res. 13, 2305-2323.

Van Dyke, M.M. and Dervan, P. (1984) Echinomycin binding sites on DNA. Science 225, 1122-1127.

Viswamitra, M.A., Kennard, O., Cruse, W.B.T., Egert, E., Sheldrick, G.M., Jones, P.G., Waring, M.J., Wakelin, L.P.G. and Olsen, R.K. (1981) The structure of TANDEM, a quinoxaline antibiotic analogue and its implication for bifunctional intercalation into DNA. Nature 289, 817-819.

Wakelin, L.P.G. & Waring, M.J. (1976) The binding of echinomycin to DNA. Biochem. J. 157, 721-740.

Wang, A.H.J., Ughetto, G., Quigley, G.J., Hakoshima, T., van der Marel, G., van Boom, J.H. and Rich, A. (1984) The molecular structure of a DNA-triostin A complex. Science 225, 1115-1121.

Ward, D.C., Reich, E. and Goldberg, I.H. (1965) Base specificity in the interaction of polynucleotides with antibiotic drugs. Science 149, 1259-1263.

Waring, M.J. and Makoff, A. (1974) Breakdown of pulse-labelled RNA and polysomes in Bacillus megaterium: actions of streptolydigin, echinomycin and triostins. Mol. Pharmacol. 10, 214-224.

Waring, M.J. and Wakelin, L.P.G. (1974) Echinomycin: a bifunctional intercalating antibiotic. Nature 252, 653-657.

Waring, M.J. (1977) Structural and conformational studies on quinoxaline antibiotics in relation to the molecular basis of their interaction with DNA. In Drug action at the Molecular Level (ed. G.C.K. Roberts) pp 167–189. Macmillan, London.

Waring, M.J. (1979) Echinomycin, triostin and related antibiotics. In Antibiotics, Vol. 5/Part 2, Mechanism of Action of Antieukaryotic and Antiviral Compounds (ed. F.E. Hahn) 173–194. Springer-Verlag, Heidelberg.

Waring, M.J. (1981) DNA modification and cancer. Ann. Rev. Biochem. 50, 159–192.

Waring, M.J. and Fox, K.R. (1983) Molecular aspects of the interaction between quinoxaline antibiotics and nucleic acids. In Molecular Aspects of Anti-Cancer Drug Action (eds. S. Neidle & M.J. Waring), Macmillan, London, pp. 127–156.

SPECIFIC GENE REGULATION BY OLIGODEOXYNUCLEOTIDES COVALENTLY

LINKED TO INTERCALATING AGENTS

Claude Hélène

Laboratoire de Biophysique, INSERM U.201, CNRS UA 481
Muséum National d'Histoire Naturelle
61, Rue Buffon - 75005 Paris, France

INTRODUCTION

In living cells, the regulation of gene expression occurs at different levels : transcription of one of the DNA strands into RNA, splicing
of pre-messenger RNAs which eliminate intron sequences after transcription of mosaic genes, post-transcriptional modifications of mRNAs (cap-
ping, polyadenylation...), transfer of mRNA from the nucleus to the cytoplasm, translation of mRNA, post-translational modifications of proteins... In most cases the regulation is achieved by proteins which bind
either DNA sequences to block or activate transcription or RNA to block,
e.g., translation (see Hélène and Lancelot, 1982, for a review). More
recently it has been shown that, in bacteria, small RNAs could play a
regulatory role similar to that of regulatory proteins by hybridizing to
mRNAs (see Green et al., 1986, for a review). The control of plasmid copy
number and immunity in bacteria is also achieved by a small RNA (RNA I)
which arises from transcription in the reverse direction as compared to
that of the primer RNA (RNA II) utilized to initiate plasmid replication.
Being transcribed from opposite strands the two RNAs are fully complementary and their association prevents replication initiation (see Tomizawa,
1986 and references therein).The role of RNAs in the regulation of mRNA
translation was first demonstrated in E.coli. The translation of mRNAs
for proteins involved in the regulation of plasmid R1 replication is
controlled by a small RNA which is transcribed in the reverse direction
with respect to the mRNA. Complete base pairing occurs between the mRNA
and the regulatory RNA which prevents ribosome from translating the mRNA
(Light and Molin, 1983). A similar situation was observed for the regulation of transposase synthesis involved in Tn 10 transposition (Simons et
al., 1983). It was then discovered that the synthesis of the OmpF protein
was controlled by a small RNA transcribed from a DNA region close to the
ompC gene. Complex formation between the regulatory RNA and the ompF mRNA
did not involve complete base pairing but ensured sufficient stability to
interfere with translation (Mizuno et al., 1984). A more recent study
showed that hybrid formation between a small RNA and the 5'- end of a
mRNA could induce premature termination of transcription, thereby controlling gene expression at another level than translation (Okamoto and
Freundlich, 1986). All the examples briefly described above involve bacterial systems. There is some evidence that a similar involvement of RNAs
as regulatory elements could occur in eukaryotic cells as well (Heywood,

1986 ; Spencer **et al.**, 1986; Williams and Fried, 1986).

The discovery of the regulatory role of small RNAs in living cells led to the development of anti-sense RNAs to artificially control gene expression (Weintraub **et al.**, 1985). The principle of these experiments is schematically represented on figure 1. A gene (or a c DNA) or part of it can be excised and inserted in a vector in the reverse orientation under the control of a strong, possibly inducible, promoter. The transcript from this "anti-sense" gene is an RNA wich is complementary to the mRNA copied from the gene itself. This hybridization blocks mRNA translation (Green **et al.**; 1986) or other post-transcriptional processes such as mRNA migration from the nucleus to the cytoplasm (Kim and Wold, 1985).

Long before the discovery of regulatory RNAs the idea of using oligonucleotides complementary to RNA sequences to alter gene expression was put forward in several laboratories (Paterson **et al.**, 1977 ; Zamecnick and Stephenson, 1978 ; Jayaraman **et al.**, 1981 ; Trudel **et al.**, 1981 ; see also Knorre and Vlassov, 1985 ; Hélène **et al.**, 1985). Since then it has been shown that oligodeoxynucleotides complementary to mRNAs could block translation either **in vitro** (Stephenson and Zamecnik, 1978 ; Blake **et al.**, 1985); or in microinjected **Xenopus** oocytes (Kawasaki 1985; Cazenave **et al.**, 1986). The application of oligodeoxynucleotides to **in vivo** studies faces two main problems : i) their penetration into living cells in culture is very limited ; ii) their sensitivity to nucleases makes their lifetime very short. Several attempts have been made to overcome these two difficulties. The phosphodiester backbone of the oligodeoxynucleotide can be changed to a phosphonate backbone. The loss of negative charges makes oligophosphonates more efficient in penetrating through the cell membranes and much more resistant to nucleases (Miller **et al.**, 1983). Attachment of oligonucleotides to polymers such as poly-L-lysine increases the efficiency of penetration and makes oligonucleotides active **in vivo** at much lower concentrations (Bayard **et al.**, 1986).

Despite these progresses there is an obvious need for developing new families of gene regulatory substances which could be used **in vivo** to control the expression of undesirable genes such as oncogenes or to prevent the development of viruses or parasites. The approach we are following to achieve this goal can be summarized as follows : i) the recognition of a nucleic acid base sequence can be easily achieved by using an oligonucleotide of complementary sequence ; ii) the stability of the mini-double helix can be increased by covalent attachment of an intercalating agent at one end of the oligonucleotide (Asseline **et al.**, 1983 ; 1984 a,b). In addition the intercalating agent endows the oligonucleotide with a higher penetrability and stability against 3' or 5'-exonucleases, depending on the attachment site ; iii) the other end of the oligonucleotide can be substituted by a reagent which may be induced to modify the target sequence following either chemical or photochemical activation (Boidot-Forget **et al.**, 1986). Specific cleavage of a mRNA or a viral RNA target or chemical modification of the bases at the binding site of the oligonucleotide prevent translation of the mRNA ; iv) the backbone of the oligonucleotide can be modified in such a way as to make it more resistant to nucleases.

SPECIFICITY OF OLIGONUCLEOTIDE TARGETING TO UNIQUE SEQUENCES

The minimum size that an oligonucleotide should have in order to recognize a single specific sequence in a genome can be calculated assuming a random distribution of the bases. The frequency of occurrence of a sequence containing n nucleotides (a adenines, g guanines, t thymines and c cytosines with $n = a + g + t + c$) in a genome containing N base pairs is given by equation (1)

$$(1) \qquad (f/2)^{(a+t)} \times ((1 - f)/2)^{(g+c)} \times 2 N$$

where f is the fraction of A.T base pairs in the genome. From equation

(1) it can be calculated that a minimum length of 12 nucleotides is requested to find the complementary sequence only once in the E. coli genome ($N \approx 4 \times 10^6$; $f = 0.5$). For the human genome ($N \approx 4 \times 10^9$; $f = 0.6$) this minimum length varies from 15 to 19 depending on the oligonucleotide base composition.

The minimum length that the synthetic oligonucleotide should have can be further reduced. For example, let us assume that the oligonucleotide is targeted to a messenger RNA. In a given cell type, at a given time, only a small fraction of the genome is expressed as mRNA. If it is assumed that about 0.5 % of the genome is transcribed into mRNA then the complexity of the target is strongly reduced. For a human cell this leads to a reduction of N (equation 1) from 4×10^9 to 2×10^7 and consequently the minimum length n is reduced from 15-19 to 11-15. Additional constraints may be taken into account. For example, the sequence CpG is underrepresented in eukaryotic genomes due to its involvement as a signal in gene expression (as a consequence of cytosine methylation in CpG sequences).Therefore an oligonucleotide containing CpG sequences will have a lower probability of finding several complementary sequences. If the target sequence is located in a regulatory region it should have been selected during evolution in such a way as to appear only once (or a limited number of times) in the genome in order to achieve specific regulation.

OLIGONUCLEOTIDES COVALENTLY LINKED TO INTERCALATING AGENTS

The above calculations suggest that short synthetic oligonucleotides can be designed to recognize selectively a target nucleic acid. This should facilitate both the synthesis (especially if other linkages than phosphodiesters are envisaged) and the penetration accross cell membranes. However, a short oligonucleotide might not have a high enough affinity towards its target sequence if the number of base pairs involved in complex formation is too small.Consequently the biological process, e.g., translation would not be affected. There are different ways of increasing the affinity of the oligonucleotide for its complementary sequence. We have chosen to covalently link an intercalating agent to one of the oligonucleotide ends. Intercalating agents are polycyclic aromatic molecules which insert their planar aromatic ring between two consecutive base pairs of double-stranded DNA. They bind much more weakly to single-stranded structures. If the linker between the oligonucleotide and the intercalating agent is appropriately chosen, intercalation can occur in the mini-double helix formed when the oligonucleotide is bound to its complementary sequence. The free energy of binding of the composite molecule (ONBI for OligoNucleotide-Bridge-Intercalator) can be approximated as the sum of the free energy for binding the oligonucleotide to its complementary sequence (ΔG_{ON}) and that of intercalation (ΔG_I), corrected for an entropy term taking into account the restricted configurational space available to the intercalating agent when it is covalently linked to the oligonucleotide. The linker is assumed to play no role in the binding process. If it is involved in the interaction then an additional free energy term should be included in equation (2).

$$(2) \qquad \Delta G_{ONBI} = \Delta G_{ON} + \Delta G_I - T\Delta S_m$$

Since ΔS_m in equation (2) is positive the association constant for the ONBI ($K_{ONBI} = \exp - \Delta G_{ONBI}/RT$) should be at least the product of the association constants for the oligonucleotide and the intercalating agent

$$(3) \qquad K_{ONBI} = \alpha \, K_{ON} \times K_I$$

with $\alpha = \exp(T\Delta S_m/RT)$

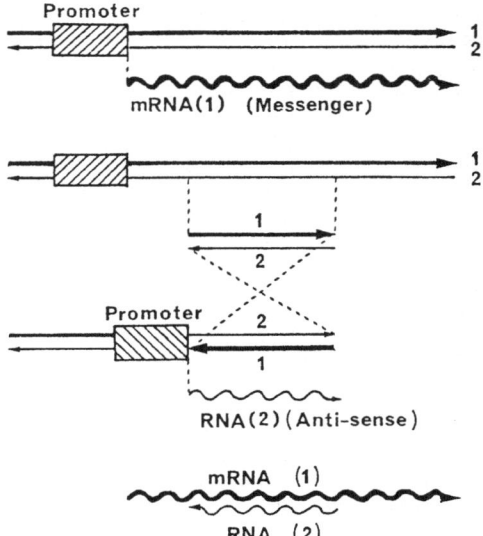

Figure 1 : Scheme for the design of an anti-sense RNA (see text). The promoter controlling the transcription of the inverted gene fragment can be different from that of the gene itself.

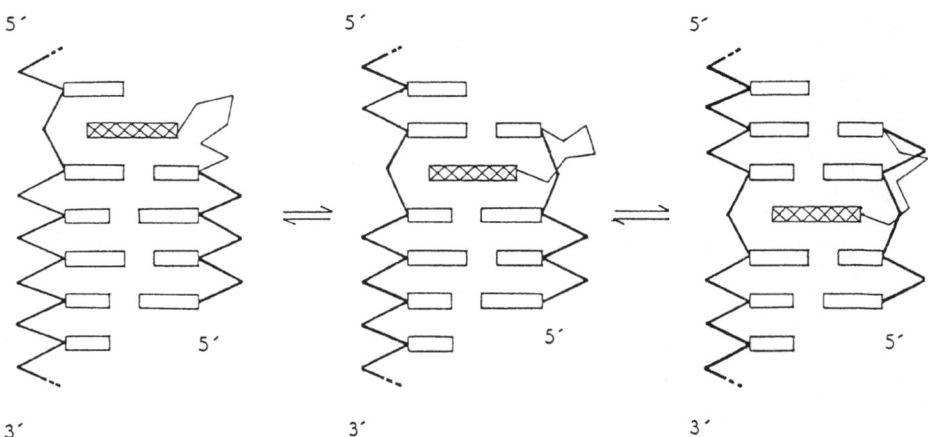

Figure 2 : Models for different types of complexes formed by an oligo-deoxynucleotide covalently linked to an intercalating agent with its complementary sequence. The hatched rectangle represents the intercalating agent. The equilibrium between these different structures depends on the base sequence of the oligo-nucleotide, on the site of attachment (3' or 5' end) of the acridine ring, on the length of the linker and on the nature of the base which follows the mini-double helix on the target sequence. For example, the stacking interaction between adenine and acridine is quite strong. If an adenine is located on the target nucleic acid next to the last base pair (on the acridine side) the equilibrium will be shifted towards the structure where the acridine stacks on top of the last base pair and interacts with the adenine (left part of the figure).

130

However the presence of the linker might not allow the intercalating agent to adopt the same intercalation geometry as it would if it was free in solution. Also the local double helical structure might be distorted in such a way as to increase ΔG_{ON} and ΔG_I. Therefore equation (3) would lead to an overestimation of the association constant for the ONBI. The results presented below show that equation (3) represents a good approximation.

BINDING OF ONBI TO COMPLEMENTARY SEQUENCES

The intercalating agent we have used in most of our studies in an acridine derivative : 2-methoxy, 6-chloro, 9-amino acridine. It exhibits no specificity of binding with respect to A.T. and G.C. base pairs. Therefore it can be covalently linked to oligonucleotides of different sequences without perturbing the specificity of binding to the complementary sequence. A polymethylene linker was used to tether the acridine ring (via its 9-amino group) to the 3'- or 5'- phosphate group of the oligonucleotide (Asseline et al., 1986). Other linkers and other intercalating agents have been developed more recently (unpublished results).

The binding of the ONBI to its complementary sequence was followed by several spectroscopic techniques, including absorption, fluorescence, circular dichroism and nuclear magnetic resonance. The results can be summarized as follows :

    i)   an hypochromism is induced in the visible absorption band of the acridine derivative upon binding. A shift of the maximum to longer wavelengths is observed which depends on the base sequence of the oligonucleotide (Asseline et al., 1983 ; 1984 a,b ; Hélène et al. 1986).

    ii)   the fluorescence quantum yield of the acridine derivative is increased when the terminal base pairs are all A.Ts. Guanine and G.C. base pairs strongly quench the acridine fluorescence. Fluorescence excitation spectra reveal several environments (at least two) for the acridine ring in the complexes (Asseline et al., 1984 a,b ; Hélène et al., 1986). The fluorescence of the complexes is highly polarized due to partial immobilization of the acridine ring. Fluorescence anisotropy decays reveal that the acridine ring retains a local mobility which is quite similar to that of acridine intercalated in DNA (Hélène et al., 1986).

    iii)   an induced circular dichroism is observed as a result of the asymmetric environment of the acridine ring in the complexes (Hélène et al., 1986).

    iv)   proton and phosphorus magnetic resonance studies indicate that the acridine ring is intercalated between the terminal base pairs of the mini-double helix (Lancelot et al., 1985, 1986 ; Lancelot and Thuong, 1986).

The results of spectroscopic studies led to the conclusion that different complexes could form which are characterized by different locations of the acridine ring, as summarized in figure 2.

Upon increasing the temperature, complexes dissociate according to a cooperative process. Base pairing and intercalation are disrupted in a single step. From the concentration dependence of the temperature of half-dissociation (T_m) it is possible to derive thermodynamic parameters according to equation (4)

$$(4)\qquad 1/T_m = \Delta S/\Delta H + (2.3R/\Delta H)\log C_m$$

where C_m is the free ONBI concentration at temperature T_m. Table 1 gives ΔG values obtained for the binding of $(T_p)_8Et$ and $(Tp)_8m6Acr$ to poly(rA).

In (Tp)$_8$Et an ethyl group has been attached to the 3'-phosphate of an octathymidylate to mimic the hexamethylene linker (m$_6$) used to tether the acridine ring (Acr) in (Tp)$_8$m$_6$Acr. The results show that the acridine strongly stabilizes the complex by providing an additional binding energy due to its interaction with A.T base pairs.

TABLE 1 : Thermodynamic parameters for the binding of two octathymidylates to poly(rA) at 273 K in a pH 7 buffer containing 10 mM Na cacodylate and 0.1 M NaCl. T_m refers to the temperature of half dissociation of the complex at 10^{-5} M total oligonucleotide concentration

	(Tp)$_8$-Et	(Tp)$_8$ m$_6$-Acr
ΔG (kcal. mole^{-1})	- 8.8	- 14.3
K (M^{-1})	10^7	2.4×10^{11}
T_m (°C)	10.3	38.2

INHIBITION OF TRANSCRIPTION AND TRANSLATION BY OLIGONUCLEOTIDES COVALENTLY LINKED TO INTERCALATING AGENTS

The sequence complementary to the oligonucleotide should be accessible in the target nucleic acid in order for the ONBI to form a complex. This is not the case in double-stranded DNA except if base pairs are locally disrupted either upon binding of proteins or enzymes, e.g., RNA polymerase, or upon introduction of a torsional stress as in supercoiled DNA. If the ONBI target is a messenger RNA, complex formation will be made easier if the target sequence is not engaged in secondary or tertiary structures.

Inhibition of transcription initiation

When RNA polymerase binds to a promoter it first forms a "closed" complex which then isomerizes to an "open" complex characterized by 10-12 open base pairs. Oligonucleotides covalently linked to an acridine derivative were directed against the transcribed strand in the open complex formed by E. coli RNA polymerase with the promoter of the β-lactamase gene (Figure 3).These ONBI of short length (hexa to nonanucleotides) were shown to inhibit transcription initiation provided the target sequence did not extend into the double-stranded region. For example, the oligonucleotide covering region -4 to +2 was more efficient than that complementary to the region -3 to +3 (+1 is the transcription start site ; see figure 3). Control experiments showed that an ONBI inhibiting β-lactamase transcription did not have any effect on the lac gene promoter. During the course of studies using a transcription-translation in vitro assay (see below) it was discovered that some ONBIs could have a non-specific effect on transcription, most probably as a result of binding to RNA polymerase. These ONBIs were characterized by a high A + T content. The binding site of RNA polymerase on promoters involves an (A + T)-rich sequence (the so-called "Pribnow" or "TATA box"). The attachment of an intercalating dye to an (A + T)-rich oligonucleotide might enhance binding to RNA polymerase as compared to the unsubstituted oligonucleotide, thereby inhibiting transcription via competition with promoter binding.

Inhibition of messenger RNA translation
The possibility of blocking gene expression at the level of translation was tested in several systems (Figure 3)

i) gene 32 from bacteriophage T4 was chosen as a prokaryotic example (Toulmé et al., 1986). This gene codes for a protein which binds selectively to single-stranded nucleic acids. The synthesis of this protein was investigated in an E. coli acellular system where a gene 32 mutant carried by a plasmid was first transcribed and then translated. An ONBI directed against a repeated sequence upstream of the Shine-Dalgarno ribosome binding site inhibited gene 32 protein synthesis. It was shown that the ONBIs used in this study [d(TTTAA)$_n$ m$_5$Acr with n = 2 and 3] were able to bind to RNA polymerase and blocked transcription in a non-specific way. If transcription was carried out in the absence of the ONBIs and the ONBIs added when translation started then a specific effect could be observed on gene 32 mRNA <u>translation</u>.

ii) translation of rabbit β-globin mRNA was investigated both in an **in vitro** acellular system (wheat germ extract) and in microinjected **Xenopus** oocytes (Cazenave et al., 1986). An ONBI (11 nucleotides in length) complementary to the β-globin mRNA upstream from the initiation codon was synthesized (Figure 3). Using a mixture of α- and β-globin mRNAs its was demonstrated that this ONBI specifically inhibited β-globin synthesis in the wheat germ extract without any effect on α-globin synthesis (the ONBI target sequence was not present on α-globin mRNA). In microinjected oocytes the synthesis of β-globin was also inhibited in a concentration-dependent process. The ONBI concentration inhibiting 50 % of β-globin synthesis was about 0.5 μM (internal concentration) as compared to 0.2 μM in the **in vitro** wheat germ extract. An undecanucleotide with the same base sequence as the ONBI was also tested in both assays. It was unable to block β-globin synthesis in microinjected oocytes even at a 10 times higher concentration than the ONBI concentration required to block 100 % protein synthesis. In the wheat germ extract the undecanucleotide had an inhibitory effect on β-globin synthesis. It should be noted that the wheat germ extract contains an RNase H activity which hydrolyzes the RNA strand in RNA-DNA hybrids (Minshull and Hunt, 1986). This might explain the high efficiency of oligodeoxynucleotides in this system because mRNA hydrolysis will obviously block translation. In both the wheat germ extract and microinjected oocytes the effect of the ONBI was sequence specific . AN ONBI with no complementary sequence in β-globin mRNA had no inhibitory effect.

OLIGODEOXYNUCLEOTIDES COVALENTLY LINKED TO INTERCALATING AGENTS AS ANTI-VIRAL, ANTI-PARASITIC AND ANTI-TUMORAL AGENTS

In order to test the efficiency of ONBIs as potential anti-viral, anti-parasitic and anti-tumoral agents, several systems were investigated.

Inhibition of influenza virus

The influenza virus genome is segmented and contains eight different RNAs coding for different proteins involved in viral life. The 3'-end sequences of these eight RNAs are identical over 12 bases except for the fourth one in type A viruses (figure 3). An ONBI seven nucleotides in length was able to block the cytopathic effect of the virus using MDCK cells in culture. A type B virus was used as a control. The sequences at the 3'-ends of the eight RNAs of type B viruses are identical over 12 bases except for a degeneracy at the eleventh position. The type A and type B sequences are however different at four positions over the 12 nucleotide-long 3'-terminal sequence. The heptanucleotide-containing ONBI which is fully complementary to the type A viral RNAs has three mismatches with the type B viral RNAs. Therefore the complex should be highly destabilized. As a matter of fact this ONBI had no effect on the cytopathic effect of a type B virus. This experiment demonstrated that the ONBI

a β-LACTAMASE GENE

β-LACTAMASE MRNA

```
                                                                    Met  Ser
  • 1
TGAGACAATAACCCTGATAAATGCTTCAATAATATTGAAAAAGGAAGAGTATGAGT
ACTCTGTTATTGGGACTATTTACGAAGTTATTATAACTTTTTCCTTCTCAGACTCA
```

OPEN REGION IN PROMOTER-RNA POLYMERASE COMPLEX

```
            -1 +1
      ACAATAACCCTGA
5'────AG            TAAA────3'
3'────TC            ATTT────5'
      TGTTATTGGGACT
```

```
5'C C C T G A∿∿Acr
5'C C T G A T∿∿Acr        OLIGONUCLEOTIDES COVALENTLY LINKED TO ACRIDINE
5' T A A C C C T G A∿∿Acr   ⋆ AND COMPLEMENTARY TO THE TRANSCRIBED STRAND
```

b GENE 32 FROM BACTERIOPHAGE T4

```
                              SHINE-
         REPEATED SEQUENCE    DALGARNO
                              SEQUENCE     Met Phe Lys Arg
5' ...G U U A A U U A A A U U A A U U A A A A A G G A A A U A A A A A U G U U U A A A C G T
```

```
        Acr∿∿∿ A A T T T 5'         OLIGONUCLEOTIDES COVALENTLY LINKED TO
     Acr∿∿∿A A T T T A A T T T 5'    ACRIDINE AND COMPLEMENTARY TO THE REPEATED
Acr∿∿∿A A T T T A A T T T A A T T T 5'   SEQUENCE OF GENE 32 MRNA
```

c β-GLOBIN MRNA

```
      1           10        8    50          60
5'G^M P P P A C A C U U G C U U U ...A A A C A G A C A G A A U G G U G C A U ....
```

```
Acr∿∿∿T G T C T G T C T T 5'
```

Figure 3 : Target sequences and oligodeoxynucleotides covalently linked
         to intercalating agents used in the **in vitro** studies reviewed
         in the text
         a. β-lactamase promoter region (Hélène **et al.**, 1985 ; Saison-
            Behmoaras **et al.**, unpublished results).
         b. gene 32 mRNA from bacteriophage T4 (Toulmé **et al.**, 1986)
         c. β-globin mRNA (Cazenave **et al.**, 1986).
         d. influenza virus mRNA, common sequence at the 3'-end (unpu-
            blished results)
         e. trypanosome mRNA, 5'-end mini-exon (Cornelissen **et al.**,
            1986 and unpublished results)

d

GENOME : 8 DIFFERENT RNAs

CONSERVED SEQUENCE AT THE 3'-END OF THE 8 RNAs

```
3'U C G C U U U C G U C C A U...      TYPE A (A/PR/8/34)
          U
          | | | | | | | |
       5'A A A G C A G→Acr            SYNTHETIC OLIGONUCLEOTIDE
          | | | |
    3'U C G U C U U C G C U U C...     TYPE B (B/HK/8/73)
                        G
                  •       •   •  •
```

e

TRYPANOSOMES

CONSERVED SEQUENCE AT THE 5' END OF ALL mRNAS
(MINI-EXON)

```
T. VIVAX    5'A A A G C T T T A T T A G A A C A G T T T C T G T A C T A T A T T G...
                        10          20          30

T. BRUCEI   5'A A C G C T A T T A T T A G A A C A G T T T C T G T A C T A T A T T G...

                       3'T G C G A T A A T A A T 5'
OLIGONUCLEOTIDES {  Acr→T G C C A T A A T 5'
                       3'T A A T A A T C T T G T 5'
```

Figure 3 : Continued

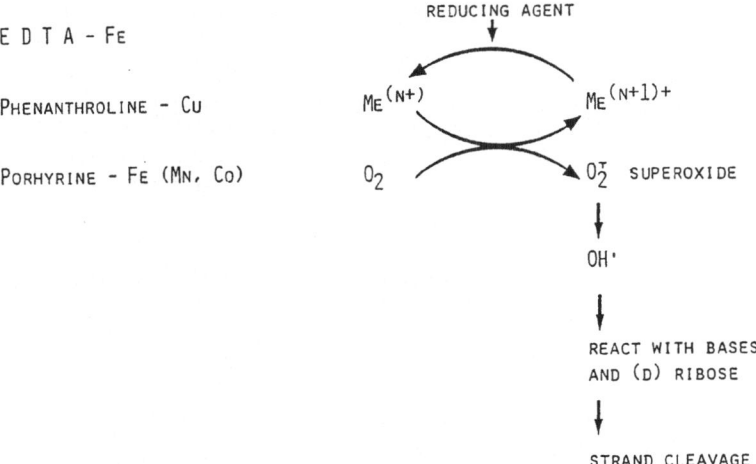

Figure 4 : Reaction pathways for the cleavage of nucleic acid chains by metal complexes.

was acting at the level of the viral RNAs, not on a cellular component
essential for viral development.

Inhibition of DNA replication of an oncogenic virus, SV40

The simian virus SV40 is a DNA-containing virus which replicates in
simian CV1 cells in culture. We investigated the effect of several ONBIs
of sequence $(Tp)_8$, $(Tp)_7Ap$ and $(Tp)_6ApTp$. They are all complementary to
an (A + T)-rich region located in the promoter of the T-antigen and in
the origin of replication which are **overlapping in** this region of the
viral genome. When CV1 cells infected with SV40 were treated with one of
the three ONBIs the cytopathic effect of the virus was strongly inhibited. An oligodeoxynucleotide of sequence $(Tp)_8$ lacking the covalently
linked acridine did not have any inhibitory effect nor did a shorter ONBI
containing only four thymines. This last experiment demonstrated that the
observed effect was not due to the acridine derivative which could have
been released upon nuclease digestion of the ONBI. Other ONBIs which did
not possess complementary sequences in the SV40 genome were also shown to
be devoid of any effect on the SV40 cytopathic effect.

Using an immunofluorescence assay the ONBI $(Tp)_8m_6Acr$ was shown to induce
a decrease in the synthesis of T-antigen by about 25 %. Concomitantly
more than 90 % of SV40 DNA synthesis were inhibited. The decrease of
T-antigen concentration did not seem sufficient to explain this large
effect on replication. The ONBI could block DNA synthesis by hybridizing
with the open region at the replication origin. Experiments on SV40 replication using an **in vitro** system should demonstrate whether such an
hybridization blocks the initiation of replication.

Inhibition of trypanosome mRNA translation

Trypanosomes are characterized by an unusual property of their RNA
transcripts. All mRNAs possess a common 35-nucleotide sequence at their
5'-end. This sequence originates from a mini-exon (encoded in tandem
repeats) which is attached at the end of pre-mRNA either via a trans-
splicing event or by acting as a primer during gene transcription (Figure
3). Oligodeoxynucleotides complementary to the 35 nucleotide-long sequence (or part of it) inhibited the translation of all mRNAs in an **in
vitro** assay. (Cornelissen **et al.,** 1986 ; Walder **et al.,** 1986). ONBIs as
short as 9 nucleotides behaved similarly and blocked translation of all
trypanosome mRNAs. They had no effect on RNAs which did not contain the
complementary sequence, e.g., Brome Mosaic Virus RNA. An ONBI without any
target sequence in the 35-nucleotide mini-exon did not inhibit trypanosome mRNA translation.

There are several trypanosome strains with different species spe-
cifity. Each strain has a 35-nucleotides mini-exon sequence in all of its
mRNAs but there are several point mutations which differentiate one
strain from another. An ONBI complementary to the region [2-10] of the
mini-exon of **Trypanosoma brucei** has two mismatches when **Trypanosoma vivax**
is used instead. These mismatches which occur within a nonanucleotide
sequence destabilize the complex. As a matter of fact this ONBI inhibited
translation of **Trypanosoma brucei** mRNA but had no effect on **Trypanosoma
vivax** mRNA translation in an **in vitro** assay.

Inhibition of oncogene expression

Cell transformation is associated either with a deregulated expression of cellular genes or expression of mutated genes. The myc gene falls
into the first category. In the premyelocytic cell line HL60 the myc gene
is amplified about 30 times. In **Burkitt lymphoma** it is translocated in
the vicinity of highly expressed immunoglobulin genes. The ras genes fall
into the second category. Mutations, usually in the 12th or 61st codons,

alter the functions of the ras gene products which leads to cell transformation.

We synthesized ONBIs directed against the myc and ras genes. **In vitro**
experiments showed that translation of myc mRNA in a wheat germ extract
could be inhibited by short ONBIs directed against a region close to the
AUG initiation codon (A. Darveau, unpublished results) . No effect was
observed on TMV RNA translation. Experiments are presently under way to
test the effect of these ONBIs on transformed cells in culture. One might
expect that a reduction in oncogene expression level would reverse the
transformed to a normal phenotype.

OLIGONUCLEOTIDES COVALENTLY LINKED TO INTERCALATING AGENTS AS VECTORS FOR SITE-DIRECTED MODIFICATIONS OF NUCLEIC ACIDS

In all the examples described above the oligodeoxynucleotide binds
to a complementary sequence in a specific way. The intercalating agent
provides an additional binding energy which stabilizes the complex but
the system retains reversibility. Due to the increased residence time of
the ONBI on its target sequence the biological process is inhibited.
However when the ONBI dissociates the biological system can resume its
activity and synthesize the final product. In order to make the system
irreversible the target sequence should be specifically altered in such
a way as to prevent recognition by the biological system. This can be
achieved by attaching a reactive group at the end of the ONBI or by using
the intercalating agent itself as a reactive group.

Site-directed cleavage of nucleic acids (Figures 4 and 5)

Several chelating agents, such as EDTA, phenanthroline and porphyrin
derivatives, can be attached at the 3'- or 5'-end of an oligodeoxynucleotide. An intercalating agent can be covalently linked at the remaining
free end to stabilize the complex formed with the complementary sequence.
The metal complexes EDTA-Fe, (Phenanthroline)$_2$-Cu and Porphyrin-Fe (Mn,
Co) can generate radical species, especially OH· radicals, in the presence of molecular oxygen and a reducing agent (a thiol derivative, ascorbic
acid...). The radicals attack the ribose or deoxyribose and induce cleavage reactions in the phosphodiester backbone. These reactions work not
only on DNA but also on RNA and can be used to cleave mRNA thereby preventing translation in an irreversible way (Boutorin **et al.**, 1984 ; Chu
and Orgel, 1985 ; Dreyer and Dervan, 1985 ; Boidot-Forget **et al.**, 1986 ;
Sigman, 1986)

Site-directed chemical modifications of nucleic acids

Site-directed chemical modification of nucleic acids can be achieved
in different ways. An oligonucleotide can be used to bring a reactive
group close to the region which is to be modified. Cross-linking of the
oligonucleotide to the target sequence has been previously described
(Knorre and Vlassov, 1985) . We have recently shown that photochemical
modifications could also be induced in the target sequence (figure 6).
When acridine derivatives such as 2-methoxy,6-chloro,9-amino acridine
or proflavine are stacked with a guanine base or a G.C base pair their
fluorescence is quenched, most probably as a result of electron-transfer
reactions in the excited state. Upon visible-light excitation, photooxi-
dation reactions are induced which lead to base modifications in the
target sequence near the end of the complementary oligonucleotide where
the photosensitizer is attached.

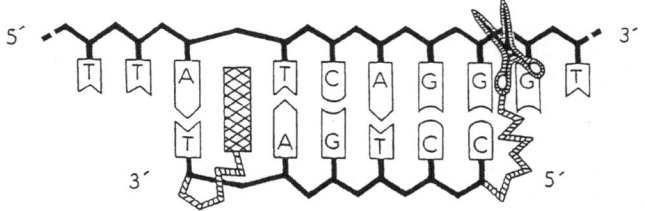

Figure 5 : Artificial site-specific endonuclease : schematic representation of an oligonucleotide covalently linked to an intercalating agent (hatched rectangle) and to a nucleic acid-cleaving
reagent (represented by the pair of scissors).

PHOTOSENSITIZED OXIDATION REACTIONS

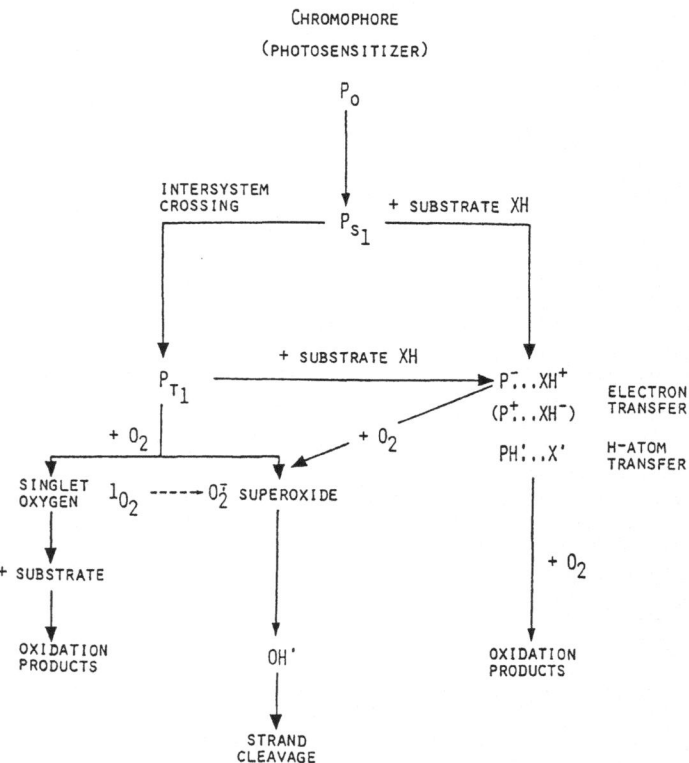

Figure 6 : Reaction pathways for photosensitized oxidation reactions in
nucleic acids. The substrate XH represents either a base or a
sugar. If the photosensitizer (P) interacts with the bases
prior to light excitation, reactions can take place from the
short-lived singlet state (P_{S1}). The longer lived triplet
state (P_{T1}) can also be involved. When the photosensitizer
does not interact in the ground state with a nucleic acid
component, then the reactions will take place from the triplet
state (P_{T1}) whose lifetime may be long enough to allow for
encounter between the excited photosensitizer and the sensiti-
zed molecule (either substrate or oxygen).

CONCLUSIONS

The synthesis of new families of molecules involving oligodeoxynucleotides has been described. An intercalating agent is attached at one (or both) end(s) of the oligonucleotide. It provides an additional binding energy which stabilizes the specific complex formed by the oligonucleotide with its complementary sequence. A nucleic acid-cleaving reagent can be attached at the other end of the oligonucleotide and used to cleave the phosphodiester backbone of the target sequence. A photoactive group can also be attached to the oligonucleotide to induce site-directed irreversible photochemical modificiations in the target sequence. The intercalating agent can itself be used as a photoactive group.

These molecules have been shown to inhibit DNA transcription or mRNA translation **in vitro**. They are also active in microinjected **Xenopus** oocytes and block the cytopathic effect of viruses in cells in culture. These results provide the basis for the rational design of gene-specific inhibitors which could be used not only as tools for cellular biology but also as anti-viral, anti-parasitic or anti-tumoral agents. The development of cellular applications requires additional steps to be solved. The penetration of these molecules across the plasma membrane of living cells should be increased. Their resistance to nuclease action should be improved. This raises challenging questions to the chemist and the biochemist. Their resolution could provide new tools for molecular and cellular biologists who cannot use the powerful techniques of bacterial genetics when studying the relationships between gene expression and phenotypic behavior in most eukaryotic systems. The development of therapeutical applications still requires more general problems to be solved such as, e.g., disponibility in biological fluids or cell targeting. The data which have been collected both **in vitro** and **in vivo** pave the way for the rational conception of substances exhibiting a high selectivity in their mode of action on genetic information.

REFERENCES

Asseline, U., Thuong, N.T., and Hélène, C., 1983, C.R. Acad. Sci. Paris, 297 (III) : 369-372.

Asseline, U., Delarue, M., Lancelot, G., Toulmé, F., Thuong, N.T., Montenay-Garestier, T., and C. Hélène 1984 a, Proc. Nat. Acad. Sci. USA, 81 : 3297-3301

Asseline, U., Toulmé, F., Thuong, N.T., Delarue, M., Montenay-Garestier T., and Hélène C. 1984 b, EMBO J. 3 : 795-800.

Asseline, U., Thuong N.T., and Hélène C., 1986, Nucleosides and Nucleo tides, 5 : 45-63

Bayard, B., Bisbal, C., and Lebleu, B., 1986, Biochemistry, 25 : 3730-3736.

Blake, K.R., Murakami, A., and Miller, P.S., 1985, Biochemistry, 24 : 6132-6138.

Boidot-Forget, M., Thuong, N.T., Chassignol, M., and Hélène, C., 1986 C.R. Acad. Sci. Paris, 302 (II) : 75-80

Boutorin, A.S., Vlassov, V.V., Kazakov, S.A., Kutiavin, I.V., and Podyminogin, M.A., 1984, FEBS Setters, 172 : 43-46

Cazenave, C., Loreau, N., Toulmé J.J., and Hélène, C., 1986, Biochimie, 68 : 1063-1069

Chu, B.C.F., and Orgel, L.E., 1985, Proc. Nat. Acad. Sci. USA, 82 : 963-967

Cornelissen, A.W.C.A., Verspieren, M.P., Toulmé, J.J., Swinkels, B.W., and Bost, P., 1986, Nucl. Ac. Res., 14 : 5605-5614.

Dreyer, G.B., and Dervan, P.B., 1985, Proc. Nat. Acad. Sci. USA, 82 :968-972

Green, P.J., Pines, O., and Inouye, M., 1986, Ann. Rev. Biochem., 55 : 569-597.

Hélène, C., and Lancelot, G., 1982, Prog. Biophys. Mol. Biol., 39 : 1-68.

Hélène,C., Montenay-Garestier, T., Saison,T., Takasugi, M., Toulmé, J.J., Asseline, U., Lancelot, G., Maurizot, J.C., Toulmé, F., and Thuong, N.T., 1985, Biochimie, 67 : 777-783.

Hélène, C., Toulmé, F., Delarue, M., Asseline, U., Takasugi, M., Maurizot, M., Montenay-Garestier, T., and Thuong, N.T., 1986, in "Biomolecular Stereodynamics III",. Sarma R.H. and Sarma M.H., Eds, Adenine Press, pp. 119-130.

Heywood, S.M., 1986, Nucl. Ac. Res., 14 : 6771-6772.

Jayaraman, K., McParland, K., Miller P., and Ts'o, P.O.P., 1981, Proc. Nat.Acad. Sci. USA, 78 : 1537-1541.

Kawasaki, E.S., 1985, Nucl.. Ac. Res., 13 : 4991-5004.

Kim, S.K., and Wold, B.J., 1985, Cell, 42 : 129-138.

Knorre, D.G., and Vlassov, V.V., 1985, Prog. Nucl. Ac. Res. Mol. Biol., 32 : 291-320

Lancelot, G., Asseline, U., Thuong, N.T., and Hélène, C., 1985, Biochemistry 24 : 2521-2529

Lancelot, G., Asseline, U., Thuong, N.T., and Hélène, C., 1986, J. Biomol. Struct. Dyn., 3 : 913-921.

Lancelot, G., and Thuong, N.T., 1986, Biochemistry, 25 : 5357-5363.

Light, J., and Molin, S., 1983, EMBO J., 2 : 93-98

Miller, P.S., Agris, C.H., Blake, K.R., Murakami, A., Spitz, S.A., Reddy, P.M., and Ts'o, P.O.P., 1983, in "Nucleic Acids : The Vectors of Life", Pullman, B., and Jortner, P., Eds, D. Reidel, pp.521-535.

Minshull, J., and Hunt, T., 1986, Nucl. Ac. Res., 14 : 6433-6451.

Mizuno, T., Chou, M.Y., and Inouye, M., 1984, Proc. Nat. Acad. Sci. USA, 81 : 1966-1970

Okamoto, K., and Freundlich, M., 1986, Proc. Nat. Acad. Sci. USA, 81 : 5000-5004

Paterson, B.M., Roberts, B.E., and Kuff, E.L., 1977, Proc.Nat.Acad.Sci. USA, 74 : 4370-4374.

Simons, R.W., Hoopes, B.C., McClure, W.R., and Kleckner, N., 1983, Cell, 34 : 673-682

Sigman, D.W., 1986, Accounts Chem. Res., 19 : 180-186

Spencer, C.A., Gietz, R.D., and Hodgetts, R.B., 1986, Nature 322 : 279-281

Stepheson, M.L., and Zamecnick, P.C., 1978, Proc. Nat. Acad. Sci.USA, 75 : 285-288

Swinkels, B.W., and Bost, P., 1986, Nucl. Ac. Res., 14 : 5605-5614.

Tomizawa, J., 1986, Cell, 47 : 89-97

Toulmé, J.J., Krisch, M.M., Loreau, N., Thuong, N.T., and Hélène, C., 1986, Proc. Nat. Acad. Sci.USA, 83 : 1227-1231.

Trudel, M., Dondon, J., Grunberg-Manago, M., Finelli, J., and Buckingham, R.H., 1981, Biochimie, 63 : 235-240.

Weintraub, H., Izant, J.G., and Harland, R.M., 1985, Trends in Genetics, 1 : 22-25

Williams, T., and Fried, M., 1986, Nature, 322 : 275-279

PHARMACOLOGY OF DNA BINDING DRUGS

Bernard Lambert and Jean-Bernard Le Pecq

Laboratoire de Pharmacologie Moléculaire (UA n°147 CNRS, U n°140 INSERM) Institut Gustave-Roussy, Rue Camille Desmoulins, Villejuif 94805 Cedex, France

INTRODUCTION

Many compounds from various sources have been found able to bind to nucleic acids. Among those are numerous basic dyes used to stain the chromatin in the cell nucleus, several antibiotics, plant alcaloïds and various chemicals synthetized during the last fifty years. Several of these compounds have useful pharmacological properties and are used in human and veterinary medicine for the treatment of parasitic diseases (rev. Van den Bossche, 1978). and cancers (rev. Le Pecq, 1978 ; Pratt and Ruddon, 1979). All these chemicals act by killing or preventing specifically the growth of target cells. A limited list of the most important derivatives is given in Table 1 with their corresponding structures shown in Figure 1.

The interaction of these compounds with nucleic acids have been studied by numerous scientists and recently reviewed (Wilson and Jones, 1981; Waring, 1981; Zimmer and Wähnert, 1986).

The structure of the complexes formed by several of these chemicals with DNA is well characterized since drug-oligonucleotide mixed crystals have been obtained and studied by X rays diffraction (rev. Waring, 1981).

These agents belong to two classes :

i/ intercalating agents :

These compounds have at least one aromatic ring, the size of which is close to that of a base pair. In the bound state, the aromatic ring is intercalated in between two adjacent base pairs.

ii/ groove binding agents :

These derivatives have generally an elongated shape and lie in one of the two grooves of the double helix.

One could have thought that the detailed characterization of the mechanisms of interaction of these drugs with DNA would be paralleled with detailed pharmacological studies leading to the understanding of the mechanisms of their pharmacological action. Such understanding is required for a rational design of these drugs. Unfortunately this has not been the case for most of these drugs.

Table 1. Main DNA binding drugs.

Drugs	Pharmacological properties

Intercalating agents :

9 aminoacridine	mutagenic
proflavine	"
quinacrine	antimalarial
chloroquine	"
ethidium	antitrypanosomal
lucanthone	antischistosomal
hycanthone	"
tilorone	interferon inducer
mAMSA	anticancer
daunomycin	"
adriamycin	"
actinomycin D	"
9 OH N-methyl ellipticinium	"
mitoxantrone	"
bleomycins	"
Bifunctional intercalators :	
echinomycin	"
ditercalinium	"

Non intercalating agents :

netropsin	antiviral
distamycin	"
chromomycin	anticancer
berenil	antitrypanosomal
hydroxystilbamidine	antileishmanial

The proper study of the pharmacological action of a drug implies that two basic questions are answered :

1°/ What is the primary molecular target of the drug ?

2°/ What are the physiological consequences of the interaction of the drug with its target ?

Generally, it is quite difficult to give a clear and unambiguous answer to the first question. Two methods are generally used for this purpose :

i/ Structure-activity relationship studies.

One will look, for instance, for a strong correlation between binding constant to a supposed target macromolecule and pharmacological activity.

ii/ Analysis of mutant or variant cells either resistant or more susceptible to the drug.

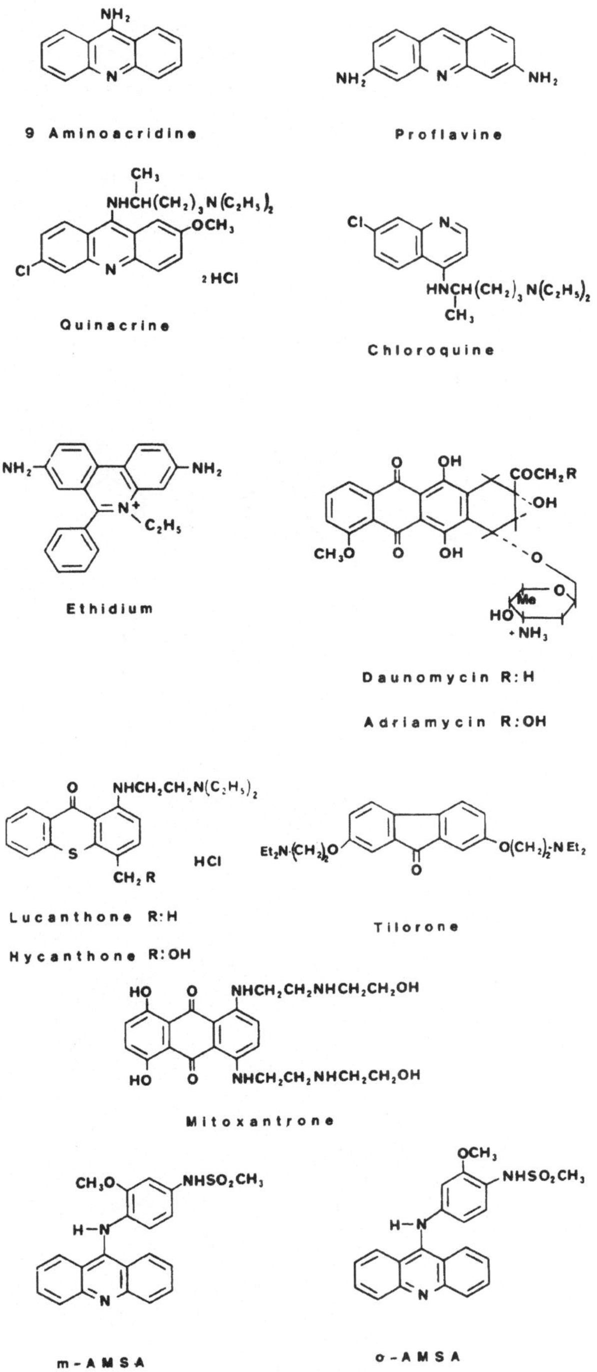

Figure 1 : Main DNA binding drugs.

Intercalating drugs

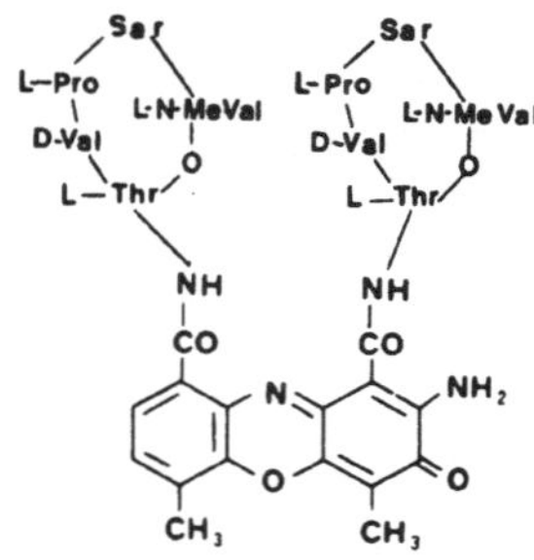

Actinomycin D

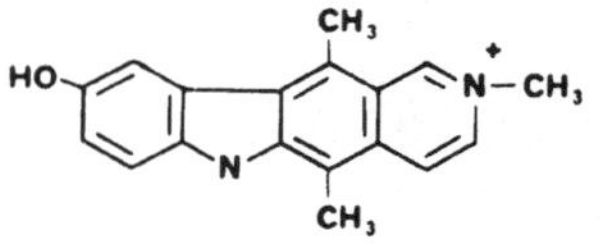

9 OH N methyl ellipticinium

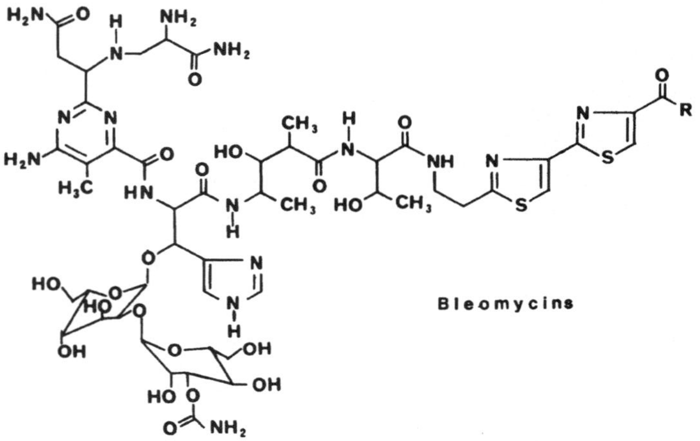

Bleomycins

Bifunctional intercalators

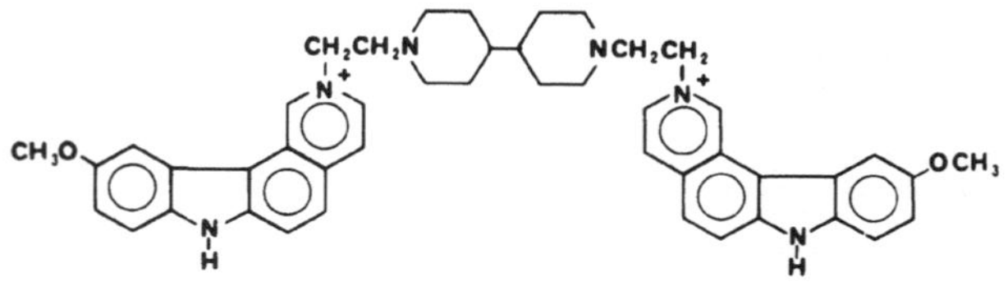

Ditercalinium

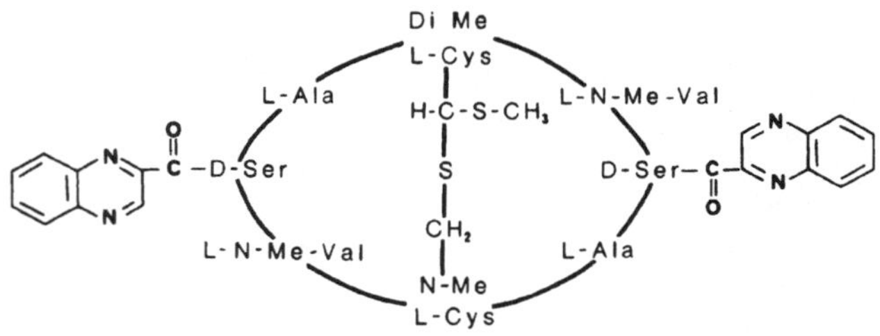

Echinomycin

Figure 1 : Main DNA binding drugs (continued).

Non-intercalating agents

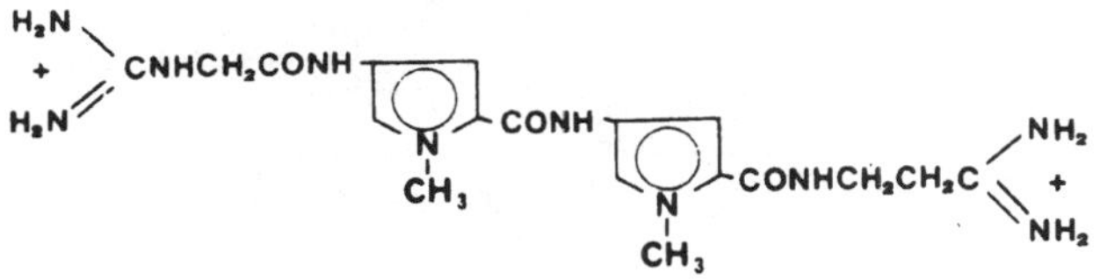

Netropsin

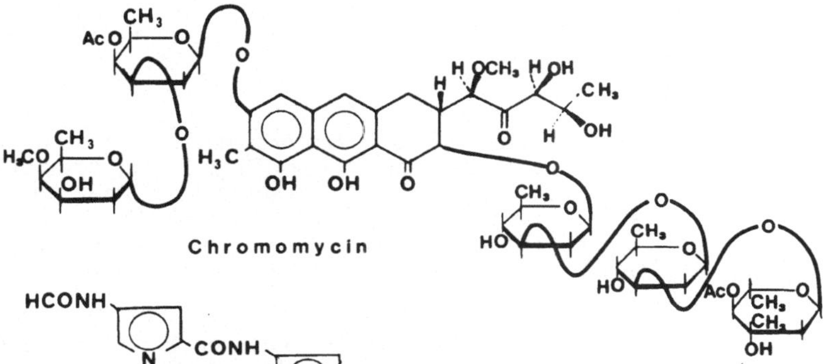

Chromomycin

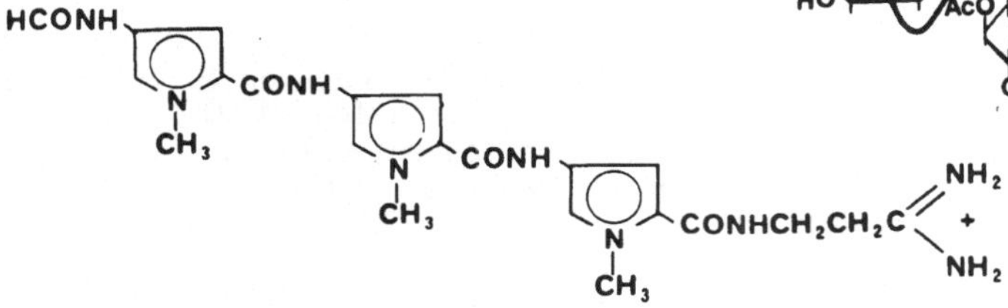

Distamycin A

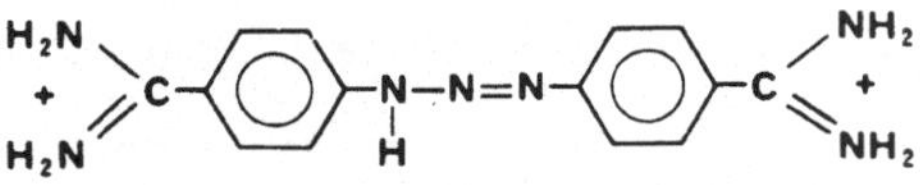

Berenil

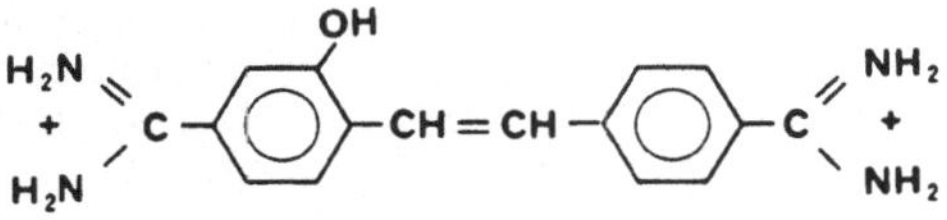

Hydroxystilbamidine

Figure 1 : Main DNA binding drugs (continued).

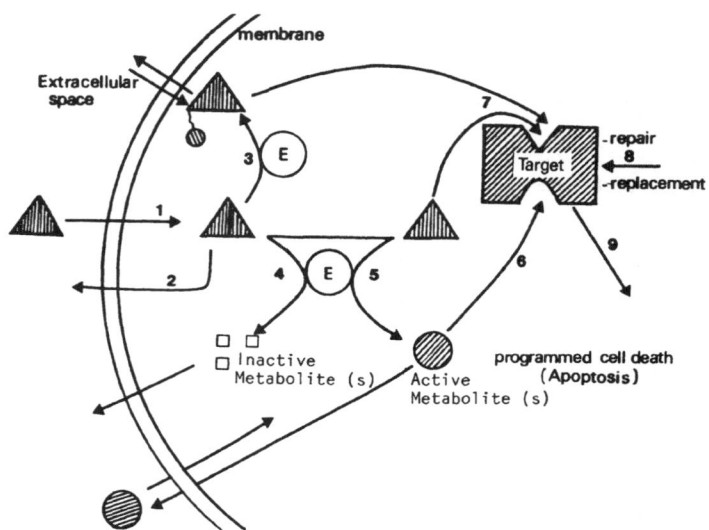

Figure 2 : Schematic representation of the factors which can lead to
specific cell toxicity: membrane permeability (1-2), metabolic
activation or inactivation (4-5), transport alteration of
modified drug (3), target interaction with drug (6) or modified drug (7), repair of replacement of inactivated target
(8), induction of a programmed cell death (apoptosis).

In that case, one will study whether the decrease or the increase of
activity is paralleled by a corresponding variation of the interacting
ability of the drug with its modified supposed molecular target.
Therefore, for DNA binding drugs, one must study and try to answer
the two following questions :

1°/ Is DNA (or RNA) the primary target of "DNA binding agents" ?

2°/ What are the biological effects resulting from the binding to
DNA of a drug inside the cell ?

Finally, in that case, one must also explain by which mechanisms
such a DNA binding drug can kill a cell and particularly by which mechanisms it can kill selectively a given type of cell such as bacteria
parasites, or tumor cells.
Antibiotics kill or prevent the growth of bacteria cells in vivo
without affecting significantly the cells of the infected host, because
most of them interact specifically with a structure inside bacteria which
is not found as such in the cells of the host. Prokaryotic ribosomes, or
enzymes such DNA topoisomerase (gyrase), dihydrofolate reductase... have
structures sufficiently different from that of their counterparts in
eukaryotic cells to bind specifically the antibiotics. The specificity of
the antibiotic action is in that case clearly associated with the specificity of the interaction between the drug and its target.
In the case of antiparasitic and antitumor DNA binding drugs, the
situation is totally different. The structure of DNA is supposed to be
very similar in cells of all living organisms. Therefore, if DNA is
really the drug target, one cannot imagine that the specificity of the
toxic action toward a cell results simply from a specific interaction of
the drug with the DNA of that cell. However, as discussed previously (Le
Pecq, 1982), this specificity could originate indirectly from diffe-

146

rential accessibility, membrane permeability or transport metabolism, etc... as schematized in Figure 2.

I - DNA, THE TARGET OF DNA BINDING DRUGS : A CONTESTED DOGMA

Until very recently, it was generally accepted that the primary pharmacological target of DNA binding drugs was indeed DNA, although no convincing experiments had ever documented such a belief. It was just observed that the binding of these drugs to DNA inhibited in vitro the DNA polymerases and DNA dependent RNA polymerases at concentrations much higher than those leading to toxic action in tissue culture. A common belief was that toxicity arose from unscheduled DNA arrest. Tumor cells were thought to be more sensitive because of their higher rate of cell division.

However, when the pharmacological properties of these drugs were studied more precisely, several observations were made which showed that the mechanism of action of these drugs were more complicated than previously thought. Let us recall some of these observations :

i/ Structure-activity relationship

It was observed in several series of DNA binding drugs that strong DNA binding was required to get biological activity but that this property alone was not sufficient (Le Pecq et al, 1974). A minor chemical modification of a drug which does not modify in any way its DNA binding characteristics can totally suppress its biological activity and decrease its cytotoxicity by several orders of magnitude. One of the most striking illustrations of this phenomenon is given by the comparison of the properties of mAMSA and oAMSA (structure Figure 1) (Denny et al, 1983). These two compounds differ only by the position of a methoxy group on a phenyl ring which is probably not involved in the DNA interaction. The DNA binding characteristics of the two compounds are very similar. But only mAMSA is cytotoxic and has antitumor action.

ii/ Cells are killed at doses which do not block DNA, RNA or protein synthesis

Charcosset et al (1985) studied the effects of ellipticine derivatives on two cell lines derived from Lung chinese hamster. A parental cell line (DC3F) was found very sensitive to these drugs and a variant cell line (DC3F/9OHE) was made resistant by adaptation to 9-hydroxy ellipticine (Salles et al, 1982). The membrane permeability of the variant cell line is not modified and the drug enters the two different cells at the same rate (Charcosset et al, 1983). It can be seen in Figure 3 that sensitive cells are killed at doses which do not induce any inhibition of macromolecule synthesis. Contrastingly, in the resistant cell line, cells died at doses which markedly inhibited both DNA and RNA synthesis. These data strongly suggest that, in sensitive cells, the cytotoxicity does not result from DNA synthesis arrest. However, in the resistant cell line, DNA arrest might well contribute to the toxicity. This experiment indicates therefore that the mechanism of action of a drug might differ depending on the cell line studied.

iii/ Most of the DNA binding drugs block cells in the G2 part of the cell cycle

The effect of drugs on cell cycle can be very easily studied by flow-cytofluorometry. It is observed, for most of the drugs, that exposed cells go through S phase where DNA synthesis occurs without being arrested. Block occurs only in the G2 phase after DNA synthesis is com-

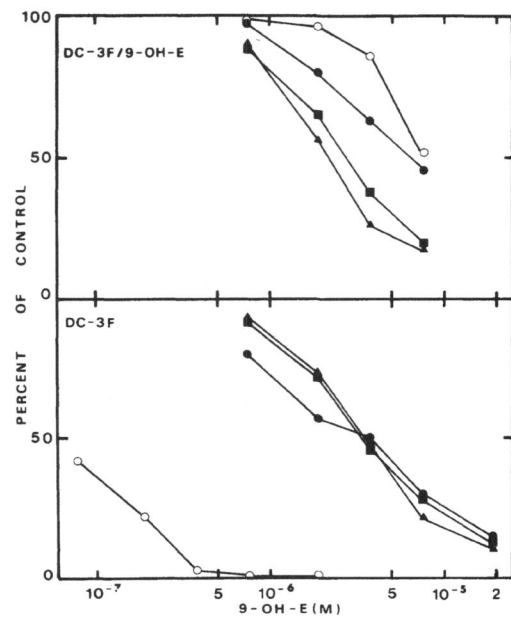

Figure 3 :

Comparison of the cytotoxic effect of 9-hydroxyellipticine measured by determination of the cloning efficiency on semi solid medium O——O with macromolecular synthesis inhibition,
▲——▲ DNA,
●——● RNA,
■——■ protein,
on sensitive (DC-3F-lower) and resistant (DC-3F/9-OH-E, upper).

Reproduced from Charcosset et al, 1985, with permission.

pleted (rev. Rao, 1979). These experiments show clearly that DNA syn- thesis is not prevented in vivo in the presence of the drugs at pharmaco- logical doses. Experiments where DNA synthesis inhibition was measured in non synchronized cells using labelled thymine incorporation have been very misleading. These experiments showed an inhibition of thymine incor- poration in presence of drug. This inhibition resulted certainly from the arrest of the cell in G2 but not from a direct inhibition of DNA syn- thesis.

iv/ Adriamycin can kill cells from without

Tritton and Yee (1982) performed a very provocative experiment. They attached adriamycin to large particles which cannot enter cells. Such adriamycin loaded particles were toxic to cells. After this observation, it was suggested that membrane could be the target of these drugs.

All these data clearly showed that DNA binding alone could not explain the pharmacological activity of DNA binding drugs. Several inter- calating agents are widely used in cancer chemotherapy. Therefore, their pharmacological properties have been more extensively studied. Recently, important results have been obtained which indicate that the DNA topo- isomerase II-DNA complex is probably the primary target of these drugs.

II - DNA TOPOISOMERASE II, A TARGET FOR DNA MONOINTERCALATING AGENTS

It was first observed that DNA intercalating agents such as adria- mycin, ellipticine derivatives, mAMSA induced DNA strand breaks in mamma- lian cells (Paoletti et al, 1979; Ross et al, 1978,1979; Ross and

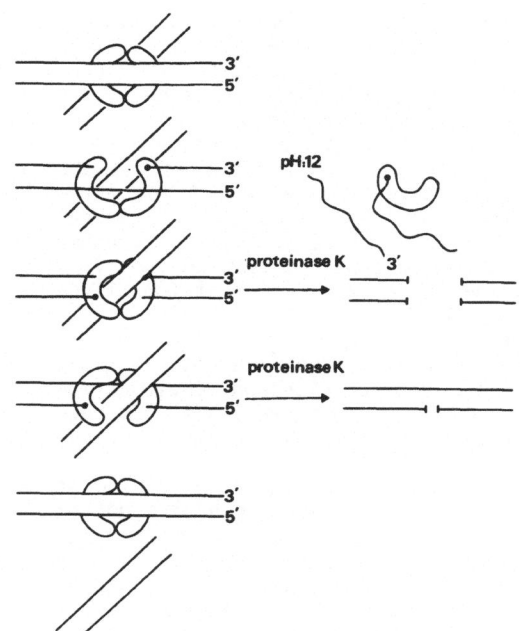

Figure 4 :

Schematic representation of DNA topoisomerase II mechanism of action. Interuption of the enzymatic process at various steps by DNA intercalating agents can lead to hidden protein DNA associated breaks which are revealed by sodium dodecyl sulfate or proteinase K treatment.

Bradley, 1981; Zwelling et al, 1981,1982). Proteins were found associated at the breaks. These breaks occured well at doses which corresponded to the pharmacological action. Studies with purified DNA topoisomerase II showed that, indeed, the DNA strand cleavage reaction could be reproduced in vitro (Nelson et al, 1984; Tewey, et al, 1984). Interestingly, mAMSA, an active compound, could induce DNA breaks in vitro as oAMSA, an inactive analogue could not. Later, it was confirmed that in vivo, DNA topoisomerase II was indeed associated with the DNA breaks in treated cells. DNA extracted from these cells could be precipitated by antibodies directed against DNA topoisomerase II (Mindford et al, 1986; Yang et al, 1985). Recently, a strong correlation between topoisomerase II linked DNA breaks, both in vitro and in vivo, and drug cytotoxicity for a variety of mAMSA analogues was reported (Rowe et al, 1986a).

DNA topoisomerase II (Wang, 1985) catalyses the passage of a double stranded DNA chain across another one. In covalently closed circular DNA, this changes the topological linking number by step of two. The supposed mechanism of action of the enzyme is illustrated in Figure 4. The enzyme gets first inserted in the double helix forming covalent links between tyrosine of the protein and a 5'phosphate terminus of DNA. The enzyme allows the concerted passage of a double stranded DNA through the protein like a boat passing through a lock. DNA intercalating agents can block the process at various steps as illustrated in Figure 4. Single strand or double strand breaks with a topoisomerase subunit linked to the 5'end are obtained. No swiveling can occur at those breaks in constrained structures (Pommier et al, 1984). The breaks are hidden and are only revealed after sodium dodecyl sulfate or proteinase K treatments. Depending on the intercalating agent, one observes mainly single strand breaks (mAMSA), double strand breaks (N-methyl 9-OH-ellipticinium) or a mixture of the two types of breaks (anthracyclines). Ethidium bromide inhibits the enzymatic reaction but does not promote the DNA cleavage (Tewey et al, 1984). Inhibition occurs very likely at an early step, before strand cleavage takes place. Interestingly, ethidium inhibits the formation of breaks caused by other intercalating agents in vitro and reduces the cytotoxicity of these drugs in vivo (Rowe et al, 1985; Tewey et al, 1984; Riou et al, 1986b). Therefore it seems likely that active compounds are those which are able to interact both with DNA and the protein within the enzyme substrate complex to cause an abortive termination of the enzymatic process at different intermediate steps. Such a model is in agreement with the results of the structure activity relationship discussed above. Surprisingly, intercalating agents promote breaks at specific points along the DNA chain. The cleavage points are different for different intercalating agents (Tewey et al, 1984).

Another series of antitumor drugs, the epipodophyllotoxin derivatives seem to have a very similar mechanism of action (Chen et al, 1984; Ross et al, 1984). Those compounds do not interact with DNA. Their action might be mediated by direct interaction with the enzyme alone. Interestingly, ethidium inhibits both in vitro and in vivo the action of epipodophillotoxin (Rowe et al 1985).

Recent studies (Yang et al, 1985 ; Pommier et al, 1986b ; Riou et al, 1986a,b; Rowe et al, 1986b) show that in vivo the DNA breaks do not occur at random in the genome after the cells have been treated with intercalating agents or epipodophillotoxin. SV40 chromosome was cleaved inside infected cells at specific sites. One of the major site was mapped in the DNase I hypersensitive region (Yang et al, 1985). In tumor cells where the c myc protooncogene is very actively transcribed, intercalating agents induce cleavage in this gene at locations close to that observed in vitro with purified DNA topoisomerase II. Furthermore, the major sites were also observed in that case in the DNase I hypersensitive regions (Riou et al, 1986a,b).

These results, as recently discussed (Ross, 1985), open quite new perspectives for the understanding of the action of intercalating agents which appears to be modulated by the accessibility and structure of DNA in specific genes. These factors could well play a crucial role in determining the specificity of cytotoxic action in malignant cells. Definite proof that DNA topoisomerase is indeed involved in the action of intercalating agents might well soon be obtained using variant cells made resistant to these agents. Already, it has been observed that in the chinese hamster lung cell line (DC3F/90HE) resistant to ellipticine derivatives, intercalating agents do not induce protein associated DNA strand breaks. In addition, these cells are resistant to epipodophyllotoxin derivatives (Pommier et al, 1986a). On-going studies suggest that a protein associated with DNA topoisomerase might be altered in the variant cell (Pommier et al, 1986b). Furthermore, a cell line selected for resistance to epipodophyllotoxin has been studied. A qualitative change of DNA topoisomerase II in this cell seems able to confer at the same time resistance to epipodophyllotoxin and to intercalating agents (Glisson et al, 1986a,b).

Finally, an important problem remains without solution. It is not understood at all how the protein associated DNA breaks induce cell death. Surprisingly, it is observed that, after drug removal, the DNA breaks disappear rapidly (Zwelling et al, 1981; Ross and Smith, 1982; Pommier et al, 1984). It has been suggested that the DNA breaks could induce sister chromatide exchanges or other DNA rearrangements which could be the true cause of cell death (Pommier et al, 1985). But this hypothesis remains for the present conjectural.

DNA topoisomerase II from trypanosome and plasmodium have recently been purified (Douc-Rasy et al, 1986; Riou et al, 1986c). The enzyme from trypanosome appears extremely interesting. It differs markedly from other known DNA topoisomerase II because it does not need ATP. This suggests

that its structure is sufficiently different from that of the enzyme of
mammalian cells to constitute a specific target for chemotherapy of parasitic diseases. DNA topoisomerase II is an important target for antibiotic action. It could well be that, in the future, this enzyme would
represent also a useful target for antiparasitic and anticancer drugs.
This would then open the way for a rational design of these drugs.

III - MECHANISM OF ACTION OF THE ANTITUMOR BIFUNCTIONAL INTERCALATORS
      DITERCALINIUM

Recently, a new class of synthetic intercalating agents, the poly-
functional intercalators have been prepared (rev. Le Pecq and Roques,
1986).
Molecules with two or three intercalating rings linked by chains of
appropriate length and structure to allow the DNA intercalation of each
subunit have been obtained. Those molecules bind to DNA with an extremely
high binding constant (up to $10^{14}M^{-1}$ for trisintercalating molecules
(Laugaa et al., 1985)).
Among these molecules, several dimeric molecules deriving from 7H-
pyridocarbazole such as ditercalinium (figure 1) elicited strong antitumor activities (Roques et al., 1979; Pelaprat et al., 1980). As for the
other DNA binding drugs, high DNA affinity was required to obtain biological activity, but this property alone was not sufficient (Pelaprat et
al., 1980).
Studies on the mechanism of action of these molecules (Bendirdjian
et al., 1984; Esnault et al., 1984; Markovits et al., 1986), indicated
that the mechanism of action of ditercalinium was completely different
from that of other antitumor drugs.

i/ In vitro DNA topoisomerase II action was inhibited when the drug
was bound to DNA, but no protein associated breaks were registered. This
was observed both with the enzyme of mammalian cells (Markovits et al.,
1986) and with the enzyme of trypanosome and plasmodium (Douc-Rasy et
al., 1984; Riou et al., 1986c).
In vivo no protein associated breaks were registered in the nuclear
DNA of treated mammalian cells although a compaction of the chromatin was
noticed (Markovits et al., 1986).

ii/ Ditercalinium induced a delayed cytotoxicity

Mammalian cells were still able to grow for five to six generations
after drug treatment before getting arrested. Treated cells gave rise to
small abortive clones when plated in semi-solid medium.

iii/ No G2 arrest was observed when treated cells were analysed by
     flow-cytofluorometry

The mechanism of action of ditercalinium seemed quite difficult to
elucidate. Important information was obtained when the action of this
molecule was studied in E. coli.
Wild-type E. coli cells are totally resistant to this compound. An
E. coli strain extremely sensitive to ditercalinium was selected after
mutagenesis (Lambert and Le Pecq, 1982). For this strain, cells were
killed at a concentration close to 10 ng/ml.
This mutation was later identified and located in the pol A gene[1].
The pol A cells reverted with high frequency to a resistant phenotype
when selected in presence of drug. All the cells which acquired the
resistant phenotype after reversion had still the pol A mutation. These
cells had therefore acquired an additional mutation which suppressed the
sensitive phenotype conferred by pol A. 50 % of these revertants were

extremely UV sensitive. It was observed that a strain having the two mutations pol A and uvr A was completely drug resistant. In pol A strain, the SOS system was efficiently induced as seen by β galactosidase synthesis induction in the pol A (λ(sfi A::lac Z) (Huisman and d'Ari, 1981) and by important filamentation of the cells.

These results indicate that the non covalent DNA ditercalinium complex is recognized by the uvrABC repair system in E. coli. In the case of monointercalating agents, the non covalent DNA complexes are not recognized by DNA repair enzymes. Formation of a covalent bond between the drug and DNA is required to activate the repair system (Lambert et al., 1986). The fact that the "DNA lesion" which is recognized by the uvrABC system is induced by a drug which binds reversibly to DNA has unexpected consequences and could lead to a futile DNA repair process[1].

It remains to be established whether such a phenomenon is involved in the determination of toxicity in mammalian cells.

IV - CONCLUSION

Important information on the mechanism of the pharmacological action of DNA binding drugs is still lacking. This is particularly noticeable for the non intercalating agents which have not been studied yet in details. However, remarkable results have already been obtained in the case of monointercalating agents implying the DNA topoisomerase II as a possible target.

Most of the anticancer drugs act at the level of DNA. Although they are very different molecules forming various kinds of covalent adducts or reversible complexes with DNA, a common feature, as far as the mechanism of the cytotoxic action is concerned, is emerging. Almost all these molecules lead directly or indirectly to the formation of DNA cleavage inside the cells.

- Bleomycins, a series of antibiotics isolated by Umezawa and collaborators (rev. Umezawa, 1976) form reversible intercalating DNA complexes (Povirk et al., 1979). When bound to DNA, they cause the cleavage of the N glycosidic bonds between deoxyribose and bases by catalysing, at the level of DNA, complex reactions leading to the formation of free radicals in presence of oxygen.

- alkylating agents form covalent adducts with DNA. These adducts are recognized as lesions by the repair enzymes which excise the modified part of DNA. Cleavage of DNA is an obligatory step in the repair process.

- monointercalating agents interfere with the functioning of the DNA topoisomerase II and induce the formation of protein associated strand breaks.

- ditercalinium, a recently developped bifunctional intercalator, forms a reversible complex with DNA. Nevertheless this complex is recognized by the DNA repair enzymes. DNA breaks are formed during the repair process. The fact that the pol A uvr A strain is completely resistant to ditercalinium indicates that the DNA binding of a drug with a very high

---

[1]Lambert B. and Le Pecq J-B. (unpublished results).

affinity in vivo has no biological effect by itself. It is only because this complex is recognized by the uvr ABC system that a toxic effect is elicited.

Other DNA binding drugs could induce unusual DNA structures which could be recognized by various enzymatic systems. For instance, Echinomycin has been shown to cause the formation of Hoogsteen base pairs in DNA (Quigley et al., 1986). The detailed characterization of the complexes of these drugs with DNA at atomic resolution appears therefore of extreme importance.

It must be pointed out also that nuclear DNA might not be the only possible target for DNA binding drugs. Mitochondrial DNA could also be a relevant target. A recent report underlines frequent alterations of mitochondria in carcinoma cells (Chen et al., 1985). Furthermore, it has been reported that mitochondrial topoisomerases have distinct properties and drug sensitivities (Fairfield et al., 1985).

It remains to understand how the formation of DNA breaks after drug treatment can lead to cell death in mammalian cells and to enhanced cytotoxicity in parasites and cancer cells. It must also be understood why closely related compounds able to induce similar types of DNA breaks are devoided of useful pharmacological properties. Results discussed in this paper show that DNA cleavage does not occur at random in the genome. The reasons and the physiological consequences of the preferential cleavage of specific regions of the genome can now be analysed. It is possible that an increased sensitivity to cleavage of genes like oncogenes expressed in malignant cells contribute to the specificity of action of intercalating antitumor drugs. As discussed above, indirect factors such as cellular metabolism of the drugs can contribute to the specificity of the cytotoxic action. For instance, it has been established that Bleomycin specificity is determined in part by its rate of enzymatic inactivation inside the cells (Umezawa, 1976).

The complete understanding of the action of these drugs requires the characterization of all the reactions in which the drug is involved from the time it is injected to the patient up to the time it is excreted. Reaction with DNA is probably a crucial step but it is obviously not the only one.

Results concerning monointercalating agents suggest that the drug target is neither the DNA nor the protein but indeed the DNA-enzyme complex. In this respect, it is worth mentioning the results of Arai and Kornberg (1981) concerning the dnaB protein from E. coli. They observed that poly(dT), dnaB protein or poly(dT)-dnaB protein complex in absence of ATP could not bind ethidium. Contrastingly, the ternary complex ATP-dnaB protein-poly(dT) had a very high affinity for ethidium. This result suggests that in the ternary complex, a unique structure was generated which is specifically recognized by ethidium. Such a situation could well be encountered in other DNA protein complexes. Studying interaction of drugs with DNA alone is probably a biased approach. The study of the mechanism of action of DNA binding agents could permit to identify DNA binding proteins of pharmacological interest. The design of molecules able to specifically recognize DNA-protein complexes would be extremely interesting.

Acknowledgments

The authors are very grateful to Dr A. Jacquemin-Sablon for fruitful discussions and permission to reproduce results of his laboratory.

153

REFERENCES

Arai,K.I., and Kornberg, A., 1981, Mechanism of dnaB protein action III allosteric role of ATP in the alteration of DNA structure by dnaB protein in priming replication,
J. Biol. Chem., 256:5260-5266.

Bendirdjian, J-P., Delaporte, C., Roques, B. P., and Jacquemin-Sablon, A., 1984, Effects of 7H-pyridocarbazole mono and bifunctional DNA-intercalators on chinese hamster lung cells in vitro,
Biochem. Pharmacol., 33:3681-3688.

Charcosset, J-Y., Salles, B., and Jacquemin-Sablon, A., 1983, Uptake and cytofluorescence localization of ellipticine derivatives in sensitive and resistant chinese hamster lung cells,
Biochem. Pharmacol., 32:1037-1044.

Charcosset, J-Y., Bendirdjian, J-P., Lantieri, M-F., and Jacquemin-Sablon, A., 1985, Effects of 9-hydroxyellipticine on cell survival, macromolecular synthesis, and cell cycle progression in sensitive and resistant chinese hamster lung cells,
Cancer Res., 45:4229-4236.

Chen, G.L., Yang, L., Rowe, T.C., Halligan, B.D., Tewey, K.M., and Liu, L.F., 1984, Non intercalative antitumor drugs interfere with the breakage-reunion reaction of mammalian DNA topoisomerase II,
J. Biol. Chem., 259:13560-13566.

Chen, L.B., Weiss, M.J., Davis, S., Bleday, R.S., Wong, J.R., Song, J., Terasaki, M., Shepherd, E.L., Walker, E.S., and Steele, Jr. G.D., 1985, in: "Cancer Cells 3.-growth factors and transformation",
J. Feramisco, B. Ozanne and Ch. Stiles eds.
Cold Spring Harbor laboratories publ. pp.433-443.

Denny, W.A., Baguley, B.C., Cain, B.F., and Waring, M.J., 1983, Antitumor acridines, in: "Molecular aspects of anti-cancer drug action",
Neidle S. and Waring M.J. ed., Verlag Chemie Publ., p.1-34.

Douc-Rasy, S., Kayser, A., and Riou, G., 1984, Inhibition of the reactions catalysed by a type I topoisomerase and a catenating enzyme of Trypanosoma cruzi by DNA-intercalating drugs. Preferential inhibition of the catenating reaction,
The EMBO J., 3:11-16.

Douc-Rasy, S., Kayer, A., Riou, J.F., and Riou, G., 1986, ATP-independent type II topoisomerase from trypanosomes,
Proc. Natl. Acad. Sci. USA, in press.

Esnault, C., Roques, B.P., Jacquemin-Sablon, A., and Le Pecq, J.B., 1984, Effects of new antitumor bifunctional intercalators derived from 7H-pyridocarbazole on sensitive and resistant L1210 cells,
Cancer Res., 44:4335-4360.

Fairfield, F.R., Bauer, W.R., and Simpson, M.V., 1985, Studies on mitochondrial type I topoisomerase and on its function,
Biochim. Biophys. Acta, 824:45-57.

Glisson, B., Gupta, R., Hodges, P., and Ross, W., 1986a, Characterization of acquired Epipodophyllotoxin resistance in a chinese hamster ovary cell line : Loss of drug-stimulated DNA cleavage activity,
Cancer Res., 46:1934-1938.

Glisson, B., Gupta, R., Hodges, P., and Ross, W., 1986b, Cross resistance to intercalating agents in a epipodophyllotoxin-resistant hamster ovary cell line : Evidence for a common intracellular target,
Cancer Res., 46:1939-1942.

Huisman, O., and D'Ari, R., 1981, An inducible DNA replication cell division coupling mechanism in E.coli,
Nature, 290:727-799.

Lambert, B., and Le Pecq, J.B., 1982, Isolement et caractérisation de souches d'E.coli sensibles à des toxiques hydrophiles et/ou chargés,
C. R. Acad. Sci. Paris, 294:447-450.

Lambert, B., Laugaa, Ph., Roques, B.P., and Le Pecq, J.B., 1986, Cytotoxicity and SOS inducing ability of ethidium and photoactivable
analogs on E.coli ethidium sensitive (Ebs) strains,
Mutation Res., in press.

Laugaa, Ph., Markovits, J., Delbarre, A., Le Pecq, J.B., and Roques,
B.P., 1985, DNA tris-intercalation : First acridine trimer with DNA
affinity in the range of DNA regulatory proteins. Kinetics studies,
Biochemistry, 24:5567-5575.

Le Pecq, J.B., 1978, "Chimiothérapie anticancéreuse", Hermann Publ.
Paris.

Le Pecq, J.B., 1982, Spécificité d'action des substances antitumorales,
J. Pharmacol., 13:53-75.

Le Pecq, J.B., Dat-Xuong, N., Gosse, Ch., and Paoletti, C., 1974, A new
antitumoral agent : 9 hydroxyellipticine. Possibility of a rational
design of anticancerous drugs in the series of DNA interalating :
drugs,
Proc. Natl. Acad. Sci. USA, 71:5078-5082.

Le Pecq, J.B., and Roques, B.P., 1986, DNA binding and biological properties of bis- and trisintercalating molecules, in: "Mechanisms of DNA
damage and repair", Simic M.G., Grossman L. and Upton A.C. ed.,
Plenum Press Publ. New York, p.219-230.

Markovits, J., Pommier, Y., Mattern, M.R., Roques, B.P., Le Pecq, J.B.,
and Kohn, K.W., 1986, Effect of the bifunctional antitumor intercalator Ditercalinium on DNA in Mouse leukemia (L1210) cells and on
L1210 DNA topoisomerase II,
Cancer Res., in press.

Mindford, J., Pommier, Y., Filipski, J., Kohn, K., Derrigan, D., Mattern,
M.R., Michaels, S., Schwartz, R., and Zwelling, L., 1986, Isolation
of interalator-dependent protein linked DNA strand cleavage activity
from cell nuclei and identification as topoisomerase II,
Biochemistry, 25:9-16.

Nelson, E.M., Tewey, K.M., and Liu, L.F., 1984, Mechanism of antitumor
drug action : Poisoning of mammalian DNA topoisomerase II on DNA by
4'-(9-acridinylamino)-methane sulfon-m-anisidide,
Proc. Natl. Acad. Sci. USA, 81:1361-1365.

Paoletti, C., Lesca, C., Cros, S., Malvy, C., and Auclair, C., 1979,
Ellipticine and derivatives induce breakage of L1210 DNA in vitro,
Biochem. Pharmacol., 28:345-350.

Pelaprat, D., Delbarre, A., Le Guen, I., Roques, B.P., and Le Pecq, J.B.,
1980, DNA intercalating compounds as potential antitumor agents. 2.
Preparation and properties of 7H-pyridocarbazole dimers,
J. Med. Chem., 23:1336-1343.

Pommier, Y., Mattern, M.R., Schwartz, R.E., and Zwelling, L.A., 1984a,
Absence of swiveling at sites of intercalator-induced protein-associated deoxyribonucleic acid strand breaks in mammalian cells
nucleoïds,
Biochemistry, 23:2922-2927.

Pommier, Y., Schwartz, R.E., Kohn, K.W., and Zwelling, L.A., 1984b,
Formation and rejoining of deoxyribonucleic acid double-strand
breaks induced in isolated cell nuclei by antineoplastic intercalating agents,
Biochemistry, 23:3194-3201.

Pommier, Y., Zwelling, L.A., Kao-Shan, C.S., Whang-Peng, J., and Bradley,
M., 1985, Correlation between intercalator-induced DNA strand breaks
and sister chromatid exchanges, mutations and cytotoxicity in
chinese hamster cells,
Cancer Res., 45:3143-3147.

Pommier, Y., Schwartz, R.E., Zwelling, L., Kerrigan, D., Mattern M.,
Charcosset, J.Y., Jacquemin-Sablon, A., and Kohn, K., 1986a, Reduced

formation of protein-associated DNA strand breaks in chinese hamster cells resistant to topoisomerase II inhibitors,
Cancer Res., 46:611-616.

Pommier, Y., Kerrigan, D., Schwartz, R., Swack, J.A., and McCurdy, A., 1986b, Altered DNA topoisomerase II activity in chinese hamster cells resistant to topoisomerase II inhibitors,
Cancer Res., 46:3075-3081.

Povirk, L.F., Hogan, M., and Dattagupta, N., 1979, Binding of Bleomycin to DNA : Intercalation of the bithiazole rings,
Biochemistry, 18:96-101.

Pratt, W., and Ruddon, R., 1979, "The Anticancer Drugs", Oxford University Press Publ. New York.

Quigley, G.J., Ughetto, G., Van der Marel, G.A., Van Boom, J.H., Wang, A.H.J., and Rich, A., 1986, Non-Watson-Crick G.C and A.T base pairs in a DNA-antibiotic complex,
Science, 232:1255-1258.

Rao, P.N., 1979, G_2 arrest induced by anticancer drug, in: "Effects of drugs on the cell nucleus", H. Busch, S.T. Crooke and Y. Daskal ed., Academic Press New York, p.475-490.

Riou, J.F., Multon, E., Vilarem, M.J., Larsen, Ch., and Riou, G., 1986a, In vivo stimulation by antitumor drugs of the topoisomerase II induced cleavage sites in c-myc protooncogène,
Biochem. Biophys. Res. Comm., 137:154-160.

Riou, J.F., Vilarem, M.J., Larsen, C.J., and Riou, G., 1986b, Characterization of the topoisomerase II-induced cleavage sites in the c-myc protooncogene : in vitro stimulation by the antitumoral intercalating drug mAMSA,
Biochem. Pharmacol., in press.

Riou, J.F., Gabillot, M., Philippe, M., Schrevel, J., and Riou, G., 1986c, Purification and characterization of plasmodium berghei DNA topoisomerase I and II : Drug action inhibition of decatenation and relaxation, and stimulation of DNA cleavage,
Biochemistry, 25:1471-1479.

Roques, B.P., Pelaprat, D., Le Guen, I., Porcher, G., Gosse, Ch., and Le Pecq, J.B., 1979, DNA bifunctional intercalators. Antileukemic activities of new pyridocarbazole dimers,
Biochem. Pharmacol., 28, 1811-1815.

Ross, W.E., 1985, DNA topoisomerases as targets for cancer therapy,
Biochem. Pharmacol., 34:4191-4195.

Ross, W.E., and Bradley, M.O., 1981, DNA double-strand breaks in mammalian cells after exposure to DNA intercalating agents,
Biochim. Biophys. Acta, 654:129-134.

Ross, W.E., and Smith, M.C., 1982, Repair of deoxyribonucleic acid lesions caused by adriamycin and ellipticine,
Biochem. Pharmacol., 31:1931-1935.

Ross, W.E., Glaubiger, D.L., and Kohn, K.W., 1978, Protein-associated DNA breaks in cell treated with adriamycin and ellipticine,
Biochim. Biophys. Acta, 519:23-30.

Ross, W.E., Glaubiger, D.L., and Kohn, K.W., 1979, Qualitative and quantitative aspects of intercalator-induced DNA strand breaks,
Biochim. Biophys. Acta, 562:41-50.

Ross, W., Rowe, T., Glisson, B., Yalowich, J., and Liu, L., 1984, Role of topoisomerase II in mediating epipodophyllotoxin-induced DNA cleavage,
Cancer Res., 44:5857-5860.

Rowe, T., Kupfer, G., and Ross, W., 1985, Inhibition of epipodophyllotoxin cytotoxicity by interference with topoisomerase-mediated DNA cleavage,
Biochem. Pharmacol., 34:2483-2487.

Rowe, T., Chen, G., and Liu, L., 1986a, DNA damage by antitumor acridines mediated by mammalian DNA topoisomerase II,
Cancer Res., 46:2021-2026.

Rowe, T., Wang, J.C., and Liu, L., 1986b, In vivo localization of DNA topoisomerase II cleavage sites on Drosophila heat shock chromatin,
Molecular and Cellular Biology, 6:985-992.

Salles, B., Charcosset, J.Y., and Jacquemin-Sablon, A., 1982, Isolation and properties of chinese hamster lung cells resistant to ellipticine derivatives,
Cancer Treat. Rep., 66:327-338.

Tewey, K., Chen, G., Nelson, E., and Liu, L., 1984, Intercalative antitumor drugs interfere with the breakage-reunion reaction of mammalian DNA topoisomerase II,
J. Biol. Chem., 259:9182-9187.

Tritton, T.R., and Yee, G., 1982, The antitumor agent adriamycin can be actively cytotoxic without entering cells,
Science, 217:248-250.

Umezawa, H., 1976, Bleomycin : Discovery, chemistry and action,
Gann, 19:3-36.

Van den Bossche, H., 1978, Chemotherapy of parasitic infections,
Nature, 273:626-630.

Wang, J.C., 1985, DNA topoisomerases,
Ann. Rev. Biochem., 54:665-697.

Waring, M.J., 1981, DNA modification and cancer,
Ann. Rev. Biochem., 50:159-192.

Wilson, W.D., and Jones, R.L., 1981, Intercalating drugs : DNA binding and molecular pharmacology,
Adv. Pharmacol. Chemother., 18:177-222.

Yang, L., Rowe, R.C., Nelson, E.M., and Liu, L., 1985, In vivo mapping of DNA topoisomerase II-specific cleavage sites on SV40 chromatin,
Cell, 41:127-132.

Zimmer, Ch., and Wähnert, U., 1986, Nonintercalating DNA-binding ligands : specificity of the interaction and their use as tools in biophysical, biochemical and biological investigations of the genetic material,
Prog. Biophys. Molec. Biol., 47:31-112.

Zwelling, L., Michaels, S., Erickson, L., Ungerleider, R.S., Nichols, M., and Kohn, K.W., 1981, Protein associated deoxyribonucleic acid strand breaks in L1210 cells treated with the deoxyribonucleic intercalating agents 4'-(9-acridinylamino)methanesulfon-m-anisidide and adriamycin,
Biochemistry, 20:6553-6563.

Zwelling, L., Michaels, S., Kerrigan, D., Pommier, Y., and Kohn, K., 1982, Protein-associated deoxyribonucleic acid strand breaks produced in mouse leukemia L1210 cells by ellipticine and 2-methyl-9-hydroxyellipticinium,
Biochem. Pharmacol., 31:3261-3267.

ON THE NATURE AND SPECIFICITY OF DNA-PROTEIN

INTERACTIONS IN THE REGULATION OF GENE EXPRESSION

Peter H. von Hippel and Otto G. Berg*

Institute of Molecular Biology and Department of Chemistry
University of Oregon
Eugene, Oregon 97403 USA

INTRODUCTION

In this paper we describe a number of approaches to the nature and specificity of DNA-protein interactions involved in the regulation of gene expression at the transcriptional level. We discuss primarily the binding of "single-specific-site" regulatory proteins to DNA targets. This involves consideration of several aspects of the specificity of DNA-protein interactions, including: (i) the combinatorial specification of the number of base pairs required to define a unique binding site in a genome of given size; (ii) the structure of DNA and protein binding sites, including structural complementarity and steric aspects; (iii) the energetics of the binding interaction, including both specific and nonspecific binding; (iv) the thermodynamics and kinetics of the overall interaction, as determined by the net binding free energy of specific complex formation and the effects of competing sites; and (v) equilibrium binding selection, which determines the actual level of saturation of the specific (regulatory) target under various environmental conditions. These aspects are all interdependent, and a coherent picture of the specificity of such interactions can only be obtained by considering them all in context. Such an approach has been set forth by us previously in von Hippel and Berg (1986), and portions of this overview are taken directly from that treatment. In conclusion, we consider how these ideas modulate the evolutionary "design" of regulatory proteins, as well as the formulation of purification procedures and binding assays for these proteins, and how these approaches may apply in vivo.

SINGLE PROTEIN BINDING TO A REGULATORY DNA SITE

Regulation of transcription at the DNA level clearly must involve, as at least an initial step, the binding of the regulatory protein to a DNA target site. This binding may comprise the entire process, as in (e.g.) E. coli lac repressor binding to its operator target. On the

---

*Present address: Department of Molecular Biology, The Biomedical Center, Box 590, S-751 24 Uppsala, Sweden.

other hand a regulatory protein may operate as an activator, changing the vicinal affinity, and thus the activity, of a "primary" regulatory protein such as RNA polymerase. And finally, of course, the protein may itself be a primary regulatory protein, for which binding serves only as the first of a series of sequential functional steps (e.g., polymerase interacting with promoter to form a closed polymerase-promoter complex).

The general problem of DNA-protein interaction specificity is best described in functional terms; i.e., in terms of the degree of saturation of a regulatory target site on a particular chromosome. The lac operon of E. coli provides a useful illustration. This operon occurs once per bacterial genome. Depending on the physiological state of E. coli, and thus on its level of replication in proportion to the rate of cell division, the average bacterium may contain one or several (up to 3 or 4) copies of the lac operon. Each operon contains one operator site, which serves as the specific binding target for lac repressor. In wild-type cells there is an average of ten to thirty copies of the lac repressor protein. In functional terms the central protein-nucleic acid interaction that defines this system is the competitive (with RNA polymerase) binding of lac repressor to lac operator. The repressor and polymerase binding sites (operator and promoter, respectively) overlap (Dickson et al., 1975); thus when repressor is bound the promoter is occluded and transcription is inhibited. The lac operon in wild-type E. coli is expressed at $\sim 10^{-3}$ of the induced (constitutive or lac$^-$) level. In binding terms this means that the ratio of free to repressor-complexed operator sites in vivo is $\sim 10^{-3}$. A detailed thermodynamic analysis of this system has been presented elsewhere (von Hippel et al., 1974; von Hippel, 1979).

The degree of specificity of the interaction can best be appreciated when one realizes that the in vivo system contains $\sim 10^7$ DNA binding sites that can, in principle, compete for lac repressor (each base-pair of the chromosome comprises the beginning of a potential competing binding site). Thus the total concentration of potential DNA binding sites (D_T) greatly exceeds that of repressor molecules (R_T), which in turn exceeds that of operator sites (O_T); i.e., $D_T \gg R_T > O_T$.

The effective specificity of the system is measured in terms of the fractional saturation of the operator site with repressor. Obviously this will relate to the free concentrations of the various species, and will thus depend, in large measure, on ratios of specific to nonspecific binding constants.

LEVELS OF SPECIFICITY

Binding Site Specification

We first consider the system in terms of absolute specificity; i.e., we assume a protein that can absolutely (and only) discriminate between the four "information elements" of DNA. These are the four canonical nucleotide residues (A, T, G, and C) in single-stranded DNA and the four types of base pairs (A·T, T·A, G·C and C·G) in double-stranded DNA. The latter will be our primary focus here. (An A·T pair can be discriminated from a T·A pair because of the polar (5' → 3') character of the sugar-phosphate backbones of the individual DNA chains.)

A specific binding site is defined as a specific sequence of base pairs. A conditional probability approach (von Hippel, 1979) can be used to determine n, the minimum length of a sequence of recognition elements (base pairs) required to specify a site, so that the expected frequency ($P_n \cdot 2N$) with which that site reappears at random within the genome is less than unity. An example of this approach is presented in Table 1. For E. coli DNA, this minimal length is ~12 base-pairs, assuming a double-stranded sequence within a genome of overall composition A = T = G = C; see Figure 1.

Table 1. Conditional Probability Approach to DNA Sequence Specification

Position indices:	1	2	3	4	5	6	7	8	
(5'-)	A	T	G	C	G	T	T	C	(-3')
(3'-)	T	A	C	G	C	A	A	G	(-5')
Probabilities in each position:	$P_{1,A}$	$P_{2,T}$	$P_{3,G}$	$P_{4,C}$	$P_{5,G}$	$P_{6,T}$	$P_{7,T}$	$P_{8,C}$	

For $N \gg n$, the probability of occurrence of this particular sequence at a specific position i in the genome is:

$$P_n = (P_{1,A})(P_{2,T})(P_{3,G})(P_{4,C})(P_{5,G})(P_{6,T})(P_{7,T})(P_{8,C})$$

For genomes containing equal numbers of A·T, T·A, G·C, and C·G base pairs, $P_A = P_T = P_G = P_T = 0.25$, and the probability of occurrence is:

$$P_n = (P_A)(P_T)^3(P_G)^2(P_C)^2 = (0.25)^8 = 1.526 \times 10^{-5}$$

The expected frequency of random occurrence of this particular sequence in an entire genome of N base pairs is:

$(P_n)(2N) = (1.526 \times 10^{-5})(2 \times 10^7) = 305$ for a genome of 10^7 base pairs (potential binding sites). (Each base pair represents the beginning of a potential binding site; the factor 2 enters because a binding site in double-stranded DNA can be read 5' → 3' along either strand.)

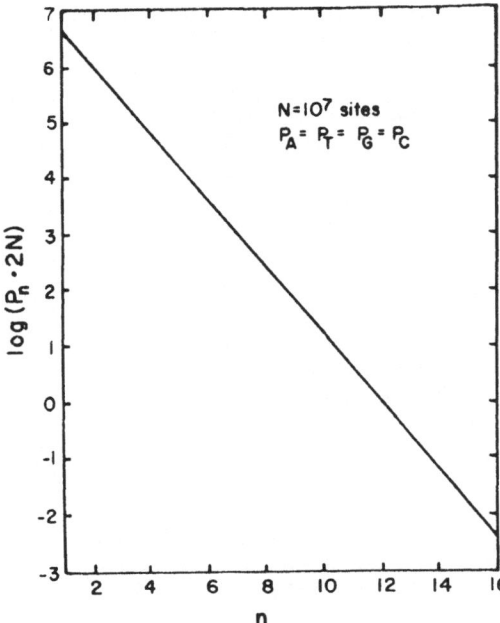

Figure 1. Plot (as a function of n, the specific site size, in base pairs) of the logarithm of $P_n \cdot 2N$, the expected frequency of random occurrence of a specific base pair sequence of length n within a genome containing N = 10^7 sites, for a DNA in which the individual probabilities of A·T, T·A, G·C and C·G base pairs are all equal ($P_A = P_T = P_G = P_C$). (Taken from von Hippel, 1979.)

161

This approach assumes that the overall sequence of the genome can be treated as chemically (though obviously not genetically) random,* and that every base-pair is fully specified (in terms of base-pair type). Unspecified loci can, of course, interrupt the overall sequence in defined positions, but these loci will not count toward n. Similarly, specification only at the level of Pu·Py (purine·pyrimidine) (vs. Py·Pu) base-pairs can occur; such loci are weighted less in establishing n. (For further details of this approach, see von Hippel, 1979.)

The Chemical Basis of Recognition

Primary Sequence Recognition Mechanisms. The primary molecular mechanism that can unambiguously recognize and discriminate individual base pairs in double-stranded DNA is complementary hydrogen bonding. Just as Watson-Crick base-base interactions involve the articulation of complementary hydrogen bond donor and acceptor groups between the central "faces" of the bases to discriminate "right" from "wrong" base-pairing in the template recognition events involved in replication and transcription, so also hydrogen bond donor and acceptor groups of amino acid residues of the protein binding site will discriminate base-pairs by complementary hydrogen bonding with the acceptors and donors of the major and minor grooves of the double-helix (Yarus, 1969;

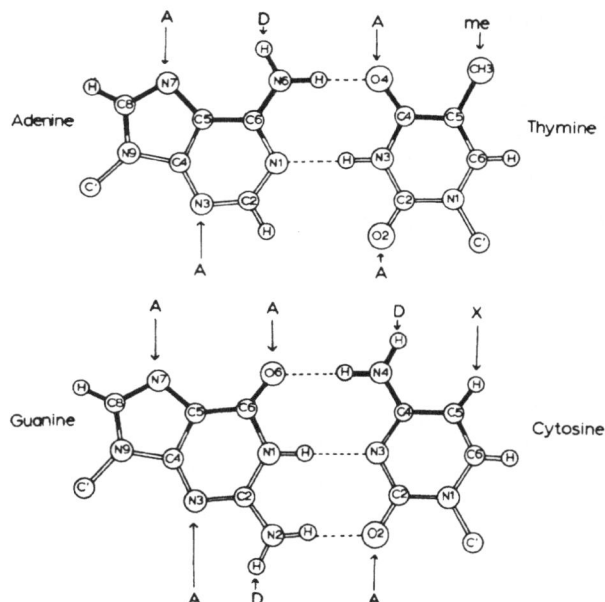

Figure 2. Molecular models of an A·T and G·C base pair, showing the functional groups in the major and minor grooves that may be important for protein recognition. A = hydrogen-bond acceptor; D = hydrogen-bond donor; me = the methyl group of thymine; X = the readily modified 5-position of cytosine.

---

*This assumption is supported (for sequences above the tetranucleotide level) both by comparison of calculated versus observed restriction enzyme cuts in viral genomes of defined size (von Hippel, 1979), and by the frequencies of occurrence of specific sequences tabulated by computer search of genomic libraries.

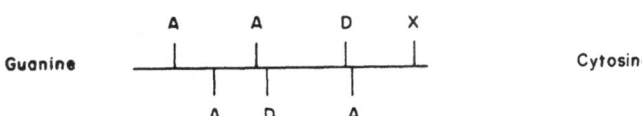

Figure 3. Schematic ("stick-figure") representations of the functional groups of an A·T and a G·C base pair that may be important in protein recognition. (Taken from Woodbury and von Hippel, 1981.)

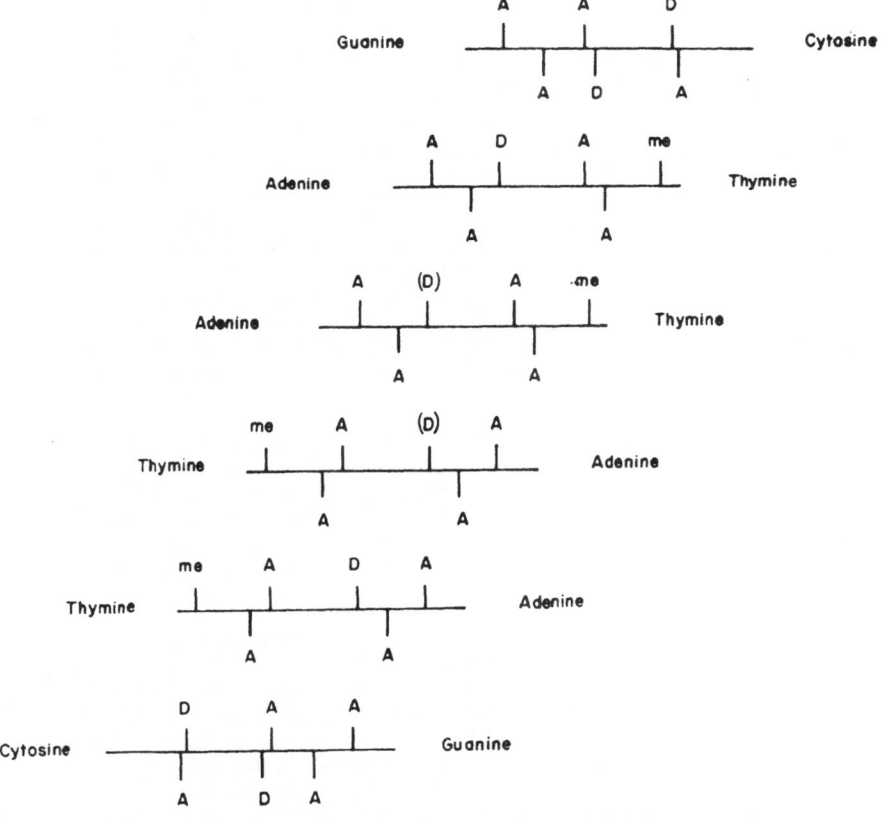

Figure 4. Schematic representation of the canonical-recognition base-pair sequence for EcoRI endonuclease and methylase. All primary functional groups are shown as in Figure 3. (D) represents hydrogen-bond donors in this sequence that are subject to methylation by the EcoRI modification enzyme. (Taken from Woodbury et al., 1980.)

von Hippel and McGhee, 1972; Seeman et al., 1976). These functional groups are shown in Figure 2 (modified after Seeman et al., 1976) and in a useful "stick-figure" representation in Figure 3 (see Woodbury et al., 1980; Woodbury and von Hippel, 1981; von Hippel et al., 1982; Ohlendorf et al., 1982). A stick figure representation of the sequence recognized by the EcoRI restriction and modification enzymes is shown in Figure 4. Clearly all these hydrogen donor and acceptor interactions are not utilized in any single protein-DNA acid binding interaction, nor are the functional groups in both the major and the minor grooves used in recognizing any one base pair, nor will the DNA always remain in the undistorted double-stranded B conformation. Nevertheless, Figure 4 does show the repertoire of possibilities from which a selection of recognition elements can be made by a binding protein.

Secondary Sequence Recognition Mechanisms. When the issue is not the absolute identification of a particular sequence of base pairs by a regulatory protein, other recognition mechanisms, at a "lower level" of specification, can also come into play. Thus DNA "regions" can be discriminated at the level of strandedness (e.g., single- vs. double-stranded), groove geometry and secondary structure (e.g., B- vs. Z-form DNA), etc., based on differences in protein binding affinity. The origins of these affinity differences can be steric, conformational or electrostatic, and may reflect structural consequences of regional differences in base-pair composition. Examples include the preferred binding of tetra-alkylammonium ions to dA·dT sequences in the (major) groove of B-DNA (Melchior and von Hippel, 1973), the preferred binding of the antibiotics netropsin and distamycin in the minor grooves of poly(dA·dT) and of dA·dT-rich sequences of B-form DNA (Kopka et al., 1985), the nonspecific binding of E. coli lac repressor to double-stranded DNA (deHaseth et al., 1977; Revzin and von Hippel, 1977), the preferential binding of phage T4-coded gene 32 protein to single-stranded DNA and RNA (Kowalczykowski et al., 1981) and the differential sensitivities of gene-specific eukaryotic DNA regions to endonucleases (Weintraub and Groudine, 1976). Such interactions obviously provide a measure of regional binding specificity. However they do not carry enough structural information to provide a mechanism for the primary recognition of specific base-pair sequences by proteins. We expect that primary recognition will be dominated by hydrogen bonding interactions.

Affinity

Accepting the notion that specific target site recognition does indeed involve the "reading", by a regulatory protein, of a specific array of hydrogen bonding donors and acceptors in the major and minor grooves of the DNA double-helix, we next consider the question of quantitative specificity or discrimination, which can also be termed "the problem of the other sites".

This problem exists because discrimination between (e.g.) "right" and "wrong" base-pairs cannot be absolute. Rather there is some finite level of affinity of the protein for the "correct" site, and some lower (but non-zero) and progressively decreasing affinity for other sites with decreasing degrees of homology with the correct one. To the extent that the great preponderence of "wrong" sites can compete with the regulatory target for protein and thus reduce the free protein concentration, the effective affinity of the protein for the correct sites will also be reduced.

This idea is shown schematically in Figure 5, which provides a specific, but over-simplified, illustration of the notion that recognition is not absolute. Figure 5 shows the (noncooperative) titration by

164

(e.g.) repressor of a specific (e.g.) operator site, labeled O_1, followed, at higher free protein concentrations, by the titration of a somewhat heterogeneous group of related operator sites located on other genes and labeled "O_2, O_3, O_4...". Figure 5 points out that this pair of titrations defines a "window of specificity" in terms of free protein concentration; at very low protein concentration there is no specific binding because the free protein concentration is below the affinity "threshold" for site O_1. At high protein concentrations repression is no longer specific only for O_1. Only at concentrations between the two binding isotherms do we see what we define as _specific_ functional binding of this particular regulatory protein.

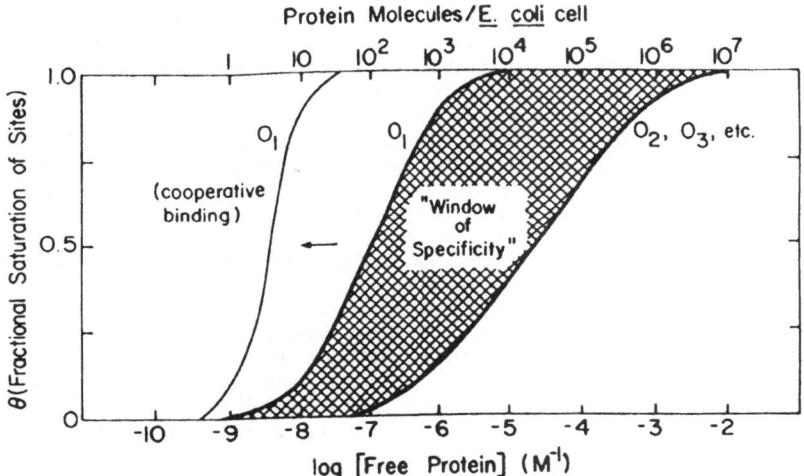

Figure 5. Titration of a specific operator site O_1 and of pseudospecific sites O_2, O_3, O_4... with repressor. Fractional saturation (θ) is plotted as a function of the logarithm of the _free_ protein (repressor) concentration. The numbers at the top of the graph indicate the number of protein molecules per E. coli cell (assumed volume = 10^{-15} liters) that these concentrations represent. The cross-hatched area between the two titration curves defines the "window of specificity" for this binding system. Cooperative binding of repressor to O_1 can strengthen the effective binding (shift it to the left) _and_ sharpen the binding isotherm (see text).

Representations such as Figure 5 also make it clear that while the window of specificity spans a finite protein concentration range, its position and size will depend on the affinity of the protein for sites O_1 and sites O_2, O_3, etc. Mutations in either the protein or the nucleic acid sites, or changes in the composition of the cellular environment, can narrow the specificity window, widen it, or obliterate it all together. In addition, binding of other proteins manifesting positive cooperativity to an extended operator site can both sharpen the specific titration and strengthen the effective binding (see Figure 5), thus also widening the window of specificity.

Sequence-Specific Binding Free Energy. We can attempt to estimate the favorable binding free energy expected per correctly positioned hydrogen bond donor-acceptor pair between protein and nucleic acid from first principles. Since the functional groups of both the protein and the nucleic acid binding sites will be involved in hydrogen bonding with water molecules when the complex is dissociated, for illustrative purposes we assume an average (differential) contribution of ~ -0.5 kcal/mole for each correctly formed protein to nucleic acid hydrogen bond. Assuming an average of one to two hydrogen bonded recognition events per base-pair*, this gives us a range of favorable specific binding free energies (for a protein with a recognition site size (n) of 12 base pairs) of ~ -6 to ~ -12 kcal/mole protein bound.

These are not large numbers, and it is important to recognize that much more favorable free energy is likely to be lost per mispaired position than is gained per proper recognition event. This follows because the interaction of a protein "recognition" functional group with an "incorrect" base pair can result in the total loss of at least one hydrogen bonding interaction; i.e., a protein hydrogen bond donor may end up "facing" a nucleic acid donor, or an acceptor may be "buried" facing an acceptor. In either case at least one hydrogen bond that was broken in removing the protein and nucleic acid donor (or acceptor) groups from contact with the solvent is not replaced, and an unfavorable contribution of as much as +5 kcal/mole may be added to the binding free energy unless the protein-DNA complex can adjust its overall conformation somewhat to minimize this problem.** This phenomenon illustrates the principle that generally applies to recognition interactions that are based on hydrogen bond donor-acceptor complementarity in water; i.e., correct acceptor-donor interactions may not add much to the stability of the complex, but incorrect hydrogen bond complementarities are markedly destablizing. Thus differential specificity of this type can usually be attributed to the unfavorable effects of incorrect contacts.

Non-Sequence-Specific Binding Free Energy. If, as discussed above, the main determinants of specificity are the unfavorable contributions of "wrong" base pairs, specific binding will also require a large

---

*As Seeman et al. (1976) have pointed out, we expect generally that two protein-nucleic acid hydrogen bonds will be required to recognize one base-pair in isolation. But one specifically positioned hydrogen bond may suffice in the context of a base-pair "stack", since the hydrogen bonding groups of both the protein and nucleic acid are held in fixed positions relative to the rest of the nucleic acid sequence and protein binding site in such structures.

**Recently, Tronrud et al. (1986) have obtained a direct measure of this maximum destabilization effect by comparing two enzyme-substrate complexes, in one of which a proper hydrogen bond is formed and in the other of which a functional group of the (otherwise identical) substrate has been altered from a hydrogen bond donor to an acceptor, thus leaving two acceptor groups "facing" one another. Crystallographic analysis show that the two ES complexes involved are entirely isomorphous, and the locus of the putative hydrogen bond is entirely shielded from water. Binding constant measurements show that the stability of the two complexes differs by ~4 kcal/mole, close to the value expected for the uncompensated loss of one hydrogen bond.

nonspecific contribution to the binding free energy to achieve sufficient binding affinity. Such nonspecific interactions usually involve a large electrostatic component, due mostly to the displacement of condensed counterions from DNA phosphate groups by positively charged protein side-chains (see Record et al., 1976). For example, for lac repressor binding to the operator site we estimate a total standard free energy of binding of approximately −17 kcal/mole under physiological salt concentrations. This interaction involves the formation of 7-8 nonspecific charge-charge interactions between protein and DNA (Record et al., 1977; Winter and von Hippel, 1977) and numerous base-pair-specific recognition interactions (Goeddel et al., 1978). The binding of lac repressor to nonspecific DNA involves ~11 charge-charge interactions and no base-pair specific interactions (deHaseth et al., 1977; Revzin and von Hippel, 1977), resulting in a standard free energy of binding of approximately −7 kcal/mole under the same conditions.

Conformational Change of the Regulatory Protein to a Totally Nonspecific Binding Mode. The above discussion suggests that at the rate at which the specific binding free energy decreases with misplaced (noncomplementary) hydrogen-bonding contacts, more than 3 to 5 "incorrect" base-pairs in the lac operator sequence may result in complete dissociation of a repressor-pseudo-operator complex. Instead we find that under these conditions lac repressor "isomerizes" to a binding mode where the interaction free energy is totally electrostatic and involves no sequence-dependent components. The same behavior may also characterize E. coli RNA polymerase (deHaseth et al., 1978) and T4-coded DNA polymerase (Fairfield et al., 1983). One of the advantages of this nonspecific binding mode for lac repressor (and perhaps for other genome-regulatory proteins as well) may involve the ability of the protein to "slide" over the surface of the DNA molecule in a one-dimensional diffusion process in this binding mode, thus facilitating translocation to the regulatory target site (Winter et al., 1981; Berg et al., 1982).

Distortions of the Protein and/or the DNA Target Site. In concluding these remarks on affinity, it is important to stress that neither the protein nor the DNA sites involved in binding are totally rigid. Thus both partners in complex formation can (and will) distort to optimize sterically sensitive binding interactions, within the limits of energetically available conformations. We have already seen an example of this in the isomerization of lac repressor to a totally electrostatically bound, non-sequence-dependent binding mode in the presence of an excess of unfavorable hydrogen bonding interactions. DNA sites can also change their local conformations. Thus, for example, hydrogen bonding of DNA base-pairs to a potentially complemetary protein matrix could either be improved or degraded by a local conformational change of the DNA that modifies the relative positions, directions and exposures of the hydrogen bonding functional groups located in the grooves of the DNA double-helix.

It is important to emphasize that such distortions from optimal solution conformations are associated with a thermodynamic cost. The free energy required to maintain the optimal (distorted) binding conformation of either partner must be subtracted from the favorable free energy of the binding interaction. Beyond a certain point this thermodynamic cost exceeds the free energy gained as a consequence of the interaction, and complex formation no longer occurs.

Equilibrium Selection

How are binding sequences designed to provide sufficient specificity in vivo? In the example of repressor-operator binding, the effective (functional) selection is determined by the fractional saturation of the operator site(s). And this, in turn, is controlled both by the affinities discussed above __and__ by the concentrations and availabilities of repressors, operators and "pseudo-operator" and nonspecific DNA binding sites.

These overall competitive binding reactions can best be described by a coupled equilibrium model, which describes the binding of (e.g.) repressor to the entire distribution of binding sites that show some affinity for it, and ultimately control the free repressor concentration and thus the final level of repression manifested by the system. Clearly this is not just a "nuisance"; as shown earlier (von Hippel et al., 1974; von Hippel, 1979) the overall level of repression of the __lac__ system, as well as the ability of inducer to derepress the system and the effects of various repressor and operator mutations can only be understood within the context of such coupled equilibrium calculations. The __quantitative__ treatment of such systems, including the handling of binding to various classes of "pseudo-specific" sites (i.e., sites that differ from the specific site in only a few base-pairing positions and to which the regulatory protein will bind with approximately the same conformation that it uses to bind to the specific site) and completely nonspecific sites, is described in von Hippel and Berg (1986).

CONCLUSIONS, APPLICATIONS, FURTHER DEVELOPMENTS AND MORE COMPLICATED PROBLEMS

In this paper we have summarized a series of approaches to the specificity of protein-nucleic acid interactions involved in the regulation (at the DNA level) of biological function. The ultimate manifestation of regulatory specificity must be the degree to which a specific biological process (e.g., the transcription of a particular gene or the translation of a particular protein) is "turned up" or "turned down" as a consequence of a specific DNA-protein interaction. Even though a given process of gene expression may reflect the end-product of a lengthy series of pre-equilibrium, steady-state or kinetically controlled steps, over an appreciable range of rates and concentrations most regulatory processes will reflect directly the equilibrium extent of specific DNA target saturation by the relevant binding protein(s), and it is this level of specificity with which we are concerned here.

Nomenclature

In order to discuss this problem with precision, we have attempted to differentiate the various levels of specificity with an appropriate terminology. Thus specification (or "information") refers to the length (in base pairs) of the sequence actually involved in specifying the target binding site (the term "information content" will be reserved for its precise statistical meaning; see Schneider et al., 1986, and Berg and von Hippel, 1986). Recognition is defined by the physico-chemical mechanisms that actually control the specificity of the interactions. Discrimination (or selectivity) refers to the thermodynamics of the interactions involved, and is determined by the differences in affinity of the protein for the various DNA targets over which the protein is distributed. Thus the discrimination ratio for pairs of binding sites is a ratio of binding constants. Finally, selection (in the equilibrium case), or the final level of biological expression (which reflects regulatory site saturation) is determined by the effective binding relation for the whole system of proteins and DNA binding sites. This hierarchy of specificity considerations makes it possible to examine a number of issues related to functional specificity.

Evolutionary "Design" of Regulatory Proteins

As indicated above, in order to saturate a regulatory DNA site adequately, the absolute binding affinity of the protein must be high enough to permit fairly complete titration of the site at the level of free protein concentration present in the cell. Operating under conditions in which the concentration of free protein is effectively determined by the concentrations and affinities for protein of the "other" DNA binding sites of the genome, the minimum specific binding constant required depends on the various selectivity ratios involved and can be calculated using eq. (5) of von Hippel and Berg (1986). The maximum value of the specific binding constant is also constrained, since the dissociation time of the protein must be set below the cell cycle times needed to achieve (e.g.) DNA replication, organelle duplication and cell division (though we note that the cell may devise special allosteric mechanisms, such as inducer binding, to lower the binding constant into a biologically acceptable range). Regulatory protein and DNA binding site "design" may cope with such problems in various ways.

(i) Proteins may be composed of (loosely or tightly associated) identical subunits, with two or more subunits binding in tandem to regulatory DNA binding sites of repeating base-pair sequence (examples include E. coli lac repressor, lambda CI or cro repressor, etc.). The advantages of this procedure include the fact that genetically coded protein units can be small and still can recognize a large site; i.e., a site specified by the totality of the tandem sequences. Additional levels of regulation of net binding affinity can be introduced by increasing or decreasing the inter-protein-subunit binding affinity, and thus the effective (at the subunit level) degree of protein binding cooperativity.

(ii) The overall protein may also be designed to bind with intermolecular cooperativity to regulatory sites; i.e., the affinity of the protein for a subsite of an (e.g.) operator can be increased (by favorable protein-protein interactions or by local DNA-lattice deformation) by the binding of another protein of the same type (homoprotein cooperativity) or of a different type (heteroprotein cooperativity) to a contiguous site. (For a treatment of the lambda phage lytic lysogenic regulatory switching system, which clearly operates this way, see Ptashne et al. (1980). The general principles of such systems, in the context of the ribosome assembly problem, have recently been formulated by von Hippel and Fairfield, 1985, and Fairfield and von Hippel, in preparation.)

(iii) The possibility of at least two protein binding conformations also introduces an additional regulatory variable. Thus, as indicated in the preceding section, a protein can bind specifically to the regulatory site (and to pseudosites differing from the canonical sequence in only a very few base-pair positions), and can assume another (totally nonspecifically binding) conformation that becomes favorable for binding when the number of incorrect (and thermodynamically unfavorable) base-pairs exceeds some number j (see von Hippel and Berg, 1986). This additional degree of conformational freedom at the protein level permits the establishment of a constant discrimination ratio for the majority of the available sites on the genome. Furthermore, if this general nonspecific affinity is totally electrostatic and if the salt dependence of the specific and nonspecific binding constants differ, this discrimination ratio can be manipulated by small changes in the ionic environment. In addition, of course, such totally electrostatic nonspecific affinity of the regulatory protein for the overall genome may permit the facilitated location of regulatory sites (see above).

(iv) Small molecule ligands that can bind to the regulatory protein and alter its affinity for the regulatory DNA target (e.g., inducers, as in the lac system), provide another dimension of regulatory control, and can be included by expanding the system of components involved in the overall coupled equilibrium system (see von Hippel et al., 1974; von Hippel, 1979; O'Gorman et al., 1980).

Nonspecific DNA Binding and the Formulation of Purification Procedures and Binding Assays for Regulatory Proteins

The existence of nonspecific DNA binding of regulatory proteins considerably complicates the design of experiments to purify and assay these proteins, especially if the specific and nonspecific binding affinities vary differently with salt concentration, because then the selectivity ratios of the protein become a function of this variable. Most protein purifications and in vitro assays are conducted at low salt concentrations, generally because protein concentrations are limited and tighter binding is achieved at low salt. The consequence of this, however, is often that the competition of nonspecific DNA sites for protein (which is generally very intense at low salt) completely "swamps" binding to the putative (and often correctly identified) specific sites, and renders worthless the classical purification paradigm of using specific DNA sites (e.g., in a column) to isolate the regulatory protein from an impure extract. In the same way an excess of nonspecific DNA, attached to an isolated DNA fragment containing the site of interest, can completely suppress specific binding under conditions where the selectivity ratio for specific binding is too low. It is clearly important to take these considerations into account in formulating purification and assay procedures, since, if one does not set up the experiment within a "window" of salt concentration at which specific site selection is possible, the experiment is doomed to fail. The "take-home" lesson may be that salt' concentration must be a variable in the design of a purification or assay procedure for any protein for which the specific or nonspecific binding parameters depend on this factor.

Binding Selectivity and Affinities in vivo

In addition to developing insight into protein-DNA interaction mechanisms and possibilities, the final objective of studies of this sort must be to understand the situation in vivo in both prokaryotic and eukaryotic cells. To define in vivo levels of selection it is necessary to extrapolate discrimination ratios established by in vitro experiments to in vivo conditions. This requires that the ionic environment, genome size, concentration of total and of free regulatory protein, and the degree of availability (to the regulatory protein) of various competing DNA sites, must all be known for the system at hand. Approaches to various aspects of this problem are under development in our laboratory and elsewhere.

ACKNOWLEDGEMENTS

The research from our laboratory that is reviewed here has been supported by USPHS Research Grants GM-15792 and GM-29158 (to PHvH), and by partial salary support from the Swedish National Science Research Council (to OGB).

REFERENCES

Berg, O. G., and von Hippel, P. H., 1986, J. Mol. Biol., in press.

Berg, O. G., Winter, R. B., and von Hippel, P. H., 1982, *Trends Biochem. Sci.*, 7:52.

deHaseth, P. L., Lohman, T. M., and Record, M. T. Jr., 1977, *Biochemistry*, 16:4783.

deHaseth, P. L., Lohman, T. M., Record, M. T. Jr., and Burgess, R.R., 1978, *Biochemistry*, 17:1612.

Dickson, R. C., Abelson, J. N., Barnes, W. M., and Reznikoff, W. S., 1975, *Science*, 182:27.

Fairfield, F. R., Newport, J. W., Dolejsi, M. K., and von Hippel, P. H., 1983, *J. Biomol. Struct. Dyn.*, 1:715.

Goeddel, D. V., Yansura, D. G., and Caruthers, M. H., 1978, *Proc. Natl. Acad. Sci. USA*, 75:3578.

Kopka, M. L., Yoon, C., Goodsell, D., Pjura, P., and Dickerson, R. E., 1985, *Proc. Natl. Acad. Sci. USA*, 82:1376.

Kowalczykowski, S. C., Lonberg, N., Newport, J. W., and von Hippel, P. H., 1981, *J. Mol. Biol.*, 145:75.

Melchior, W. B. Jr., and von Hippel, P. H., 1973, *Proc. Natl. Acad. Sci. USA*, 70:298.

O'Gorman, R. B., Rosenberg, J. M., Kallai, O. B., Dickerson, R. E., Itakura, K., Riggs, A. D., and Matthews, K. S., 1980, *J. Biol. Chem.*, 255:10107.

Ohlendorf, D. H., Anderson, W. F., Fisher, R. G., Takeda, Y., and Matthews, B. W., 1982, *Nature*, 298:718.

Ptashne, M., Jeffrey, A., Johnson, A. D., Maurer, R., Meyer, B. J., Pabo, C. O., Roberts, T. M., and Sauer, R. T., 1980, *Cell*, 19:1.

Record, M. T. Jr., deHaseth, P. L., and Lohman, T. M., 1977, *Biochemistry*, 16:4791.

Record, M. T. Jr., Lohman, T. M., and deHaseth, P. L., 1976, *J. Mol. Biol.*, 107:145.

Revzin, A., and von Hippel, P. H., 1977, *Biochemistry*, 16:4769.

Schneider, T. D., Stormo, G. D., Gold, L., and Ehrenfeucht, A., 1986, *J. Mol. Biol.*, 188:415.

Seeman, N. C., Rosenberg, J. M., and Rich, A., 1976, *Proc. Natl. Acad. Sci. USA*, 73:804.

Tronrud, D. E., Holden, H. M., and Matthews, B. W., 1986, *Science*, in press.

von Hippel, P. H., 1979, in "Biological Regulation and Development," R. F. Goldberg, ed., Plenum, New York.

von Hippel, P. H., Bear, D. G., Winter, R. B., and Berg, O. G., 1982, in "Promoters: Structure and Function," M. Chamberlin and R. Rodriquez, eds., Praeger, New York.

von Hippel, P. H. and Berg, O. G., 1986, *Proc. Natl. Acad. Sci. USA*, 83:1608.

von Hippel, P. H., and Fairfield, F. R., 1985, *Pure and Appl. Chem.*, 57:45.

von Hippel, P. H., and McGhee, J. D., 1972, *Ann. Rev. Biochem.*, 41:231.

von Hippel, P. H., Revzin, A., Gross, C. A., and Wang, A. C., 1974, *Proc. Natl. Acad. Sci. USA*, 71:4808.

Weintraub, H., and Groudine, M., 1976, *Science*, 193:848.

Winter, R. B., Berg, O. G., and von Hippel, P. H., 1981, *Biochemistry*, 20:6961.

Winter, R. B., and von Hippel, P. H., 1981, *Biochemistry*, 20:6948.

Woodbury, C. P. Jr., Hagenbuchle, O., and von Hippel, P. H., 1980, *J. Biol. Chem.*, 255:11534.

Woodbury, C. P. Jr., and von Hippel, P. H., 1981, in "The Restriction Enzymes," J. Chirikjian, ed., Elsevier, Amsterdam.

Yarus, M., 1969, *Ann. Rev. Biochem.*, 38:841.

SEARCHING FOR THE CODE OF IDEAL PROTEIN-DNA-RECOGNITION

Norbert Lehming, Juergen Sartorius, Brigitte
von Wilcken-Bergmann and Benno Mueller-Hill

Institut fuer Genetik der Universitaet zu Koeln
Weyertal 121
5000 Koeln 41, FRG

ABSTRACT

A system is described which allows testing of specific
protein-DNA-interactions. The system consists of two mutually
compatible plasmids carrying different origins of replication
and resistance markers. One plasmid carries a lac I gene in
which the DNA-recognizing domain has been replaced by synthetic
DNA saturated with restriction sites. The other carries a lac P
Z unit in which the natural operator has been deleted and
replaced by a unique restriction site. Into this restriction
site any operator can be cloned.

Our results suggest that: 1) The innermost seven base
pairs of the lac operator can not be replaced without
diminishing repression. 2) If Tyr1 Gln2 of the recognition
helix of lac repressor is substituted by Val1 Ala2, the mutant
repressor recognizes the mutant operator TGTAAGC GCTTACA better
than the ideal or any other lac operator variant. Finally, a
model is presented which may describe the binding of an alpha
helix to the deep groove of B-DNA.

THE PROBLEM

The extremely simple repetitive structure of DNA which
offers just two clefts for recognition led already in 1972 to
the suggestion of the possible existence of a code of protein-
DNA-recognition[1]. A code would imply simple rules governing
all or most specific interactions between proteins and DNA of a
particular sequence. The proposal of such a code gained in
weight when it was shown by X-ray analysis that two suitably -
spaced recognition helices existed in dimeric lambda cro[2] and
that the DNA recognizing domains of lambda cro and various
other repressors were homologous in sequence[3-5]. Later X-ray
analysis has shown several other DNA recognizing proteins to
contain potential recognition helices[6-8]. Moreover it could be
shown that the exchange of a single amino acid either within
the sequence preceding the recognition helix[9] or inside the
recognition helix[10], or the swapping of all[11] or part[12] of the
recognition helix, changed the specificity of particular
proteins. So far, however, the existence of a code could not be

173

proven, but merely remained a possibility. Intrigued for many years by this question, we established a system which would allow our old proposal to be tested. At present, we are unable to answer the question, although the work we present provides some support for the existence of a code.

Even if a code of protein-DNA recognition existed in an ideal form it would possibly be degenerate in most existing systems, owing to selection against perfect recognition. The lac repressor-operator system which we and others have studied extensively in the past[13,14] may serve as a case in point. The apparent, albeit partial symmetry of the lac operator was not reflected by symmetry in methylation protection patterns[15]. At that time this led to the rather pessimistic notion of an asymmetric interaction between lac repressor and lac operator[15]. A search for the best possible lac operator out of "random" mouse DNA showed finally that fully symmetric ideal operator indeed existed [16]. (The same fully symmetric operator was simultaneously found in another study[17]).

We have carried out DNA protection experiments with dimethylsulfate to show that there was symmetric protection of ideal lac operator by lac repressor (figure 1). Moreover the lac system seemed well-suited for analysis because of its high specificity. In contrast to the lambda system, it contains only a single operator. Sloppy recognition is not safeguarded against by cooperativity of a second operator nearby. Specific interaction of lac repressor with asymmetric wild type lac operator decreases by a factor of 10^3 in the presence of inducer[13]. In vitro ideal lac operator binds lac repressor eight times more strongly than wild type lac operator[16,17]. We may guess that exclusively the specific interaction is increased. A last argument in favor of the use of the lac

```
                          A
      7 6 5 4 3 2 1
          ^   ^
5' T G T G A G C   G C T C A C A   3'
                                        ideal lac operator
3' A C A C T C G   C G A G T G T   5'
                      v   v


5' T G T A A X C   C X T T A C A   3'
                                        idealised gal operator
3' A C A T T X C   C X A A T G T   5'

                          B
Tyr GlN Thr Val Ser Arg Val Val AsN GlN   lac repressor


Val Ala Thr Val Ser Arg Val Ile AsN AsN   gal repressor

 1   2   3   4   5   6   7   8   9  10
```

Fig. 1.A: DNA sequence of ideal lac operator and of idealized gal operator. ^ N 7 of G which is protected by lac repressor against methylation. X signifies that various bases may be found in that particular position. B. Protein sequence of the recognition helices of lac and gal repressor. For the numbering of base pairs and amino acids see text.

system is the fact that the lac operator is situated far enough away from the lac promoter to be deleted and replaced by other synthetic operators[18].

RESULTS

We established a system which allows repression of any mutant lac operator by any mutant of the lac I (repressor) gene by construction of two plasmids (fig. 2): the first carried a lac operon of which the lac operator had been deleted and replaced by a unique restriction site. Into this restriction site we cloned all possible symmetric lac operator variants, which differed from the ideal symmetric[16,17] lac operator by a single base pair exchange within the first seven base pairs on each half of the operator (Table 1). The second plasmid carried a lac I gene whose first 70 codons had been replaced by chemically synthesized DNA (figure 2). Here we introduced as many unique restriction sites as possible to allow for rapid amino acid replacements by chemical synthesis of duplex DNA. Both plasmids carry different origins of replication and different antibiotic resistance genes to allow selection and coexistence in a single E.coli cell.

To determine the degree of specificity of recognition, we first tested all 21 synthetic symmetric operator mutants with wild type lac repressor in vivo and found that none of them could bring about full repression (table 1). Thus full conservation of sequence within this seven base pair stretch is imperative at all positions for correct recognition. The absolute specific activity of β-galactosidase in the absence of lac repressor is, however, in most mutants lower than in wild type, a phenomenon which may be due to cis and/or trans effects.

Our next task was to find a case where a change of specificity in the lac system could be brought about by a single amino acid exchange. In analogy to the cap system[10,19] we first tried to replace residue 2 (GlN) of the recognition helix by Leu to repress the lac operator variant in which G/C at position 4 had been replaced by A/T. (We use here the same numbering system for lac operator as later for lambda operator: We denote the sometimes absent central base pair of all operators as No 0 and continue then counting 1,2,3 etc. from the center to the outside). This mutant repressor did not at all repress the suggested or any other lac operator variant (data not shown). We then reasoned that the double amino acid exchange from Tyr1 GlN2 to Val1 Ala2 (we number here according to the position in the recognition helix) which changes the recognition helix of lac repressor to that of gal repressor[20] (fig. 1) might lead to repression of the same operator variant since it was found to be similar to gal operator[20,21] (fig. 1). We found indeed that Val1 Ala2 lac repressor did repress strongly and specifically Plasmid 341 which carried the particular G to A exchange (table 2). We then proceeded to synthesize and test various repressor mutants which had exchanges only in residue 1 and/or 2 of the recognition helix (table 2). Of all the mutants we synthesized, only Ala1 Ala2 did repress operator 341 as well as the Val1 Ala2 derivative. Furthermore, an exchange of residue 1 or 2 of the recognition helix exerted a noteworthy influence only on base pair 4 or 5 of lac operator.

Table 1: Symmetric <u>lac</u> operator variants diminish repression <u>in vivo</u>. The ideal <u>lac</u> operator and its variants have been cloned into the <u>Xba</u>I site of plasmid pWB300 (see fig. 2). All operators share the consensus sequence AATTGTGAGC GCTCACAATT. The host was the <u>E.coli</u> strain DC41-2 <u>(lacpro)</u>ᐃ <u>RecA GalE</u>, <u>Smr</u>. Plasmid pWB100 was used as source of <u>lac</u> repressor. To test for the state of no repression we used a variant of pWB100 which had a deletion within the <u>lac I</u> gene (codon 14 to 60). Bacteria were grown in glycerol minimal medium in the presence of proline, ampicilline and tetracycline to an OD580 of 1.0. Repression is defined as specific activity of β-galactosidase in the absence of <u>lac</u> repressor divided by the specific activity of β-galactosidase in the presence of <u>lac</u> repressor.

Plasmid pWB	Operator Sequence	Spec.Activity of β-galactosidase		Repression
		<u>lac</u> Repressor	no <u>lac</u> Repressor	
310	7654321 TGTGAGC*GCTCACA	55	38,000	700
311	TGTGAGA TCTCACA	3,300	7,500	2.2
313	TGTGAGG CCTCACA	1,400	5,200	3.7
314	TGTGAGT ACTCACA	1,100	22,000	20
321	TGTGAAC GTTCACA	1,700	4,900	2.9
322	TGTGACC GGTCACA	1,000	16,000	16
324	TGTGATC GATCACA	650	7,100	11
332	TGTGCGC GCGCACA	2,800	5,800	2.1
333	TGTGGGC GCCCACA	1,000	4,100	4.1
334	TGTGTGC GCACACA	2,800	5,300	1.9
341	TGTAAGC GCTTACA	740	3,800	5.1
342	TGTCAGC GCTGACA	730	4,500	6.2
344	TGTTAGC GCTAACA	1,700	5,800	3.4
351	TGAGAGC GCTCTCA	370	5,900	16
352	TGCGAGC GCTCGCA	660	7,000	10
353	TGGGAGC GCTCCCA	400	3,600	9
361	TATGAGC GCTCATA	2,400	6,300	2.6
362	TCTGAGC GCTCAGA	2,400	4,500	1.9
364	TTTGAGC GCTCAAA	3,600	10,000	2.8
371	AGTGAGC GCTCACT	240	1,100	4.6
372	CGTGAGC GCTCACG	400	9,000	22.5
373	GGTGAGC GCTCACC	590	6,700	11.4

The fact that residue 1 of the recognition helix seems to have little effect on base pair 3, but some effect on base pair 5 of <u>lac</u> operator (table 2) suggested that the recognition helix of <u>lac</u> repressor is oriented in the same way as the recognition helices of <u>lambda</u> cro, <u>lambda</u> repressor, and <u>cap</u> protein; the more N-terminal residues of the recognition helix recognise base pairs further away from the center of symmetry

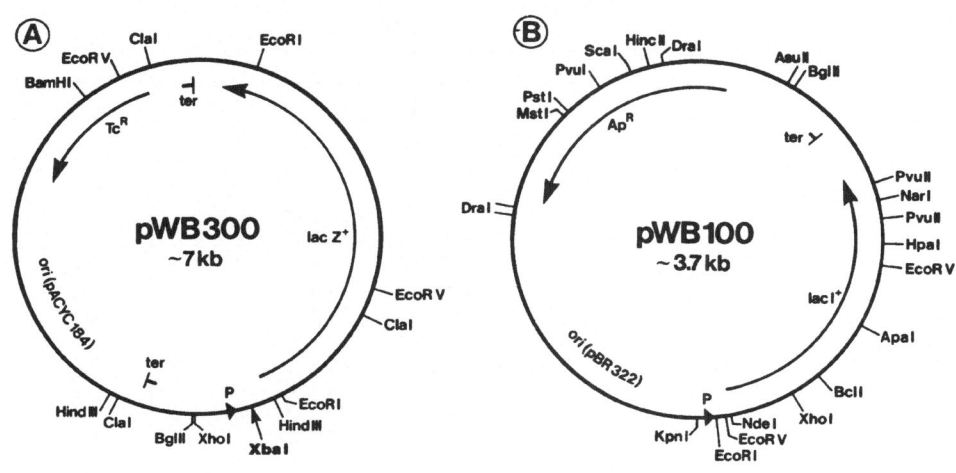

C

pBR322 GGTACCAATCGAGATAAAAAATTTAT TTGCTT TCAGGTACAATTCTTGA TATAAT ATTATCATCTAGGGAACTAGCTTTCCGGGG AATTCA
KpnI

                                        -35   promoter   -10   mRNA

EcoRI

RBS

        1   5   10   15   20
        met lys pro val thr leu tyr asp val ala glu tyr ala gly val ser tyr glN thr val ser
TAAAGGAGAT ATCA T ATG AAA CCG GTA ACG TTA TAC GAC GTC GCT GAA TAC GCC GGC GTT TCT TAC CAG ACC GTT TCT
        EcoRV NdeI   AatII   NaeI   XbaI

        25   30   35   40   45
arg val val asN glN ala ser his val ser ala lys thr arg glu lys val glu ala ala met ala glu leu asN
AGA GTG GTT AAC CAG GCT TCA CAT GTT AGC GCT AAA ACC CGG GAA AAA GTT GAA GCT GCC ATG GCT GAG CTC AAC
XbaI HpaI   XmaI   NcoI   SstI

        50   55   60   65   70
tyr ile pro asN arg val ala glN glN leu ala gly lys glN ser leu leu ile gly val ala thr ser ser leu
TAC ATC CCG AAC CGT GTT GCG CAG CAG CTG GCT GGT AAA CAA AGC TTG CTG ATC GGT GTC GCG ACC TCG AGC TTG
        MstI   PvuII   HindIII   NruI   XhoI

Table 2: Repression of lac operator variants by lac repressor variants. For the structure of plasmids, the definition of repression etc.: see legend of table 1, fig. 2 and text.

Operator	Repressor					
	WT	V_1A_2	V_1	A_2	A_1V_2	A_1A_2
ideal	700	50	110	40	15	60
332	2.1	1.8	2.3	1.2	1.8	1.8
333	4.1	1.1	1.8	1.1	1.3	1.0
334	1.9	1.3	1.3	1.3	1.5	1.4
341	5.1	100	1.6	10	26	130
342	6.2	2.0	1.9	1.0	2.6	1.1
344	3.4	9.4	7.0	1.1	8.3	20
351	16	27	13	2.8	11	60
352	10	3.3	8.3	1.7	2.1	2.7
353	9.0	16	10	1.3	11	26

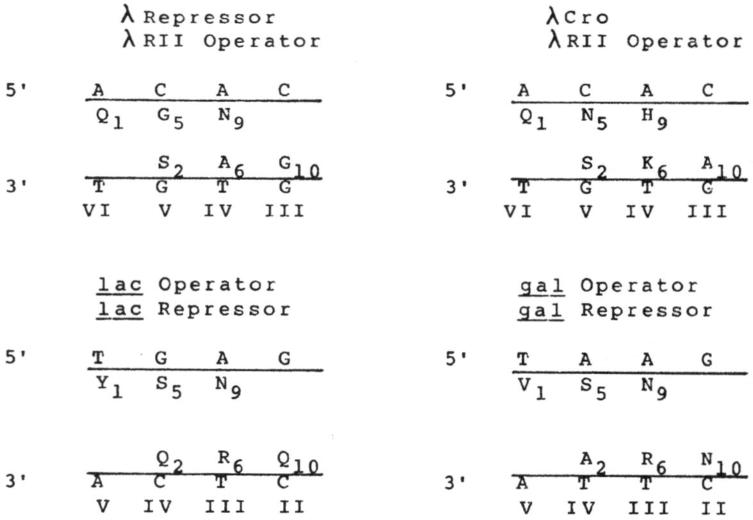

λ Repressor
λ RII Operator

5' $\underline{\begin{array}{cccc} A & C & A & C \\ Q_1 & G_5 & N_9 & \end{array}}$

3' $\underline{\begin{array}{cccc} & S_2 & A_6 & G_{10} \\ T & G & T & G \end{array}}$
    VI   V   IV   III

λ Cro
λ RII Operator

5' $\underline{\begin{array}{cccc} A & C & A & C \\ Q_1 & N_5 & H_9 & \end{array}}$

3' $\underline{\begin{array}{cccc} & S_2 & K_6 & A_{10} \\ T & G & T & G \end{array}}$
    VI   V   IV   III

lac Operator
lac Repressor

5' $\underline{\begin{array}{cccc} T & G & A & G \\ Y_1 & S_5 & N_9 & \end{array}}$

3' $\underline{\begin{array}{cccc} & Q_2 & R_6 & Q_{10} \\ A & C & T & C \end{array}}$
    V   IV   III   II

gal Operator
gal Repressor

5' $\underline{\begin{array}{cccc} T & A & A & G \\ V_1 & S_5 & N_9 & \end{array}}$

3' $\underline{\begin{array}{cccc} & A_2 & R_6 & N_{10} \\ A & T & T & C \end{array}}$
    V   IV   III   II

Fig. 3.: Possible interactions between the recognition helices of lac, gal, lambda repressor and lambda cro and their respective operator half sites. For the numbering of amino acids and base pairs see text.

of the lac operator, while the more C-terminal residues of the recognition helix recognize bases closer to the center of symmetry of lac operator. We are aware of the fact that this is in contradiction to the NMR data of Kaptein on lac headpiece - lac operator -complexes presented at this Summer School (Kaptein et al, this Volume). It could be that the lac headpieces which are not held together properly by the core of lac repressor adopt the opposite orientation on the operator-fragment which was used by Kaptein et al.

We then asked whether there were traits common to the lac and lambda systems. Playing with models does not show unambigously which residue of the recognition helix will interact with which base. Looking back at the previous failure to predict the recognition of particular base pairs from the properly predicted recognition helix of lac repressor from looking at space filling models[1], we opted this time for a different type of approach. We used symmetry and simplicity as a guideline arguing that a code, if it existed had to be simple and aesthetic. Here the following possibility got our attention. Suppose consecutive bases of one strand would be seen by the side chains of residues 1,5 and 9 and consecutive bases on the other strand would be seen by the side chains of residues 2,6 and 10. Furthermore, residues 1 and 2 as well as residues 5 and 6 or 9 and 10 would always see neighbouring bases but on opposite strands. Considering the righthandedness of the alpha-helix and of DNA, this would indicate the following interactions in the lac system (figure 3). Ala 1 or Val 1 would recognize T5 on the upper strand and Ala 2 would recognise T 4 of the lower strand. This would imply that the methylgroup of thymine is used for recognition in a purely hydrophobic interaction.

The model suggests symmetry between the N and C terminal residues of the recognition helix. In particular, residue 1 and residue 10 of the recognition helix appear to have similar properties and functions in recognition. However, that this is not so is suggested by the suppression data of these codons. Whereas Tyr 1 of the recognition helix may be not replaced by any other residue without loss of specific interaction, GlN 10 of the recognition helix may be replaced by Ser, AsN, Tyr, Leu or Lys[14]. This already indicates that the symmetry suggested by the model is not imperative. The reason for this could be that the lac recognition helices are not arranged in parallel as in the lambda cro dimer.[2,6-8] We tested whether Ser6 GlN9 (symmetric to lac wildtype GlN2 Ser5) would repress the operator carrying plasmid No 332; We found that this was not the case (data not shown). Thus simple complete symmetry is not found. However this does not exclude the possible existence of the type of interaction we envisage with some degenerated symmetry.

How would the lambda repressor-operator interaction look in our model? Evidence has recently been presented which suggests that lambda repressor and lambda cro recognize the same base pair of lambda operator through Ser 2 of their recognition helices[22]. Mutant lambda repressor and mutant lambda cro, in which residue 2 of the recognition helix was Ala, no longer shielded G No 5 of lambda operator against methylation by dimethylsulfate. It seemed remarkable that in our model GlN 2 Ser 5 recognizes a 5'G:C3' base pair in the lac system whereas Ser 2 AsN 5 (Ser 2 Gly 5) recognizes a 5'C:G3' base pair in the <u>lambda</u> cro (repressor) system. A test of these predictions is in progress.

DISCUSSION

Our results show that seven or possibly more base pairs of each half side of <u>lac</u> operator are optimal for recognition and can not be changed without disturbing recognition. The recognition helix we have tried to manipulate recognizes maximally four base pairs i.e., base pairs 3,4,5 and 6 of <u>lac</u> operator according to our model. The outmost base pair(s) and the two innermost base pairs thus have to make contact with other side chains of <u>lac</u> repressor headpiece. We recall here that the region between residue 50 and 60 of <u>lac</u> repressor has been implicated in <u>lac</u> operator contact [13] - possibly as a beta sheet in the minor groove around the symmetry center of <u>lac</u> operator[20]. If we speak about a code of protein-DNA recognition, we have to take into account that the recognition helix is only part of the recognition system.

Finally, we would like to list some of the obstacles which stand possibly in the way of a simple code and its elucidation:

● Whereas the two recognition helices of the <u>cro</u>-dimer are parallel to the DNA helix grooves, those of the <u>cap</u> and <u>lambda</u> headpiece dimers are not. It is possible, although not too likely, that the lack of parallel alignment may be corrected in the <u>cap</u> and <u>lambda</u> DNA-protein complexes.
● Our model implies that two amino acids may interact with one base pair. But the real situation may be even more complicated. The side chain of one amino acid may form hydrogen bonds with two adjacent bases of one strand. Such an interaction has been shown in the case of the restriction enzyme <u>Eco RI</u>.[23] The recognition of two adjacent bases by one single side chain increases drastically the number of possible interactions one has to consider. Such a situation would make the selection of specificity changes difficult if not impossible.
● Some of the synthetic mutants of <u>lac</u> repressor we have used to change the specific recognition of operator variants may tend to be unstable and unable to form the crucial alpha helix. We know that some missense mutants in the headpiece of <u>lac</u> repressor are readily degraded by proteolytic enzymes.[24] Such an instability may completely obscure and mislead interpretations from our type of analysis.
In summary we presently feel that the existence of a code of protein DNA recognition is still an open question. However, we feel that the data reported here allow some cautious optimism.

ACKNOWLEDGEMENTS

We thank Daniela Tils for excellent technical help and Irene and Ben Oswalt for hospitality. This work was supported by the Deutsche Forschungsgemeinschaft through grant SFB74-A2 and Bundesministerium fuer Forschung und Technologie through grant BCT 0365/2.

REFERENCES

1. Adler, K., Beyreuther, K., Fanning, E., Geisler, N., Gronenborn, B., Klemm, A., Muller-Hill, B., Pfahl, M. and Schmitz, A., How lac repressor binds DNA, Nature 237: 322 (1972).
2. Ohlendorf, D. H., Anderson, W.F., Fisher, R.G., Takeda, Y. and Matthews, B. W., The molecular basis of DNA-protein recognition inferred from the structure of cro repressor, Nature 298: 718 (1982).
3. Matthews, B.W., Ohlendorf, D. H., Anderson, W.F. and Takeda, Y., Structure of the DNA-binding region of lac repressor inferred from its homology with cro repressor, Proc.Natl.Acad.Sci.USA 79: 1428 (1982).
4. Sauer, R. T., Yocum, R. R., Doolittle, R. F., Lewis, M. and Pabo, C.O., Homology among DNA-bindig proteins suggests use of a conserved super-secondary structure, Nature 298: 447 (1982).
5. Gicquel-Sanzey, B. and Cossart, P., Homology between different procaryotic DNA-bindig regulatory proteins and between their sites of action, The EMBO J. 1: 591 (1982).
6. Pabo, C. O. and Lewis, M., The operator-binding domain of λ repressor: structure and DNA recognition, Nature 298: 443 (1982).
7. McKay, D. B. and Steitz, T., Structure of catabolite activator protein at 2.9 A resolution suggests binding to left handed B-DNA, Nature 290: 744 (1981).
8. Schevitz, R. W., Otwinowski, Z., Joachimiak, A., Lawson, D.L. and Sigler, P. B., The three-dimensional structure of trp repressor, Nature 317: 782 (1985).
9. Youderian, P., Vershon, A., Bouvier, S., Sauer, R. T., Susskind, M., Changing the DNA-binding specificity of a repressor, Cell 35: 777 (1983).
10. Ebright, R.H., Cossart, P., Gicquel-Sanzey, B. and Beckwith, J., Mutations that alter the DNA sequence specific specificity of the catabolite gene activator protein of E.coli, Nature 311: 232 (1984).
11. Wharton, R. P., Brown, E. L. and Ptashne, M., Substituting an α-helix switches the sequence-specific DNA interaction of a repressor, Cell 38: 361 (1984).
12. Wharton, R. P. and Ptashne, M., Changing the binding specificity of a repressor by redesigning an α-helix, Nature 316: 601 (1985).
13. Muller-Hill, B., Lac repressor and lac operator, Progr.Biophys.Molec.Biol. 30: 227 (1975).
14. Miller, J. H., The lac I gene: its role in lac operon control and its use as a genetic system, in, "The Operon" ed. by. J.H. Miller and W.S. Reznikoff, Cold Spring Harbor (1978).
15. Gilbert, W., Majors, J. and Maxam, A., How proteins recognize DNA sequences, in: Organisation and expression of chromosomes, Dahlem Konferenzen, Berlin (1976).
16. Simons, A., Tils, D., Wilcken-Bergmann, B.v. and Muller-Hill, B., Possible ideal lac operator: Escherichia coli lac operator - like sequences from eukaryotic genomes lack the central G C pair, Proc.Natl.Acad.Sci.USA 81: 1624 (1984).
17. Sadler, J. R., Sasmor, H. and Betz, J. L., A perfectly symmetric lac operator binds the lac repressor very tightly, Proc.Natl.Acad.Sci.USA 80: 6785 (1983).

18. Besse, M., Wilcken-Bergmann, B.v. and Muller-Hill, B., Synthetic _lac_ operator mediated repression through _lac_ repressor when introduced upstream and downstream from _lac_ promoter, The EMBO J. 5: 1377 (1986).
19. Ebright, R. H., Evidence for a contact between glutamine-18 of _lac_ repressor and base pair 7 of _lac_ operator, Proc.Natl.Acad.Sci.USA 83: 303 (1986).
20. Wilcken-Bergmann, B.v. and Muller-Hill, B., Sequence of _gal_ R gene indicates a common evolutionary origin of _lac_ and _gal_ repressor in _Escherichia coli_, Proc.Natl.Acad.Sci.USA 79: 2427 (1982).
21. Irani, M. H. Orosz, L., A control element within a structural gene. The _gal_ operon of _Escherichia coli_, Cell 32: 783 (1983).
22. Hochschild, A. and Ptashne, M., The recognition helices of λ repressor and λ cro make homologous contact with the operator, Cell 44, 925 (1986).
23. McClarin, J.A., Frederick, C.A., Wang, B.C, Greene, P., Boyer, H.W., Grable, J., Rosenberg, J.M., Structure of the DNA-EcoRI endonuclease recognition complex at 3A resolution, Science, in the press.
24. Schlotmann, M. and Beyreuther, K., Degradation of the DNA-binding domain of wild type and I^{-d} lac repressors in _Escherichia coli_, Eur. J. Biochem. 95: 39 (1979).

trp REPRESSOR, A CRYSTALLOGRAPHIC STUDY OF ALLOSTERY IN GENETIC REGULATION

P. B. Sigler, A. Joachimiak, R.W. Schevitz, C.L.Lawson, R.-G. Zhang, Z. Otwinowski and R. Marmostein

Department of Biochemistry and Molecular Biology, The University of Chicago

Chicago, IL, 60637 U.S.A.

(Abstract of the lecture)

The crystal structure of the E. coli trp repressor has been solved (1) and refined to 2.2 A. The two subunits (107 residues each) are related by an exact crystallographic dyad. Each subunit is composed of six helices, five of which intertwine about each other in a way that may make it seemingly impossible to disengage the subunits without altering their tertiary structure. The two symmetrically related L-tryptophan binding sites are formed by this interface.

Tryptophan must bind to the protein for repressor function. Tryptophan acts as an allosteric inhibitor as follows: (i) L-Tryptophan is wedged between the amino end of the C helix of one subunit and the side of the E helix of the dimer-related subunit fixing the orientation of the E helix, the most important element in recognizing the operator. (ii) The polar substituents of the bound tryptophan mold the protein's polar residues near the region of the repressor surface where the DNA backbone most closely approaches the protein. (iii) The amino group of the tryptophan mitigates the negative charge potential arising from the carboxyl terminus of the B helix.

The crystal structure of trp aporepressor (2,3) has also been solved (by molecular replacement) and partially refined to 2.1 A. The aporepressor is the unliganded inactive form of the trp repressor that is activated to the repressor upon binding two molecules of L-tryptophan per dimer. By contrasting the aporepressor and repressor structures we can now define the allosteric transition that activates this sequence-specific DNA-binding regulatory protein.

The presence or absence of bound tryptophan has essentially no effect on the overall architecture, especially the dimer interface. However, if tryptophan is not bound, the helix-turn-helix motif 'collapses' into the 'empty' binding pocket. The largest differences between the structure of the liganded active repressor and that of the unliganded inactive aporepressor occur in the D-helix (the first helix of the motif), the amino terminus of the E helix (the second helix of the motif) and the intervening turn, that is, the residues involved in the repressor/operator interaction. There is also a substantial change in the conformation of the amino-terminal segment of the molecule.

Docking studies using a canonical B-DNA for the operator's conformation show that the amino-terminal three residues of the E-helix, Ile79, Ala80 and Thr81, protrude into the major groove of the operator and make snug hydrophobic fit with the 5-6 region of the pyrimidines of the base pairs that are functionally most sensitive to mutational change. Electrostatic calculations (in collaboration with J. Warwicker, Yale) show a nearly perfect complementarity of charge potential in the complex.

We report progress on the structure analysis of <u>trp</u> pseudorepressor (2,3), the adduct formed when certain analogues of L-tryptophan (e.g. indolepropionic acid) displace L-tryptophan. Although nearly isomorphous, the pseudorepressor does not bind selectively to the operator. Similarly we report progess on the structure analysis of the crystalline <u>trp</u> repressor/operator system.

REFERENCES

1. Schevitz, R.W., Otwinowski, Z., Joachimiak, A., Lawson, C.L. & Sigler, P.B. (1985) <u>Nature 317,</u> 782-786
2. Joachimiak, A., Kelley, R.L., Gunsalus, R.R., Yanowsky, C. & Sigler, P.B. (1983) <u>Proc. Natl. Acad. Sci. U.S.A. 80,</u> 668-672
3. Joachimiak, A., Schevitz, R.W., Kelley, R.L., Yanowsky, C. & Sigler, P.B. (1983) <u>J. Biol. Chem. 258,</u> 12641-12643

STRUCTURAL STUDIES OF THREE DNA BINDING PROTEINS: CATABOLITE GENE

ACTIVATOR PROTEIN, RESOLVASE, AND THE KLENOW FRAGMENT OF DNA POLYMERASE I

T.A. Steitz, L. Beese, B. Engelman, P. Freemont,
J. Friedman, M. Sanderson, S. Schultz, G. Shields,
and J. Warwicker

Department of Molecular Biophysics and Biochemistry
Yale University, New Haven, Connecticut, USA

We have been studying the crystal structure of three proteins that interact with DNA: the E. coli catabolite gene activator protein (CAP) whose structure was determined at 2.9 Å resolution (1,2) and is now refined to 2.5 Å resolution (3), the Klenow fragment of E. coli DNA polymerase whose structure has been determined from a 3.3 Å resolution electron density map (4) and gamma-delta resolvase, a site-specific recombination protein whose structure determination is still in progress (5,6,7). In addition to working out the mechanism by which each of these proteins carries out its biological function, we have been interested in ascertaining the ways in which these proteins are similar or differ in their manner of interaction with double-stranded DNA. Both CAP and resolvase contain small DNA binding domains which confer their DNA sequence specificity (1,6). In both cases the principles of DNA sequence specific interaction appear to be very similar: side-chains from protruding alpha-helices penetrate the DNA major groove and interact with the exposed edges of base pairs. DNA polymerase, on the other hand, contains a very large cleft which is able to nearly surround a double-stranded DNA substrate.

RECENT STUDIES OF CAP STRUCTURE

CAP is a dimer of two identical 22,500 molecular weight subunits whose properties have been recently reviewed (8). In the presence of cyclic AMP (cAMP), CAP binds to specific DNA sequences in the promoter region of some operons and stimulates transcription by RNA polymerase. The structure of CAP crystallized as a complex with cAMP was the first repressor/activator structure determined and shows that cAMP binds to the larger of two domains (1,2). The smaller of the two domains is known to bind DNA (9) and contains a two alpha helical structure that is also found in other transcription repressors and activators. In fact, it was the comparison of CAP and Cro structures that demonstrated the existence of the helix-turn-helix as the common motif in DNA binding proteins (10). Recently, we have concluded that CAP binding to DNA sharply kinks the DNA and have grown two crystal forms of CAP complexed with DNA.

We have now used a newly developed algorithm (11) to recalculate the electrostatic charge potential around CAP. We find that the positive electrostatic charge potential forms a bent pathway around the DNA binding domains. If we assume that the electrostatically negative DNA binds to CAP by overlapping with the positive electrostatic regions of CAP, then we must kink or sharply bend the DNA and partially wrap the DNA around the CAP protein. In an earlier model building attempt to fit DNA to the previous electrostatic calculation (12), Weber and Steitz (13) smoothly bent the DNA to maximize the contacts between the protein and the DNA. This model, however, as was realized at that time, did not adequately account for the full length of DNA with which CAP appears to interact. Studies of phosphate ethylation interference of DNA binding indicated an interaction site size of 26 base pairs (14). We find now, however, that by kinking the DNA rather than smoothly bending it we can not only more nearly follow the positive electrostatic charge potential, but we can also more completely account for the interactions expected between the protein and the DNA. In the current model all those phosphates whose ethylation prevent DNA binding are now seen to be near protein side chains. The overall bend in the DNA predicted from this model building is in the vicinity of 140°.

The proposed kinking of the DNA is consistent with experiments of Wu and Crothers (15) who showed that the electrophoretic mobility of the CAP-DNA complex on gels was dependent on the position of the CAP site on the 203 base pair DNA fragment. These data were interpreted in terms of CAP bending the DNA. The model is also consistent with recent data from the Crothers laboratory (D.M. Crothers, private communication) that show from studies of CAP binding to various length DNA's that at least 28 to 30 base pairs of DNA are required to approach the full affinity of a 203 base pair fragment. We assume that CAP kinking of the DNA is somehow related to its function of activating the RNA polymerase. Whether this occurs indirectly via a coupling transmitted through the DNA or whether it occurs directly through a protein/protein contact or (more likely) both is not established as yet.

To experimentally determine how CAP binds DNA, we are co-crystallizing the complex of CAP with various lengths of DNA. We have been successful in obtaining at least two different crystal forms of CAP-DNA complex. In one crystal form CAP has been crystallized with a 28 base pair symmetric fragment of DNA and in the second crystal form a 29 base pair fragment of DNA corresponding to the sequence in the lac promoter region was used. The latter crystal form provides a centered lattice with unit cell dimensions of 180 x 103 x 128 Å. Co-crystallization attempts are underway with DNA's of a variety of other lengths.

KLENOW FRAGMENT OF DNA POLYMERASE I

DNA polymerase I from E. coli is 103,000 molecular weight monomer that shows three enzymatic activities: DNA polymerase, 3'-5' exonuclease and 5'-3' exonuclease (15). Klenow showed (16) that protease treatment cleaved the protein into two fragments. The larger fragment, now referred to as the Klenow fragment, retains 3'-5' exonuclease as well as the polymerase activity. The gene coding for the large fragment was cloned in a high expression vector (17) and the resulting protein crystallized and its structure determined at 3.3 Å resolution (4). Currently, crystallographic refinement of the protein coordinates is proceeding at 2.7 Å resolution and has resulted in a crystallographic R-factor of 0.25 with RMS bond length derivation from ideality of 0.01Å.

The structure has two domains. The smaller amino terminal domain binds nucleoside monophosphate in the crystal. The larger C-terminal domain contains a very large cleft into which it was possible to model build the DNA substrate of the polymerase reaction (Figure 1). These structural data interpreted in the light of biochemical studies of the protein led to the hypothesis that the larger C-terminal domain catalyzes the DNA polymerase activity, whereas the smaller amino terminal domain contains the 3'-5' exonuclease activity (4,18). This hypothesis has now been tested and has been shown to be correct.

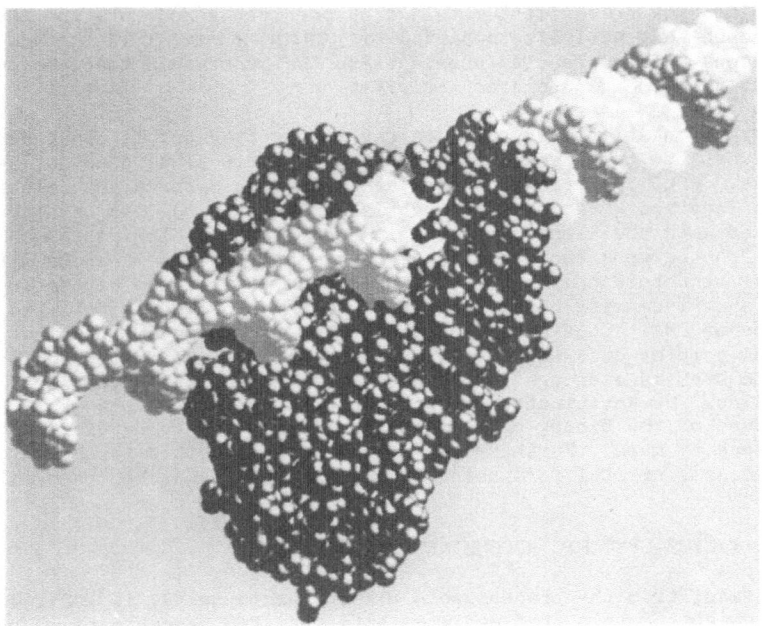

Figure 1. Space-filling model of the Klenow fragment bound to a DNA substrate. The placement of the DNA is achieved by model building. Note that the minor groove of the duplex product of DNA synthesis adjacent to the putative 3'-terminus is facing the top of the cleft where the 50 residue disordered domain is located.

The DNA encoding the large domain of the Klenow fragment was cloned in a high expression vector and the protein product purified (19). The purified large domain has the expected molecular weight of 45,000 and, as anticipated, does show DNA polymerase activity, but it possessed no measurable exonuclease activity. While the polymerase activity is reduced from that exhibited by the entire Klenow fragment, it is clear that all of the polymerase active site is present in the large domain.

Examination of amino acid side chains that surround the binding site for the nucleoside monophosphate in the small domain led to the hypothesis that the residue Asp 424 might be involved in the exonuclease activity. In a collaborative effort with Cathy Joyce and Vicky Derbyshire this residue has been changed to an alanine and the resultant protein's activity measured and the mutant protein crystallized. Preliminary studies indicate that the mutant protein has an altered exonuclase activity, though not in the manner anticipated. A 3 Å resolution difference map of nucleoside monophosphate bound to crystals of the mutant protein shows that the nucleoside monophosphate binds to the mutant protein in an identical fashion to wild type protein and there are no changes in the conformation of the protein at all. The only changes in the mutant complex are 1) the carboxylate of residue 424 is missing as expected and 2) a metal ion that binds between the carboxylate of 424 and the 5' phosphate of dTMP in the wild type complex is missing in the mutant complex. A second mutant in which 355 and 357 are changed to Ala has been crystallized and examined at low resolution. This mutant does not bind dTMP or divalent metal which presumably accounts for the lack of exonuclease activity observed by Cathy Joyce (private communication). These results substantiate the hypothesis that the small domain catalyzes the exonuclease activity.

We have been able to crystallize the Klenow fragment in the presence of a DNA substrate under conditions that the protein alone does not crystallize. HPLC analysis of the crystals and the supernatant solution from which they are grown firmly establishes that the DNA has been co-crystallized with the Klenow fragment in a one to one molar ratio. Precession photographs show that these co-crystals are isomorphous with the crystal structure that has been determined, diffract to better than 2.8 Å and show intensity changes relative to the protein crystallized in the absence of DNA. This establishes that the formation of the binary protein-DNA complex does not result in a major conformational change in the protein such as a change in the relative orientation of the large and small domains. We anticipate that we should shortly be able to examine the structure of the binary complex at high resolution using difference electron density maps. Furthermore, it should be possible to obtain exactly the same crystal form using DNA containing mismatched base-pairs.

RESOLVASE, A SITE-SPECIFIC RECOMBINATION PROTEIN

Resolvase, from the transposable element gamma delta, is a 21,000 molecular weight protein that mediates site-specific recombination between two specific duplex DNA sequences (called a res site) that are oriented the same way on closed circular DNA. Limited proteolysis of the protein using chymotrypsin gives two fragments (6). The smaller C-terminal fragment is 43 residues long and binds specifically to DNA. It also shows a sequence homology to helix-turn-helix of the DNA binding domain of CAP, Cro and lambda repressor. The larger amino terminal domain forms all the subunits contacts and is probably involved in catalyzing recombination reaction.

Large crystals have been obtained of the intact resolvase and also of the large fragment. Crystals of the native resolvase diffract only to 6 Å resolution in one direction, though to a higher resolution in another direction. However, one crystal form of the large fragment diffracts to better than 2.7 Å resolution.

In order to make heavy atom derivatives of these proteins we have, in collaboration with Nigel Grindley and Gram Hatfull, been making and crystallizing mutants of resolvase in which serine residues are exchanged for cysteine residues. We have successfully crystallized a resolvase large fragment in which serine 39 has been changed to a cysteine. It appears that mercury compounds bind to this cysteine residue and structural analysis is in progress.

Examination of the packing of subunits in the crystal leads to possible explanations of the arrangement of subunits at synapsis. In the space group P6₄22 there are twelve subunits in the crystallographic asymmetric unit arranged as three tetramers each having 222 symmetry. At synapsis the two DNA duplexes are brought together and involve twelve protein subunits. The symmmetry of the protein in this crystal is, therefore, exactly the symmetry of one possible arrangement of subunits at synapsis.

To examine the protein in the presence of the DNA attempts are being made to co-crystallize the protein with DNA corresponding to one of the three sites in the res site. We also intend to crystallize the protein with the entire 120 base-pair res site.

Acknowledgements

Research was supported by American Cancer Society grant NP-421, USPHS grant GM-22778 and National Science Foundation grant PCM-81-10880.

References

1. D.B. McKay, and T.A. Steitz, Nature 290: 744-749 (1981).
2. D.B. McKay, I.T. Weber, and T.A. Steitz, J. Biol. Chem. 257: 9518-9524 (1982).
3. I.T. Weber, and T.A. Steitz, J. Mol. Biol. 188: 109-110 (1986).
4. D.L. Ollis, P. Brick, R. Hamlin, N.G. Xuong, and T.A. Steitz, Nature 313: 762-766 (1985).
5. P.C. Weber, D.L. Ollis, W.R. Bebrin, S.S. Abdel-Meguid and T.A. Steitz, J. Biol. Chem. 157: 689-690 (1982).
6. S.S. Abdel-Meguid, N.D.F. Grindley, N. Smyth Templeton and T.A. Steitz, Proc. Natl. Acad. Sci USA 81: 2001-2005 (1984).
7. S.S. Abdel-Meguid, K. Murthy and T.A. Steitz, J. Biol. Chem., in press 1986.
8. B. de Crombrugghe, S. Busby, H. Buc, Science 224: 831-838 (1984).
9. E. Eilen, C. Pampeno, J.S. Krakow, Biochem. 17: 2469-2480 (1978).
10. T.A. Steitz, D.H. Ohlendorf, D.B. McKay, W.F. Anderson and B.W. Matthews, Proc. Natl. Acad. Sci. USA 79: 3097-3100 (1982).
11. J. Warwicker, D.L. Ollis, F.M. Richards and T.A. Steitz, J. Mol. Biol. 186: 645-649 (1985).
12. T.A. Steitz, I.T. Weber, D.L. Ollis and P. Brick, Cold Spring Harbor Symposium in Quantitative Biology 47: 419-426 (1983).
13. I.T. Weber, and T.A. Steitz, Proc. Natl. Acad. Sci. USA 81: 3973-3977 (1984).
14. H-M. Wu and D.M. Crothers, Nature 308: 509 (1984).
15. A. Kornberg, "DNA Replicatiion," Freeman, San Francisco (1980).
16. H. Klenow and I. Henningsen, Proc. Natl. Acad. Sci. USA 65: 168-172 (1970).
17. C.M. Joyce and N.D.F. Grindley, Proc. Natl. Acad. Sci. USA 80: 1830-1834 (1983).
18. C.M. Joyce, D.L. Ollis, J. Rush, T.A. Steitz, W.H. Konigsberg and N.D.F. Grindley, in: "Protein Structure, folding and Design", D. Oxender, UCLA Symposia on Molecular and Cellular Biology, pp 197-205, Alan R. Liss, Inc., New York (1986).
19. P.S. Freemont, D.L. Ollis, T.A. Steitz and C.M. Joyce, Proteins 1: 66-73 (1986).

A TWO-DIMENSIONAL NMR STUDY OF THE COMPLEX OF LAC REPRESSOR HEADPIECE WITH A 14 BASE PAIR LAC OPERATOR FRAGMENT

R. Boelens, R.M. Scheek, R.M.J.N. Lamerichs,
J. de Vlieg, J.H. van Boom* and R. Kaptein

Laboratory of Physical Chemistry
University of Groningen
The Netherlands
and
*Gorlaeus Laboratory
University of Leiden
The Netherlands

INTRODUCTION

Lac repressor of E. coli recognizes a 20 to 25 basepair DNA sequence called the lac operator.[1-4] When it binds to this region it blocks transcription of the genes coding for the lactose enzymes. Genetic and biochemical data have implicated the N-terminal part or 'headpiece' of the repressor as the DNA binding domain.[4,5] The precise way in which lac repressor recognizes its operator is, however, not known.

For other DNA binding proteins some progress has been made. Thus, crystal structures are known for a number of repressors, cI[6] and cro[7] from phage lambda and trp repressor[8] and for the catabolite activator protein (CAP).[9] A common feature of all these proteins is a helix-turn-helix structural domain, the second helix of which has been suggested to be primarily responsible for recognition of nucleic acid bases in the major groove of DNA. The most direct evidence for this comes from a low resolution (7Å) X-ray structure of a 434 repressor-operator complex.[10] Furthermore, in a few cases genetic experiments have provided evidence for direct contacts between amino acid residues belonging to the recognition helix and specific DNA bases. Thus, Ebright et al.[11] have shown that Glu 181 of CAP contacts a specific GC basepair of its binding site in the lac operon, while for lac repressor it was shown that Gln 18 makes a contact with GC 7 of the lac operator.[12] In a recent study Wharton and Ptashne[13] changed the specificity of 434 repressor to that of P22 repressor by substituting amino acid residues in the recognition helix. From these studies a picture emerges in which the helix-turn-helix structural motif contains the main determinants for specific DNA recognition. However, a detailed molecular mechanism for sequence specific DNA recognition by proteins is still lacking.

For the lac repressor-operator system a large body of genetic and biochemical data is available. Although no X-ray structure is known, detailed models for the lac repressor-operator complex have been proposed by Matthews et al.[14] and Weber et al.[15] on the basis of the supposed analogy with models for the cro-OR3 and CAP-DNA complexes, respectively. These models were further elaborated by Rein et al.[16] and Ebright et al.[17], who predicted specific protein-nucleic acid contacts.

The present work is a continuation of a study on the lac headpiece-operator interaction using the method of nuclear magnetic resonance (NMR). A two-dimensional (2D) NMR study of the headpiece alone has provided a large number of ^{1}H resonance assignments and proton-proton nuclear Overhauser effects (NOE's).[18-20] These NOE's were further translated into distance constraints, from which the three-dimensional structure could be derived.[21,22] This was accomplished using a combination of model building and structure refinement by a restrained molecular dynamics procedure, in which the NOE distance constraints were taken into account. A structure for the headpiece has been obtained, which basically consists of three α-helices folded together. The first two helices consisting of residues 6-13 and 17-25, form the helix-turn-helix supersecondary structure as had been predicted on the basis of sequence homology, while the third helix (residues 34-45) packs against the first two to form a hydrophobic core.

Regarding operator DNA, a 14 bp fragment comprising the stronger binding half of the lac operator has been synthesized for which ^{1}H resonances have been assigned.[23] In this paper we describe a 2D NOE study on the interaction between lac headpiece and this 14 bp lac operator fragment. A large number of specific resonance assignments were made, which allowed us to detect 23 NOE's between protein and DNA. All these NOE's are consistent with one major headpiece-DNA complex. The most surprising result of this study is that the orientation of the headpiece is opposite to what has been predicted in model building studies. A preliminary account of this work was given elsewhere.[24]

MATERIALS AND METHODS

Synthesis of the 14 bp DNA fragment

The single strands d(GGAATTGTGAGCGG) and d(CCGCTCACAATTCC) were synthesized by a phosphotriester method as described elsewhere.[25,26] The 14 bp DNA fragment was prepared by mixing the single strands in a 1:1 ratio. Excess of single strand material was removed by gel filtration on a Sephadex G25 column in 25 mM (NH4)HCO3. The resulting 14 bp duplex comprises the left half of the lac operator, which is known to have a stronger interaction with repressor than the right half.[27a,b]

Purification of the lac repressor headpieces 51, 56 and 59

Wild-type lac repressor, grown from the overproducing E.coli strain BMH 74-12, was purified as described previously.[28] Lac repressor solutions were stored in 1M Tris-HCl pH=8, 30% glycerol at -20° C. Headpiece 51 (HP51, the N-terminal headpiece consisting of 51 residues) was obtained by digestion with clostripain, as described previously.[28] Headpiece 56 was obtained by digestion with 2% (w/w) chymotrypsin for 3 hrs at room temperature, as described by Arndt et al..[29] Headpiece 59 was obtained by digestion with 1% (w/w) trypsin for 20 min at

18° C, [30] after which the reaction was stopped with soybean trypsin inhibitor (Sigma).

After digestion the solution was loaded on a Sephadex G50 column and eluted with 0.4 M KCl, 50 mM KPi, pH 7.5 to separate the headpiece from the core of lac repressor, intact lac repressor and the proteinases, as described previously.[28] After dialysis against 50 mM KPi, pH 7.5, 5% glycerol the headpiece solution was loaded on a small phosphocellulose column, rinsed with 50 mM KPi pH 6.5 in D_2O and eluted with 50 mM KPi, 400 mM KCl, pH 6.5 in D_2O. If necessary, the sample was concentrated on a small-volume Amicon high-pressure cell with YM2 filters (cutoff MW=1000).

Preparation of the lac repressor headpiece -operator complexes

The relative concentrations (about 2mM) of stock solutions of headpiece in 50 mM KPi, pH 6.5, 400 mM KCl in ^2H_2O and operator in 50 mM KPi, pH 6.5 in ^2H_2O were determined by obtaining 1D ^1H NMR spectra at 360 MHz and integrating the aromatic resonances (18 and 34 protons in headpiece and operator, respectively). Thereafter, both samples were mixed in a 1:1 molar ratio and the complex was concentrated in an Amicon cell, as described above, to about 5 mM. At higher concentrations aggregation causes excessive line broadening in the ^1H NMR spectra. After concentration the complex was dialysed in the same cell against 200 mM KCl, 10 mM KPi at pH 6.5 in ^2H_2O. Slight turbity, caused by small amounts of insoluble material, was removed by centrifugation and 350 µl was brought into the NMR tube. To dissociate the complex 1 M KCl (solid) was added to the NMR sample. The NMR spectra were recorded at 300 K.

^1H NMR spectroscopy

One-dimensional (1D) 360 MHz ^1H NMR spectra of free headpiece, and lac operator were obtained on a Bruker HX 360 spectrometer, equipped with an Aspect 2000 computer. The HDO line was suppressed by pre-irradiation. For the quantitative NMR measurements 128 free induction decays of both samples were added, recorded both with a long (3 s) relaxation delay to prevent saturation.

One-dimensional (1D) and two-dimensional (2D) 500 MHz ^1H NMR spectra of the lac repressor headpiece-operator complexes were obtained on a Bruker WM 500 spectrometer, equipped with an Aspect 2000 computer. All 500 MHz NMR data were transferred onto magnetic tape and processed with the '2DNMR' software package (written in Fortran 77) on a VAX 11/750 computer.

2D NOE spectra were recorded using the pulse sequence (relaxation delay -90°-t_1 -90°-t_m-90°- acquisition)n. The first two 90° pulses create a frequency labeling of all z-magnetisation components; during the mixing period t_m magnetisation is transferred between the different components, which are detected by the last 90° pulse.[31,32] To suppress all coherences generated by the first two pulses (except zero-quantum coherences) phase cycling was performed in subsequent data acquisitions according to States et al.[33] The carrier was positioned at one side of the spectrum, to avoid the appearance of an 'anti-diagonal', caused by incomplete separation of positive and negative frequencies in the Ω_1 domain. For each sample a total of 128 or 256 FID's were recorded with a spectral width of 10000 Hz, each of 2048 or 4096 points, with mixing times of 100 ms and 200 (or 250) ms, while the number of t_1

increments n was 400 to 512. Each 2D NOE measurement took about 20 - 24 hrs.

The FID's were weighted with different filters both in the Ω_2 and the Ω_1 domain, to allow for detection of different features in the 2D NOE spectra. Most assignments within the DNA fragment were performed on medium resolution enhanced spectra (filter for Ω_2 :gaussian multiplication with a line broadening of -15Hz and its optimum at 0.12 for 2048 points, filter for Ω_1 : $\pi/5$ shifted sine bell). A large number of intra- and interresidue NOE's within the headpiece were found in strongly resolution enhanced spectra (filter for Ω_2 : gaussian multiplication with a line broadening of -20 Hz and its optimum at 0.4 for 2048 points, filter for Ω_1 : $\pi/10$ shifted sine bell). Most NOE's between headpiece and DNA, however, were only observed in moderately resolution enhanced spectra (filter for Ω_2 : gaussian multiplication with a line broadening of -15Hz and its optimum at 0.08 for 2048 points, filter for Ω_1 : $\pi/2$ shifted sine bell). All spectra were phase -corrected and a baseline correction[34] was applied both in the Ω_1 domain to remove t_1 ridges and the Ω_2 domain.

2D homonuclear Hartmann - Hahn spectra of the complexes with HP59 were recorded using the pulse-sequence (relaxation delay - $90_x\circ-t_1-$ (SL_y - $60\circ_{-y}$ - $300\circ_y$ - SL_{-y} - $60\circ_y$ - $300\circ_{-y}$)$_8$ - SL_y- acquisition)$_n$, with a 2 ms spin lock pulse SL.[35] During the evolution period t_1 all magnetisation components are frequency labeled and during the spin-lock periods magnetisation is transferred between the different magnetisation components mainly through J-coupling.[35] The spin-lock field-strength corresponded to a 90° rf pulse of 35 µs. 128 FID's consisting of 2048 points were obtained, with the rf carrier in the middle. The number of t_1 increments was 400. Positive and negative frequencies for the Ω_1 domain were discriminated by the TPPI method.[36] The FID's were weighted with $\pi/3$ shifted sine bell windows in both domains. After Fourier transformation the spectra were phase corrected and baseline corrected.

RESULTS

2D NOE spectra of the headpiece - operator complex. Assignment of DNA resonances

Figure 1 shows a one-dimensional (1D) ^{1}H NMR spectrum of the complex of lac repressor headpiece 56 and the 14 basepair lac operator fragment. Indicated in the Figure are those regions which contain the resonances from protons in the DNA and the regions which contain the resonances from protons in the headpiece. This grouping of resonances is based on comparison with 1D spectra of the free constituents and is supported by titrations of the DNA fragment with headpiece 51 in which the chemical shifts of a large number of resonances were followed.[28] From this Figure it is clear that there will be large areas in the 2D spectra which can only contain cross peaks derived from DNA. Figure 2 shows a 2D NOE spectrum of the complex. The regions in this spectrum which have shown to be essential for the assignment of resonances of DNA fragments[37-39], the regions containing the contacts between the aromatic H6 or H8 protons and the aromatic protons with the ribose H1' protons, overlap only to a minor extent with resonances from the headpiece. Therefore, the sequential assignment of resonances of DNA within the complex is still straightforward and can be carried out

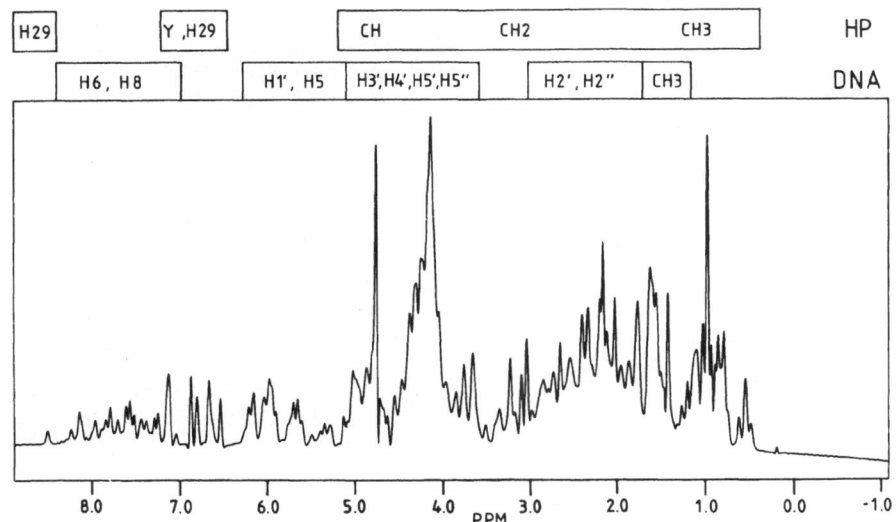

Fig. 1. 500 MHz ^{1}H NMR spectrum of the complex of <u>lac</u> repressor headpiece 56 with a 14 bp <u>lac</u> operator fragment. Experimental conditions : concentrations 5 mM in both protein and DNA in D$_2$O, 0.2 M KCl, 10 mM phosphate, 27 °C, pH 7.

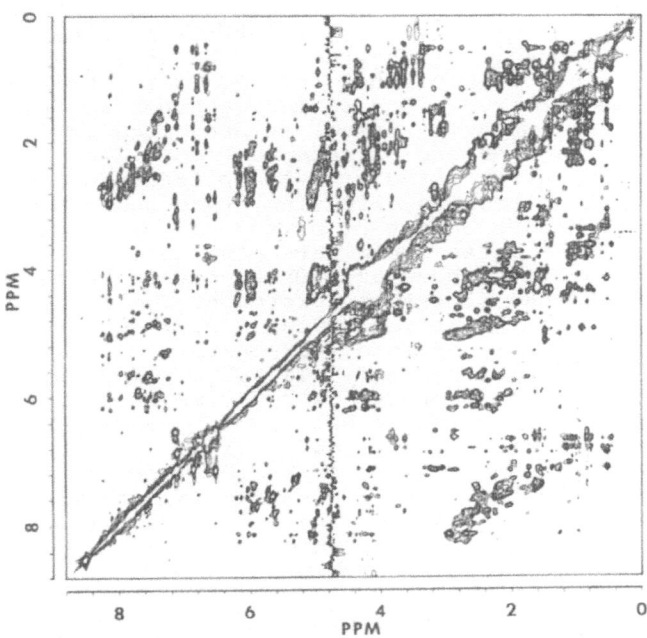

Fig. 2. 500 MHz ^{1}H 2D NOE spectrum of the complex of <u>lac</u> repressor headpiece 56 with a 14 bp <u>lac</u> operator fragment. The spectrum is a sum of two spectra taken with mixing times of 100 ms and 250 ms and was moderately resolution enhanced. For experimental conditions see Figure 1.

195

using the same strategy of sequential assignment as described for free DNA fragments.[37-39] In right-handed helical DNA a base proton is not only within NOE distance from the ribose H1' proton of the same nucleotide but also from that of its 5' neighbour. Therefore, on each row or column of the 2D spectrum corresponding to a base proton frequency, two cross peaks can be observed, one connecting it with its own ribose H1' and the other with the H1' of the neighbour (except that of the 5' terminal nucleotide). This last cross-peak identifies the frequency of a H1' proton, which in the other domain is connected by a third cross-peak to its own base-proton, and leads to a series of interconnected NOE cross-peaks as indicated in Figure 3A for the d(GGAATTGTGAGCGG) strand and in Figure 3B for the d(CCGCTCACAATTCC) strand of the 14 bp lac operator fragment in the complex.

The region in which the NOE's between the H6/H8 protons and the ribose H2' and H2" protons can be found, can, in principle, overlap with NOE's from the aromatic protons of DNA to aliphatic protons in the headpiece. However, most NOE's within this region could be explained as NOE's within the DNA and the assignment of the intra nucleotide H6/H8-H2'(i,i), H6/H8-H2"(i,i) cross peaks and the internucleotide H6/H8-H2'(i,i+1), H6/H8-H2"(i,i+1) cross peaks could be checked in the 'fingerprint' region of the ribose units, where NOE contacts between the H1', H2' and H2" of each ribose were identified. The assignments of the base protons resonances could further be verified by direct sequential contacts between all base protons and by the NOE contacts to their own H5 proton (in cytidines), their neighbouring H5 proton (3' neighbours of cytidines), the own methyl protons (in thymines) and their neighbouring methyl protons (3' neighbours of thymines). The assignment of the adenine H2 protons was obtained from NOE's to their own ribose H1' protons and/or from NOE's to the neighbouring ribose H1' protons from the opposite strand, as described previously for the free 14 bp lac operator fragment.[23]

In 2D NOE spectra of the free 14 bp fragment taken under conditions of limited spin diffusion cross-peaks are present between the H8 and H6 base protons and all ribose protons of the same nucleotide.[23] These NOE's were also observed in the complex. In addition, however, NOE's were found between the base protons and the H3' ribose proton of the 5' neighbour. Therefore, an independent sequential analysis of the resonance assignment of the base protons was possible, now using the H6/H8-H3'(i,i) and H6/H8-H3'(i,i+1) NOE's. The asignment of the resonances of the H3' protons could be verified, by the intra-residue NOE's of H3' with the H2' and H2" protons of the same nucleotide. Some H4' and H5', H5" protons could be assigned from their contacts with the base and H1' protons within the same nucleotide.[23] In this way, we were able to assign the non-exchangeable base protons and many of the deoxyribose protons of the DNA in the complex. The chemical shifts of the identified resonances are summarized in Table I.

Chemical shifts of the DNA protons in the complex

Comparison of the chemical shifts of the DNA in the complex (cf. Table I) with that of the free fragment, as was published previously [23] shows that not all resonances shift to the same extent upon complex formation. The changes in chemical shift upon complex formation for a number of resonances are summarized in Figure 4. They indicate regions in the DNA where

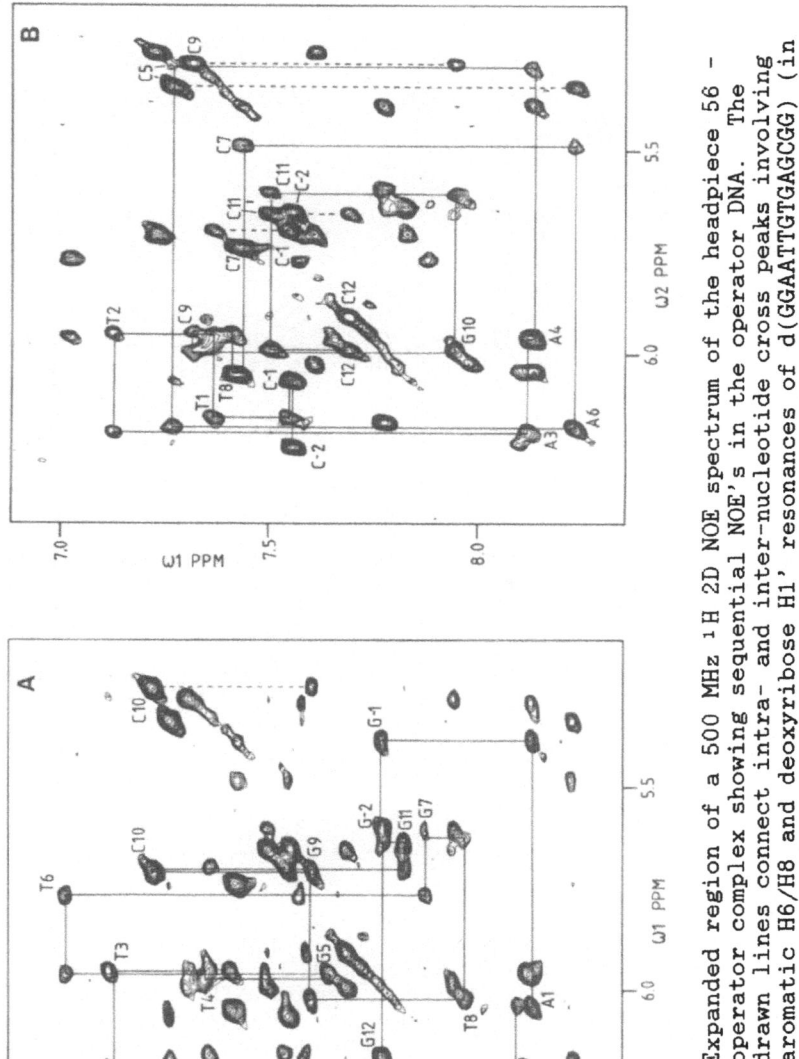

Fig. 3. Expanded region of a 500 MHz ^{1}H 2D NOE spectrum of the headpiece 56 – operator complex showing sequential NOE's in the operator DNA. The drawn lines connect intra- and inter-nucleotide cross peaks involving aromatic H6/H8 and deoxyribose H1' resonances of d(GGAATTGTGAGCGG) (in Fig. 3A) and of d(CCGCTCACAAATTCC) (in Fig. 3B). The dashed lines connect cross peaks involving cytosine H5 resonances. Intra-nucleotide cross peaks are indicated by a label. The spectrum was taken with a mixing time of 250 ms and was medium resolution enhanced (see text). For further conditions see Figure 1.

Table I
Chemical shifts (in ppm relative to DSS) of proton resonances
in the 14 bp lac operator in a 1:1 complex with lac
repressor headpiece 56.

nucleotide	GH8 AH8 CH6 TH6	AH2 CH5 TCH3	H1'	H2'	H2"	H3'	H4'
G-2	7.79		5.59	2.41	2.63	4.79	4.16
G-1	7.79		5.38	2.62	2.71	4.97	4.31
A1	8.14	7.28	6.03	2.70	2.96	5.07	4.48
A2	8.13	7.60	6.17	2.47	2.88	5.09	4.51
T3	7.14	1.25	5.90	2.00	2.56	4.85	4.22
T4	7.36	1.51	5.94	2.19	2.53	4.89	4.28
G5	7.66		5.96	2.56	2.82	4.99	4.43
T6	7.04	1.10	5.76	1.96	2.36	4.86	4.18
G7	7.89		5.61	2.70	2.85	5.00	4.36
A8	7.98	7.56	6.01	2.58	2.85	5.01	4.40
G9	7.62		5.70	2.45	2.59	4.95	4.34
C10	7.24	5.25	5.70	1.76	2.24	4.80	4.12
G11	7.84		5.63	2.62	2.70	4.96	4.31
G12	7.79		6.15	2.52	2.36	4.66	4.21
C-2	7.57	5.64	6.21	2.37	2.37	4.56	4.02
C-1	7.56	5.68	6.05	2.18	2.51	4.86	4.20
T1	7.38	1.54	6.13	2.18	2.57	4.91	4.25
T2	7.14	1.25	5.93	2.02	2.57	4.85	4.18
A3	8.12	7.51	6.17	2.58	2.94	5.02	4.45
A4	8.14	6.78	5.95	2.68	2.93	5.02	4.37
C5	7.28	5.34	5.29	1.96	2.29	4.78	4.12
A6	8.23	7.59	6.17	2.68	2.84	4.99	4.40
C7	7.45	5.72	5.48	1.82	2.21	4.79	4.07
T8	7.42	1.63	6.03	2.18	2.50	4.84	4.19
C9	7.33	5.28	5.93	2.06	2.53	4.68	4.25
G10	7.96		5.97	2.72	2.78	5.05	4.42
C11	7.52	5.65	5.60	2.12	2.43	4.86	4.13
C12	7.71	5.89	5.97	2.02	2.50	4.87	4.12

conformational changes (if any) are most likely to occur.[40] In
these terms 'hot spots' in the DNA chemical shifts are
concentrated at the basepairs GC5, GC7 and AT8. Previous
studies showed that the iminoproton of basepair AT6 undergoes a
large shift upon complex formation.[28] Around the same basepair
the DNA structure was proposed to be irregular on the basis of
the anomalous exchange observed for the iminoprotons of AT6.[41]
At this site the DNA conformation could very well change upon
complex formation, explaining the chemical shift change at GC5
and GC7. However, other trivial explanations of all chemical
shift changes could be the presence of aromatic rings of the
headpiece, the presence of charged groups near the bases or the
neutralization of phosphate backbone charges upon complex
formation.

The most prominent feature of Figure 4 is the absence of
major changes in chemical shift, except for those at the 'hot
spots', suggesting that there are no gross conformational
changes in the DNA fragment upon complexation with headpiece.

Much more reliable information on the conformational change in the DNA fragment upon headpiece binding, can be obtained from the 2D NOE spectra. Since the NOE effect is due to a magnetic dipole-dipole interaction between two protons and therefore strongly distance dependent, a change in conformation of the DNA fragment can be easily observed by the appearance or disappearance of NOE's in the complex. As was clear after the assignment of the DNA fragment in the complex and by comparison of the 2D spectra of the complex with those of the free DNA fragment, this was not the case. Therefore, the conclusion that major changes in DNA conformation upon complexation are absent, is also supported by the 2D NOE spectra. However, it should be noted that a real quantitative interpretation of the NOE effect is only possible when the NOE is measured at very short mixing times, allowing the elimination of the effects of spin diffusion.[42,43] Thus small changes in conformation of the operator upon binding of headpiece are still compatible with the NOE data.

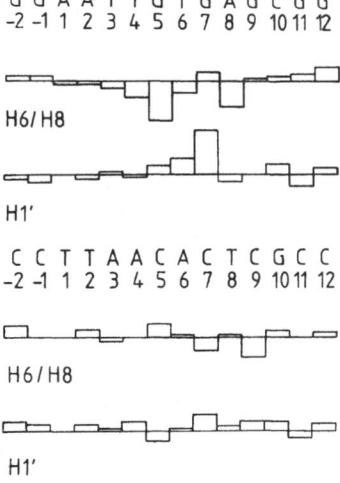

Fig. 4. Changes in chemical shift of the aromatic H6/H8 and deoxyribose H1' protons upon complex formation. The bars are the difference in chemical shift of the proton resonances in the complex minus the free operator (taken from ref. 23). The distance between the H6/H8 and H1' lines corresponds to 0.4 ppm.

Assignment of NOE's within the headpiece

The non-exchangeable protons of DNA fragments can be assigned completely from 2D NOE spectra obtained in ^2H_2O.[37-39] In contrast, the sequence specific assignment of proton resonances in proteins relies on the presence of 2D spectra, which measure the coherent interaction between spins (COSY or 2D homonuclear Hartman - Hahn spectra), recorded both in ^2H_2O and ^1H_2O for the identification of the spin-systems of the amino-acids. Furthermore, 2D NOE spectra must be recorded in ^1H_2O to find the $d_{\alpha N}$, d_{NN} and $d_{\beta N}$ connectivities between adjoining residues to assign the spin-systems sequentially to specific amino-acid residues.[44-46] Previously, this procedure was applied to the free headpiece 51 to identify all backbone protons (except those of Ile48) and most protons of the amino-acid side chains.[18-20] This has not yet been accomplished for a headpiece in a complex with the 14 bp lac operator fragment, mainly because of the complexity of the ^1H NMR spectra of the complex in ^1H_2O, where serious overlap exists between the amide protons of headpiece with the base protons and amino protons of the operator. The necessary sequential NOE contacts for the assignment of the protein ($d_{\alpha N}$, d_{NN} and $d_{\beta N}$) overlap strongly in the 2D spectra with the internal contacts in the DNA from the base and amino protons (to base, amino, H3', H4, H5', H5", H2', H2").

One-dimensional ^1H NMR spectra recorded during titrations of headpiece 51 with the 14 bp lac operator fragment showed that the complex is in fast exchange with its constituents on the NMR time-scale.[28] Therefore, another approach for the assignment of the resonances of the protons in headpiece is the tracing of resonances in titrations with the operator fragment in 1D or even 2D spectra. In this way the aromatic ring protons of His29, Tyr7,17,47 and 12, the aliphatic $C^{\alpha}H$ and $C^{\beta}H$ protons of Thr5 and the methyl protons of Leu6, Leu45 and Val9 could be assigned in the complex.[28] A two-dimensional version of this approach, monitoring cross-peaks in 2D spectra during such a titration, is hampered by the instability of free headpiece in low ionic strength solutions. In principle these problems can be overcome by titrating the 1:1 complex with salt, which causes a dissociation of the complex.[28]

At present we already made a large number of tentative assignments of cross-peaks in headpiece in the 2D NOE spectra of the complex on the basis of a comparison with 2D NOE spectra of free headpieces. It had been noted earlier that the chemical shifts of some internal residues of the headpiece, for instance Leu6, Val9, Leu45 and Tyr47, hardly change upon complex formation Therefore, it seems reasonable to assume that the folding of the headpiece in the complex is similar to that in the free state and that the chemical shift and the NOE's between the headpiece protons do not change to a large extent upon complex formation, except for those directly involved in DNA binding. Most assignments started from the previously firmly assigned protons. For instance, on a cross section of the (3,5) protons of Tyr12 NOE's were observed to protons assigned as the $C^{\alpha}H$ and $C^{\beta}H$ protons of the same residue (cf. Figure 5b). This assignment was then confirmed by NOE's and homonuclear Hartmann-Hahn cross peaks (in this case) of the $C^{\alpha}H$ of Tyr12 to the $C^{\beta}H$ protons (not shown). Such patterns could be recognized for the spin systems of Val4, Thr5, Leu6, Tyr7, Val9, Ala10, Tyr12, Ala13, Tyr17, Val24, Ala41, Met42, Glu44, Leu45, Tyr47 and Pro49 in the cross-sections at the aromatic tyrosine protons (cf. Figure 5a, 5b,

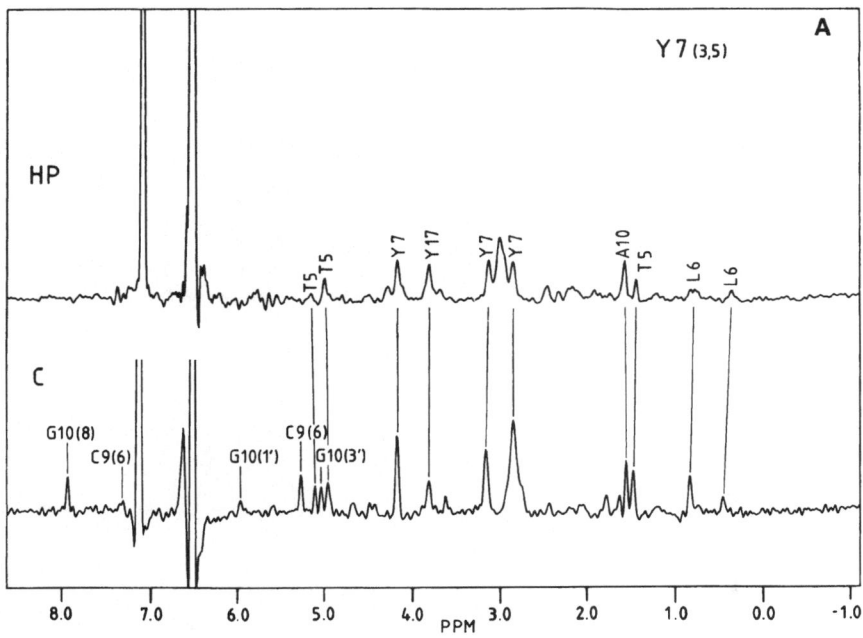

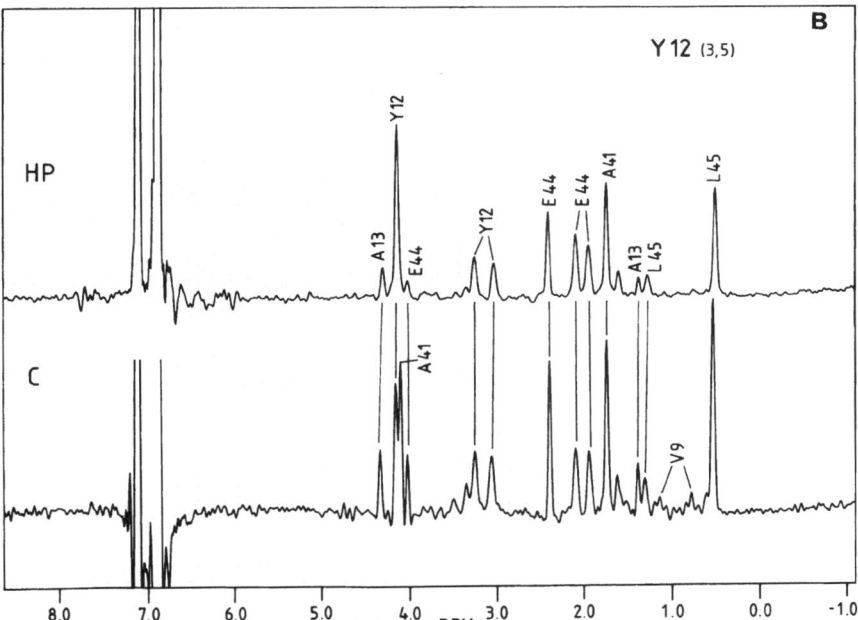

Fig. 5. Cross sections of 2D NOE spectra showing NOE's between aromatic protons of Tyr7 (in A), Tyr12 (in B), Tyr17 (in C) and Tyr47 (in D). In each case cross sections are shown for a number of residues in the free headpiece 51 (indicated by HP) and for the complex of headpiece 56 and the operator (indicated by C).

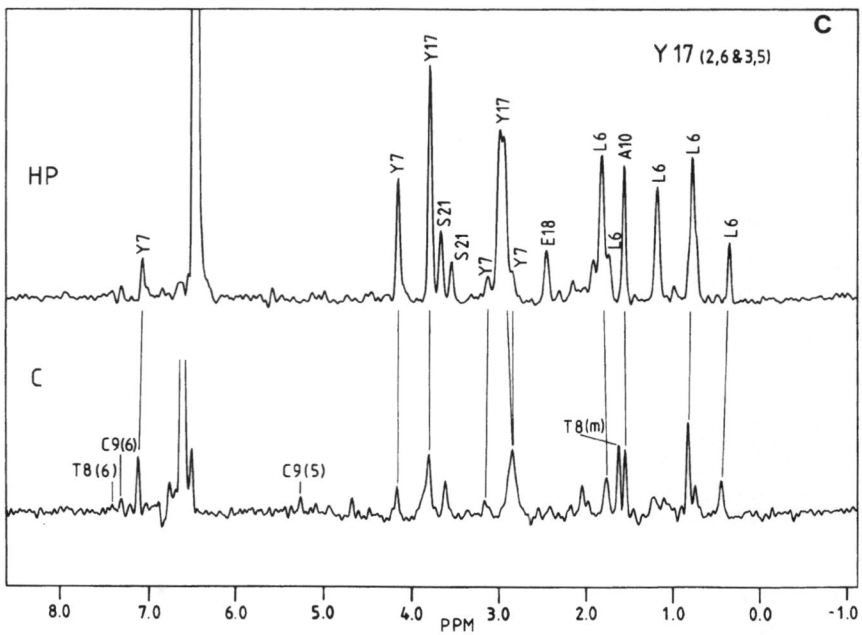

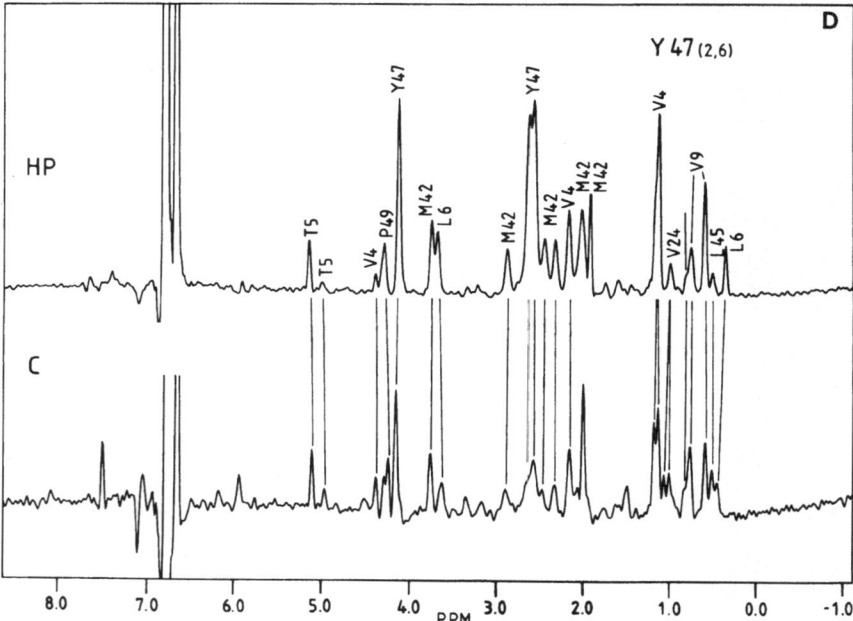

Fig. 5. (continued)
Some NOE's between protein and DNA are also indicated.
The 2D NOE spectrum of headpiece 51 was taken with a
mixing time of 200 ms and the assignment was from ref.
18,19 and 20. The 2D NOE spectrum of the complex was
taken with a mixing time of 250 ms.

5c, 5d), for His29 in a cross-section at the position of its C4
proton (not shown) and for Val4, Leu6, Asp8, Val9, Ala41, Met42,
Glu44, Leu45 and Tyr47 from the cross-sections at the position
of the methyl protons of Val9, Leu6 and Leu45 (not shown). The
remaining valines (Val15, Val20, Val23, Val30 and Val38), Ile48,
the threonines (Thr19, Thr34), the remaining alanines (Ala27,
Ala32, Ala40), the prolines (Pro3, Pro49), the serines (Ser21,
Ser28), Asn25, Asn46, Asn50 and Lys2 were assigned by comparison
of the 2D NOE pattern within a residue in the complex with that
in free headpiece. In this way we were able to assign a large
number of NOE's in headpiece in the complex. Figure 6
summarizes the NOE's observed between different residues of
headpiece in the complex and compares them with free headpiece.
Only a small number of NOE's in free headpiece were not yet
assigned in the complex. They correspond to the residues Glu11,
Arg22 and Arg35, which may have undergone too large shifts to
recognize the patterns or may overlap with other NOE's in the 2D
spectra.

Our working hypothesis that headpiece has a similar folding
in the complex as it has in the free state finds strong support
by these results, but it remains to be tested by the titration
experiment suggested above.

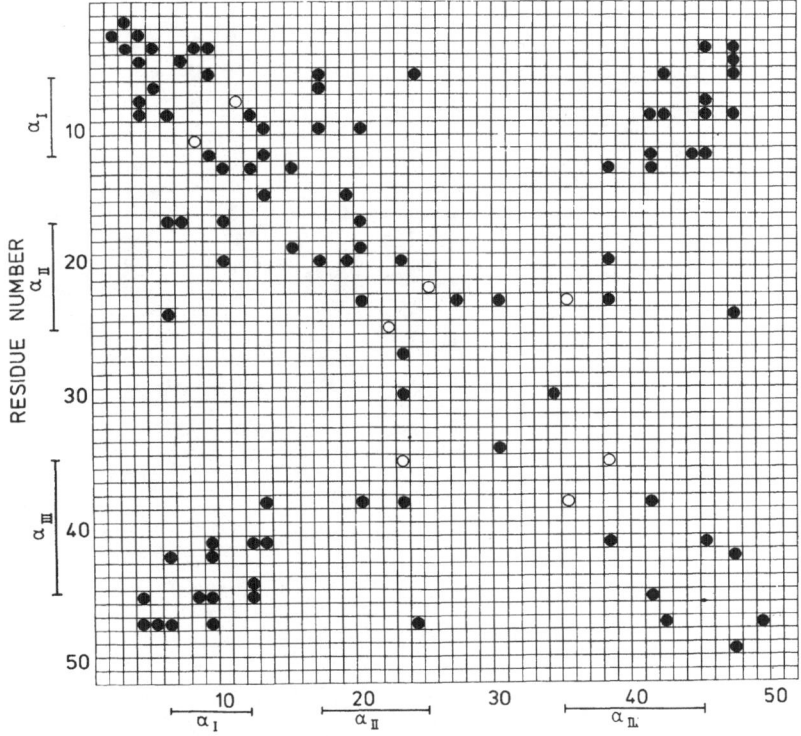

Fig. 6. NOE matrix for headpiece alone and in the complex with
        the 14 bp operator fragment on the basis of residue
        number. NOE's assigned both in the complex and in free
        headpiece are indicated by ▪; NOE's only assigned in
        free headpiece by o. The NOE's in free headpiece are
        taken from ref. 20.

NOE's between protein and DNA

Using the assignments of resonances in the 14 bp fragment
and the headpiece 56 we were able to detect cross peaks due to
NOE's between protein and DNA. These are indicated in the
Figures 7a, 7b and 7c and are summarized in Table II. Some of
the NOE's are also indicated on the cross sections. Ten of the
protein-DNA NOE's were observed at unique and unambiguously
assigned resonance frequencies. In the case protein-DNA NOE's
occur in crowded regions of the spectrum unambiguous assignment
is extremely difficult, because there is always the possibility
that an assigned resonance overlaps with another one. This
situation prevails for the remainder of the NOE's listed in
Table II. However, many of these NOE's occur between residues
where firmly assigned NOE's are also present for other proton
pairs. This is true for Leu6-C9, Tyr17-T8 and His29-A2. In a
'low resolution' picture of the complex they would not add new
information but would serve to confirm the previously found
NOE's and in fact they are in some cases much stronger in
intensity.

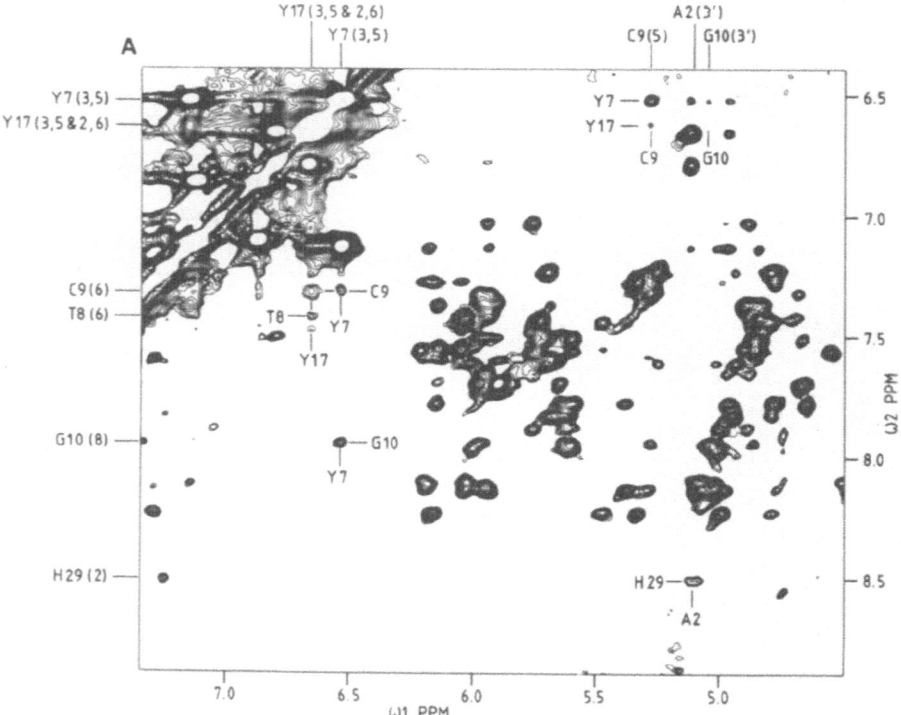

Fig. 7. NOE's between the headpiece and the operator in the
complex. A, between aromatic protons of headpiece 56
and base and ribose protons of the operator; B, between
methyl protons of headpiece 56 and base and ribose
protons of the operator; C, between the H2 proton of
His29 of headpiece 56 and protons of A2 of the operator.
The spectrum was taken with a mixing time of 250 ms and
was medium resolution enhanced. For further conditions
see Figure 1.

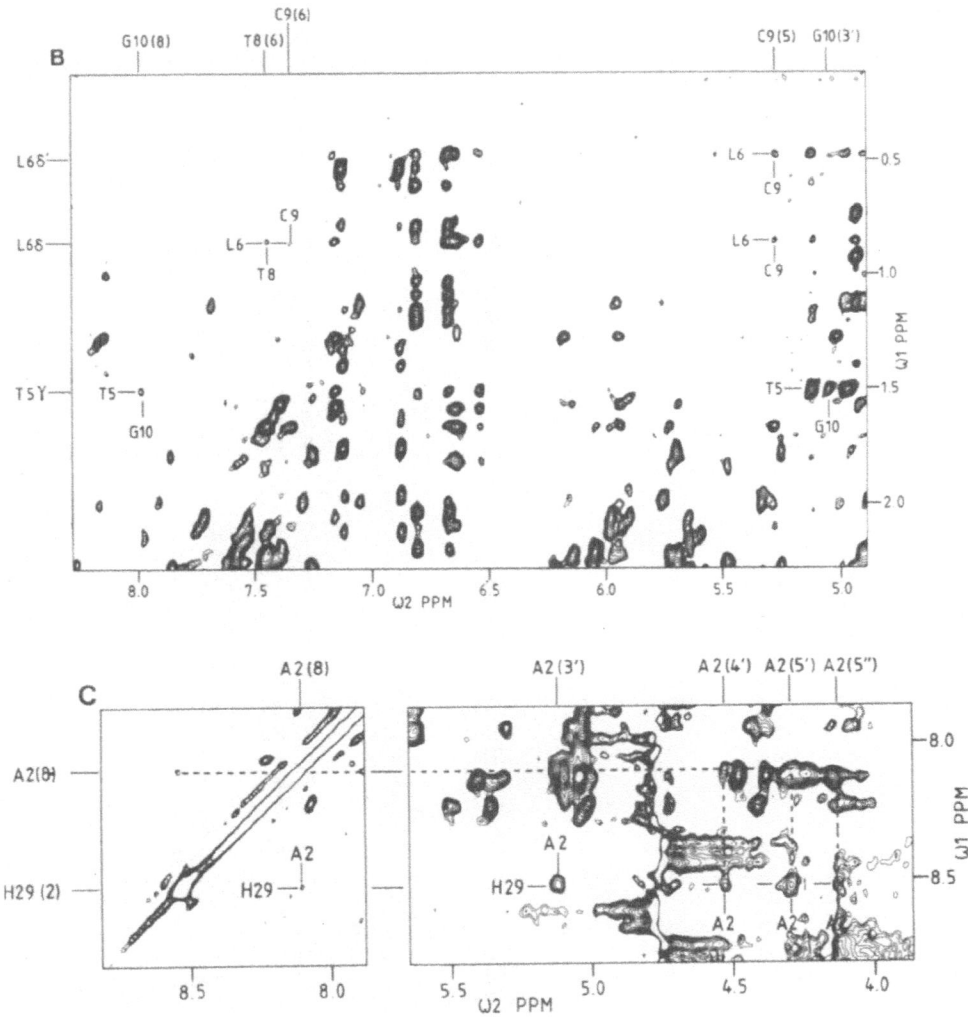

Fig. 7. (continued)

The NOE's from the base H8 and ribose H3' of G10 in DNA with the methyl group of Thr5 could in principle overlap with a NOE to another methyl group of the protein, since not all methyl resonances have yet been assigned. However, the observation of a similarly identified G10 (H3') - Thr5(C$^\delta$H$_3$) NOE in the complex of headpiece 59 with the operator, which has a slightly different chemical shift for the methyl group compared to headpiece 56, supports this assignment.

Table II

NOE's between <u>lac</u> repressor headpiece 56 and a 14 bp <u>lac</u> operator fragment.

Protein-DNA NOE				Chemical shift (ppm)
unambiguous [a]				
Tyr7	H3,5	- G10	H8	6.53 / 7.96
Tyr7	H3,5	- G10	H1'	6.53 / 5.98
Tyr7	H3,5	- G10	H3'	6.53 / 5.04
Tyr7	H3,5	- C9	H5	6.53 / 5.28
Tyr7	H3,5	- C9	H6	6.53 / 7.33
Leu6	CδH₃	- C9	H5	0.45 / 5.28
Tyr17	H3,5 + H2,6	- C9	H5	6.62 / 5.28
Tyr17	H3,5 + H2,6	- C9	H6	6.62 / 7.33
Tyr17	H3,5 + H2,6	- T8	H6	6.62 / 7.42
His29	H2	- A2	H8	8.52 / 8.10
probable [a]				
Thr5	CδH₃	- G10	H8	1.48 / 7.96
Thr5	CδH₃	- G10	H3'	1.48 / 5.04
Leu6	CδH₃	- C9	H5	0.83 / 5.28
Leu6	CδH₃	- C9	H6	0.83 / 7.33
Leu6	CδH₃	- C9	H3'	0.83 / 4.68
Leu6	CδH₃	- T8	H6	0.83 / 7.42
Tyr17	H3,5 + H2,6	- T8	CH₃	6.62 / 1.63
Ser21	CαH	- T8	CH₃	3.85 / 1.63
Ser21	CβH	- T8	CH₃	3.64 / 1.63
His29	H2	- A2	H3'	8.52 / 5.11
His29	H2	- A2	H4'	8.52 / 4.52
His29	H2	- A2	H5'[b]	8.52 / 4.28
His29	H2	- A2	H5"[b]	8.52 / 4.13

[a] the unambiguous NOE's were assigned at unique resonance frequencies, while the probable NOE's were from resonances which could overlap with resonances of other protons (see text for further discussion).

[b] H5' and H5" protons were only pair-wise assigned.

Also the methyl protons of Leu6 (δ=0.83 ppm) do not resonate at a unique frequency. Therefore these NOE's could originate from another methyl group in headpiece. However, this assignment is supported by an unambiguously assigned NOE of C9 (H5) with the other methyl group of Leu6 (δ=0.45 ppm).

A relatively strong cross peak involving the Tyr17 ring protons was found at the position of the T8 methyl group. Again, since there may be protein resonances at this position it is difficult to assign this cross peak unambiguously to a NOE involving T8 (or even to a protein-DNA NOE) in contrast to the previous ones. A cross-section (not shown) of the 2D NOE spectrum of the complex of headpiece 56 and the operator at the

resonance position of the CH₃ group of T8 (δ=1.63 ppm) shows
that most cross peaks could be explained by NOE's within the DNA
(to its own and neighbouring aromatic base and ribose protons)
except for three cross peaks at intersections with the aromatic
protons of Tyr17 and the CᵅH and one CᵝH proton of Ser 21.
Therefore, it is likely that in the complex of headpiece 56 and
the operator the T8 methyl group resonates at a unique frequency
and that these cross peaks are protein-DNA NOE's. Further
support for the assignment of the NOE of Tyr17 with the T8
methyl group comes from an experiment in which salt was added to
a complex of headpiece 59 with the operator. Figure 8a shows
the dissociation of the complex as observed from the
disappearance of the NOE between Tyr7 and G10, while Figure 8b
shows that at the same time the NOE between Tyr 17 and T8
decreases in intensity. The remaining intensity is probably due
to partial overlap from a NOE of Tyr 17 with Ala10 in the
dissociated complex. As shown in Figure 5c the cross-peak of
Tyr 17 with Ala10 in headpiece shifts to a different position
upon complex formation. In addition it should be noted that a
weak but unique NOE exists between Tyr 17 and the base H6 proton
of T8 (cf. Table II).

The DNA ribose protons, that show NOE's with the His29 H2
proton (δ=8.52 ppm) were assigned on the basis of a comparison
with a similar pattern of NOE's observable only for the H8
protons of A2 (δ=8.12 ppm). As explained above, in 2D NOE
spectra recorded under conditions with limited spin-diffusion
cross-peaks are present from the base H8 proton of A2 to all
ribose protons of the same nucleotide. In the spectrum of
Figure 7b the same combination of cross peaks to ribose protons
as observed for the A2 (H8) proton is also present at the His29
(H2) proton and therefore these cross peaks were assigned as
NOE's from His29 to the ribose protons of A2. The assignment
finds strong support from the observation of an unique (but
weak) NOE from His29 (H2) with the base proton of A2.

Not all identified NOE's indicate direct contacts between
protein and DNA. The mixing times that were used in the 2D NOE
experiments of the complex were sufficiently long (100 and 250
ms) to allow for a considerable amount of spin-diffusion and a
number of cross peaks were only weak. We estimate that the
protons between which NOE's were observed are closer than 0.6
nm. Some NOE's are relatively strong (Tyr 7 - G10 H8, Tyr7 -
G10 H3', Tyr7 - C9 H6, Tyr17 - T8 CH₃, Thr5 - G10 H3', His29 -
A2 H3') and probably correspond to proton-proton distances
within 0.4 nm. A complete quantitative analysis in terms of
distances is only possible from series of 2D NOE spectra
recorded with shorter mixing times. These experiments are now
in progress.

DISCUSSION

A structural model for the lac repressor-operator complex

The distance information obtainable from NMR spectroscopy
(NOE's) differs from that obtained by diffraction techniques
(reflections) in that in the case of NMR one knows which pair of
atoms is involved, provided that the resonances have been
assigned to specific nuclei. So, without having to solve the
complete three-dimensional structure of the protein-DNA complex,
for which obviously many more such distances are required, we
can still formulate constraints on distances between atoms on

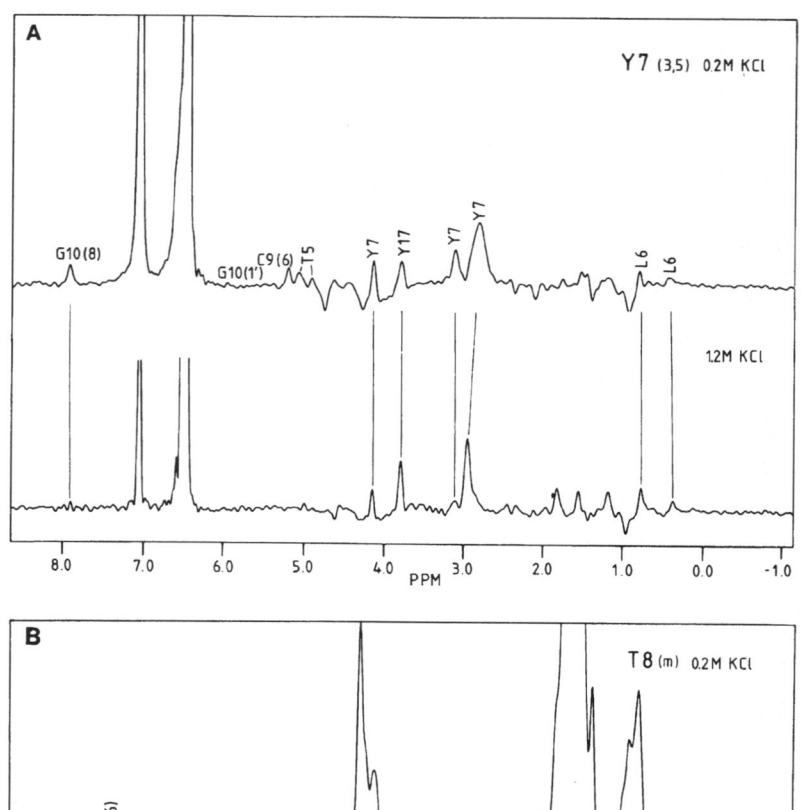

Fig. 8. Cross sections of 2D NOE spectra of the complex between
headpiece 59 and the 14 bp operator. The Figure shows
the disappearance of NOE's between protein and DNA upon
addition of salt, due to the dissociation of the complex
at high ionic strength. A, cross sections at the 3,5
protons of Tyr7 of headpiece 59, before and after
addition of salt; B, idem at the cross section at the
methyl protons of T8 of the operator. The incomplete
disappearance of the NOE between T8 and Tyr17 (in B) is
probably due to overlap of the methyl protons of T8 and
Ala10 in the dissociated complex. The 2D NOE spectra
were recorded with a mixing time of 200 ms.

the protein and on the DNA from the observed NOE's and discuss
the consequences of these constraints for structural models of
the complex. Before doing so, however, some other points need
to be discussed.

First we note that the three-dimensional structures of the
protein and the DNA do not undergo major changes upon complex
formation. In Figure 6 the NOE matrix of headpiece 51 free in
solution is compared to that of headpiece 56 in the complex.
Some 80% of the long-range NOE's found in the free protein[20]
could also be tentatively identified in the complex, as
discussed above. These include many strategic interhelical
contacts. We conclude that the overall fold of the protein and
the structure of its hydrophobic core is largely unaltered by
complex formation.

Similar arguments hold for the DNA structure: all base,
sugar H1', and many other sugar resonances could be assigned by
tracing the same cross-relaxation networks that were described
previously for the unbound DNA fragment.[23] Neither the relative
intensities of the NOE's nor the chemical shifts of the DNA
resonances are very different, so we conclude that our 14
basepair lac operator fragment undergoes only minor changes upon
binding to the protein.

Next we discuss the intermolecular NOE's. Previous studies
have shown that the complex between headpiece 51 and the 14
basepair DNA fragment is not unique: from the concentration
dependence of titration curves we concluded that weaker
complexes coexist with the major, specific complex in a fast-
exchange situation.[28] In such situations the theoretical
possibility exists that individual NOE's between protein and DNA
protons arise from different types of complexes. Realising this
we can still draw the conclusion that none of the observed NOE's
fits with any of the structural models of the lac repressor-
operator complex proposed recently. Thus both the cluster of
contacts between Tyr7,Tyr17 and basepairs 8,9 and 10, and that
between His29 and basepair 2 are incompatible with the models
proposed by Matthews et al.[14] and Weber et al.[15]

The question now arises whether there exists any plausible
model in which all experimental constraints are satisfied
simultaneously. Figure 9 is the result of some computer-aided
model building on an Evans and Sutherland picture system. We
kept the protein in a conformation compatible with all the NOE's
that were observed for headpiece 51 freely dissolved, although,
strictly spoken, only some 82% of these were actually identified
in the complex thusfar. Similarly, the DNA was kept in a rigid
B-DNA conformation. We allowed for adjustments of some protein
side chains to relieve obvious clashes between protein and DNA
atoms. In the structure of Figure 9 all the experimental intra-
and intermolecular NOE's are satisfied simultaneously. Thus, we
accept as a working hypothesis that the NOE's documented here
stem from the major, tight-binding complex; in this view other
types of complexes are too low in concentration to give rise to
observable NOE's in our experiments. In the complex of
headpiece 51 with the same operator fragment many of the same
intermolecular NOE's were also observable, albeit weaker than in
the spectra shown here. This observation suggests that a shift
in the binding equilibria has occurred in favour of the specific
complex in going from the smaller to the larger headpiece.

Even with the assumption that the observed intermolecular
NOE's are to be satisfied simultaneously there are obviously
more structures possible apart from the one we arrived at by
model building. The lack of more, and more precise distance

constraints will be felt even more strongly if we allow for some
extra uncertainty in the protein and DNA structures within the
limits imposed by the observed NOE's. Hence, at this stage we
are unable to make precise statements about the angle between α-
helices in the protein and the DNA axis. We note, however, that
in the model of Figure 9 the orientation of the second, so-
called 'recognition' helix relative to the major groove of the
DNA is roughly opposite to that observed in the 434-operator
complex[10] and proposed by Matthews et al.[14] for the cro-OR3
complex and Weber et al for the CAP-DNA complex.[15] Moreover, the
two clusters of observed NOE's are so far apart that any
structure in which this orientation is reversed, will not
satisfy the experimental constraints. In particular, the first
helix of the helix-turn-helix motif is near the pseudo two-fold
axis in our model, while it is away from this axis in the
Matthews and Weber models.

It remains an open question whether in the complex of the
complete lac repressor with its operator the orientation of the
helices is the same as we observe in our model system. It is
generally believed that the lac repressor binds with two
headpieces simultaneously to the full lac operator in such a way

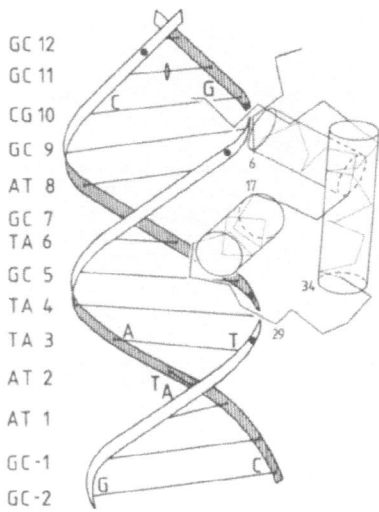

Fig. 9. Model of the lac headpiece - lac operator complex. The
figure shows schematically the orientation of the
recognition helix in the major groove of DNA using a
standard B-DNA conformation [47] for the operator
fragment. The conformation of the headpiece was taken
from ref. 22. The start of the helical regions is
indicated (helix I : residues 6-13; helix II, residues
17-25; helix III, residues 34-45). The model was built
on an Evans and Sutherland PS1 picture system. The
relative position of headpiece with respect to DNA was
adjusted as to best satisfy the NOE's of Table II. The
black dots indicate the positions of the phosphates
where ethylation with ethylnitrosourea interferes with
lac repressor binding[48].

that the protein and the DNA share an approximate two-fold axis. The interesting possibility exists that in this complex protein-protein contacts force the headpieces into an orientation different from the one we observe. An alternative possibility and the one that we consider most likely is that in different repressors the 'recognition helix' can function as such in different orientations relative to the operator.

A structural explanation of observations on mutant repressors

Recently Miller[4] compiled the results of amino acid replacements in the DNA-binding domain of the lac repressor on repressor activity. We are now in a position to interpret some of the observations made on these mutants in structural terms.

For the model of the complex shown in Figure 9 we calculated which amino acids become less accessible to water molecules upon DNA complexation. This is shown in Figure 10. Small changes in this solvent accessibility may not be significant, since Figure 9 must be regarded a low-resolution structure. We note, however, that the results of photo-CIDNP experiments on headpiece and its complexes with DNA[50] are nicely accounted for in this model: of the accessible residues Tyr7, Tyr12, Tyr17 and His29 only Tyr12 remain exposed to solvent upon complex formation. Furthermore, from phosphate ethylation interference studies[48] and methylation protection experiments[5] it is known, not only which base pairs are involved in repressor binding, but also which side of the DNA actually faces the protein in the complex. In this respect our model does not differ from the earlier proposals and is consistent with this data as well.

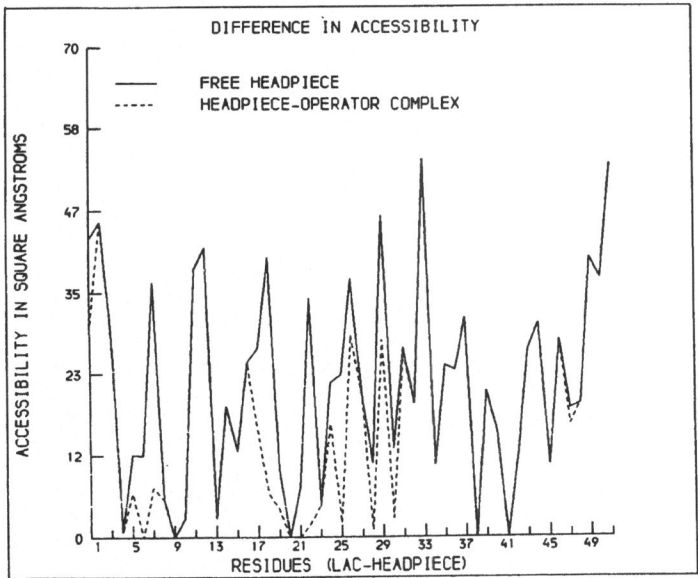

Fig. 10. Solvent accessibility of amino acids in free headpiece and headpiece in the complex. The accessibility was calculated as described by Richmond and Richards [48] and the solvent radius used was 0.14 nm. The differences in accessibility show those amino acids which are covered by the operator.

Table III

Structural role of amino acid residues in E. coli lac repressor headpiece for which I⁻ mutants are known.[4]

operator binding	tertiary structure
Thr5	Leu6
Tyr7	Val9
Tyr17	Ala10
Gln18	Ala13
Ser21	Thr19
Arg22	Val23
Asn25	Leu45
His29	Tyr47

Table III summarises our tentative interpretation of the mutation experiments reviewed by Miller.[4] The effects of replacements that result in I⁻ phenotypes are explained in one of two ways: either the mutated residue is directly involved in operator binding in our model (see Figures 9 and 10) or it is thought to be essential for the intact fold of the DNA-binding domain. Thus Thr5, Tyr7, Tyr17, Gln18, Ser21, Arg22, Asn25 and His29 are candidates for direct interaction with the DNA. For the proximity (less than 0.6 nm) of Thr5, Tyr7, Tyr17, Ser21 and His29 to DNA protons the evidence was presented in this paper; evidence for Gln18 to be directly involved in this interaction (with GC7) comes from the work of Ebright [12]; for the residues Arg22 and Asn25 the evidence is indirect and solely based on the model shown in Figure 9. On the other hand Leu6, Val9, Ala10, Ala13, Thr19, Val23, Leu45 and Tyr47 play important structural roles and replacement is thought to result in distortion of the tertiary structure and consequent loss of operator-binding capacity. Of these Leu6 is probably within .6 nm of C9 of the operator and may also be directly involved in the interaction with DNA.

CONCLUSIONS

The results from the 2D NOE study of the complex of lac repressor headpiece with a 14 bp lac operator fragment can be summarized as follows.
- A large part of the ¹H resonances of both protein and DNA could be assigned.
- From a comparison of chemical shifts and intra-protein and intra-DNA NOE's in the complex with those of the separate constituents it can be concluded that no major conformational changes occur upon complex formation. These are probably limited to adjustments of the amino acid side chains in the DNA binding region of the protein and slight bending and/or unwinding of the DNA.
- A set of 23 NOE's between protons of headpiece 56 and the 14 bp operator fragment has been identified (see Table II). Of these, 10 occur at uniquely assigned resonance positions for both protein and DNA. The remaining NOE's were assigned on the basis of pattern recognition procedures, but since they occur in crowded regions of the

spectrum, their assignment is not absolutely certain. In any case, all 23 NOE's are consistent with one protein-DNA complex, schematically depicted in Figure 9. Probably, in solution one major (specific) complex is present in fast exchange with a number of non-specific complexes[28] for which, however, we have not found any protein-DNA NOE's.
- The most significant and, indeed, surprising result is that the orientation of headpiece on the operator (in particular the orientation of the recognition helix in the major groove of DNA) is opposite to that of other repressors (cI,cro,434)[6,7,10] and CAP[9], and to that proposed on the basis of model building for <u>lac</u> repressor.[14,15] Whether this orientation is also present in complexes of the complete <u>lac</u> repressor remains to be investigated.

ACKNOWLEDGEMENTS

This work was supported by the Netherlands Foundation for Chemical Research (SON) with financial support from the Netherlands Organization for Advancement of Pure Research (ZWO).

REFERENCES

1. Bourgeois, S. & Pfahl, M. (1976) Adv. Protein Chem. <u>30</u>, 1-99
2. Miller, J.H. & Reznifoff, W. (1976) <u>The operon</u>, 2nd ed., Cold Spring Harbor Laboratory, Cold Spring Harbor, N.Y.
3. Caruthers, M.H. (1980) Acc. Chem. Res. <u>13</u>, 155-160
4. Miller, J.H. (1984), J. Mol. Biol. <u>180</u>, 205-212
5. Ogata, R.T. & Gilbert, W. (1979), J. Mol. Biol. <u>132</u>, 709-728
6. Pabo, C. & Sauer, R. (1984), Ann. Rev. Biochem. <u>53</u>, 293-321
7. Anderson, W., Ohlendorf, D., Takeda, Y. & Matthews, B. (1981), Nature <u>290</u>, 754-758
8. Schevitz, R.W., Otwinowski, Z., Joachimiak, A., Lawson, C.L. & Sigler, P.B. (1985) Nature <u>317</u>, 782-786
9. McKay, D. & Steitz, T. (1981), Nature <u>290</u>, 744-749
10. Anderson, J.E., Ptashne, M. & Harrison, S.C. (1985), Nature <u>316</u>, 596-601
11. Ebright, R.H., Cossart, P., Gicquel-Sanzey, B. & Beckwith, J. (1984) Nature <u>311</u>, 232-235
12. Ebright, R.H. (1985) J. Biomolec. Struct. Dyn. <u>3</u>, 281-297
13. Wharton, R.P. & Ptashne, M. (1985) Nature <u>316</u>, 601-605
14. Matthews, B.W., Ohlendorf, D.H., Anderson, W.F. & Takeda, Y. (1982) Proc. Natl. Acad. Sci. USA <u>79</u>, 1428-1432
15. Weber, I.T., McKay, D.B.M. & Steitz, T.A. (1982) Nucleic Acids Res. <u>10</u>, 5085-5102
16. Rein, R., Kieber-Emmons, T., Haydock, K., Garduno-Juarez, R. & Shibata, M. (1983) J. Biomolec. Struct. Dyn. <u>1</u>, 1051-1079
17. Ebright, R.H. (1986) in : <u>Protein Structure Folding and Design</u>, Dale Oxender Ed., Alan R. Liss N.Y., in press
18. Zuiderweg, E.R.P., Kaptein, R. & Wüthrich, K. (1983) Eur. J. Biochem. <u>137</u>, 279-292
19. Zuiderweg, E.R.P., Boelens, R. & Kaptein, R. (1985), Biopolymers <u>24</u>, 601-611
20. Zuiderweg, E.R.P., Scheek, R.M. & Kaptein, R. (1985), Biopolymers <u>24</u>, 2257-2277
21. Kaptein, R., Zuiderweg, E.R.P., Scheek, R.M., Boelens, R. & van Gunsteren, W.F. (1985) J. Mol. Biol. <u>182</u>, 179-182

22. Zuiderweg, E.R.P., Scheek, R.M., Boelens, R., van Gunsteren, W.F. & Kaptein, R. (1985) Biochimie 67, 707-715
23. Scheek, R.M., Boelens, R., Russo, N. & Kaptein, R. (1985) in : Structure and Motion : Membranes, Nucleic Acids and Proteins, E. Clementi, G. Corongiu, M.H. Sarma and R.H. Sarma Eds, Adenine Press, Guilderland, pp.485-495
24. Boelens, R., Scheek, R.M., van Boom, J.H. and Kaptein, R. (1987) J. Mol. Biol. 193, 213-216
25. van Boom, J.H., Burgers, P.H.J. and van Deursen, P.H. (1976) Tetrahedron Lett., 869-872
26. de Rooij, J.F.M, Wille-Hazeleger, G., van Deursen, P.H., Serdijn, J. and van Boom, J.H. (1979) Recl. Trav. Chim. Pays Bas 98, 537-548
27. a Sadler, J.R., Sasmor, H. and Betz, J.L. (1983) Proc. Natl. Acad. Sci. USA 80, 6785-6789
    b Simons, A., Tils, D., von Wilcken-Bergmann, B. and Müller-Hill, B. (1984) Proc. Natl. Acad. Sci. USA 81, 1624-1628
28. Scheek, R.M., Zuiderweg, E.R.P., Klappe, K.J.M, van Boom, J.H., Kaptein, R., Rüterjans, H. and Beyreuther, K. (1983) Biochemistry 22, 228-235
29. Arndt, K.T., Boschelli, F., Lu, P. & Miller, J.H. (1981), Biochemistry 20, 6109-6118
30. Geisler and Weber (1978) FEBS Lett. 87, 215-218
31. Jeener, J., Bachman, P., Meier, B.H. and Ernst, R.R. (1979) J. Chem. Phys. 71, 4546-4553
32. Macura, S. and Ernst, R.R. (1980) Mol. Phys. 41, 95-117
33. States, D.J., Haberkorn, R.A. and Ruben, D.J. (1982) J. Magn. Reson. 48, 286-292
34. Boelens, R., Scheek, R.M., Dijkstra, K. and Kaptein, R. (1985) J. Magn. Reson. 62, 378-386
35. Bax, A. and Davies, D.G. (1986) in: Advanced Magnetic Resonance Techniques in Systems of High Molecular Complexity, Nicolai, N., Valensin, G., Eds., Birkhauser, Basel
36. Marion,D. and Wüthrich, K. (1983), Biochem. Biophys. Res. Comm. 113, 967-974
37. Scheek, R.M., Russo, N., Boelens, R. & Kaptein, R. (1983) J. Am. Chem. Soc. 105, 2914-2916
38. Scheek, R.M., Boelens, R., Russo, N. and Kaptein, R. (1984) Biochemistry 23, 1371-1376
39. Hare, D.R., Wemmer, D.E., Chou, S.H, Drobny, G.E. & Reid, B.R. (1983) J. Mol. Biol. 171, 319-336
40. Buck, F., Hahn, K.-D., Brill, W., Rüterjans, H., Chernov, B.K., Skryabin, K.G., Kirpichnikov, M.P. and Bayev, A.A. (1986) J. Biomolec. Struct. Dyn. 3, 899-911
41. Lu, P., Cheung, S. and Arndt, K. (1983), J. Biomolec. Struct. Dyn. 1, 509-521
42. Kalk, A. and Berendsen, H.J.C. (1976) J. Magn. Reson. 24, 343-366
43. Kumar, A., Wagner, G., Ernst, R.R. and Wüthrich, K. (1981) J. Am. Chem. Soc. 103, 3654-3658
44. Billeter, M., Braun, W. and Wüthrich, K. (1982) J. Mol. Biol. 155, 321-346
45. Wagner, G. and Wüthrich, K. (1982), J. Mol. Biol. 155, 347-366
46. Wider, G. and Wüthrich, K. (1982), J. Mol. Biol. 155, 367-388
47. Arnott, S. and Hukins, D.W.L (1972), Biochem. Biophys. Res. Comm. 47, 1504-1509

48. Gilbert, W. and Maxam, A., cited by Barkley, M.D. & Bourgeois, S. (1978) in : <u>The Operon</u>, 2nd ed., J.H. Miller & W.S. Reznikoff Eds., Cold Spring Harbor Laboratory, Cold Spring Harbor, N.Y.
49. Richmond, T. and Richards, F.M. (1978), J. Mol. Biol. <u>119</u>, 537-555
50. Buck, F., Rüterjans, H. Kaptein, R. and Beyreuther, K. (1980) Proc. Natl. Acad. Sci. USA <u>77</u>, 5145-5148

DNA METHYLATION AND MISMATCH REPAIR:

MOLECULAR SPECIFICITIES

Miroslav Radman

Institut Jacques Monod, C.N.R.S.
Université Paris 7, Tour 43
75251 - Paris Cedex 05, France

INTRODUCTION

A DNA base pair mismatch can be defined as any non-complementary base pair in the DNA duplex. Such DNA duplex carrying mismatched, or unpaired, bases is called DNA heteroduplex. A DNA heteroduplex is characterized by non-identical genetic information of the two complementary strands, such that upon DNA replication a single heteroduplex molecule segregates genetically, i.e. yields mixed progeny consisting of DNA duplexes with the sequence of individual heteroduplex strands. The mismatch repair process transforms a heteroduplex molecule into a homoduplex which yields pure progeny of a single genotype.

Thus, mismatches consist of chemically normal, but mispaired or unpaired bases which can arise _in vivo_ by three different mechanisms: (i) error in DNA replication, (ii) genetic recombination by strand exchange between homologous but non-identical DNA sequences and (iii) deamination of 5-methyl-cytosine to thymine generating a G:T mismatch _in situ_. All three mismatch-generating events are rare and unpredictable, such that it is impossible to isolate mismatched DNA intermediates _in vivo_ in order to study their repair. We have attempted in our laboratory to mimic the three _in vivo_ situations by suitable _in vitro_ reconstituted DNA heteroduplexes of bacteriophage lambda and Φx174.

The availability of several hundreds of sequence mutants in the cI repressor gene of bacteriophage lambda, allowed us to construct the desired repertoire of defined mismatches and unpaired bases to test the molecular specificity of the E.coli mismatch repair system.

Below is summarized published, or in press, work from our laboratory.

MATERIALS AND METHODS

All the methods used in the preparation of heteroduplex DNA (following strand separation), transfection and the genetic analysis of the phage progeny have been described previously (Dohet et al., 1985; 1986; Radman et al., 1985). It suffices to reiterate that the

DNA heteroduplex transfection experiments are such that only one molecule enters the transfected cells which are plated immediately on agar Petri dishes and incubated to produce infective centers. The genotype of phages inside each infective center are determined by streaking on individual plates. Low fraction of mixed (mutant/wild type, c/c+) infective centers reflects efficient mismatch repair.

Since (i) each defined mispaired or unpaired base is associated with a phenotype (c or c+) and (ii) heteroduplex DNA species are pure, due to annealing of separated strands, every preference for the repair of a particular mismatched base or of a DNA strand, can be detected in these experiments.

Sequenced mutants in the cI gene of bacteriophage lambda and the E.coli strains have been described (Dohet et al., 1985; 1986; 1987; Jones et al., 1987a, 1987b).

METHYL-DIRECTED MISMATCH REPAIR

Mechanism of error correction in DNA replication

Methyl-directed mismatch repair in E.coli appears to remove mispaired and unpaired bases from the newly synthesized DNA strands, thus correcting replication errors (review, Radman and Wagner, 1986). The discrimination between the parental and the newly synthesized strand is based on the transient undermethylation of GATC sequences in nascent strands (Radman et al., 1980; Pukkila et al., 1983). Full methylation of the GATC sequences inhibits mismatch repair in heteroduplex DNA (Pukkila et al., 1983; Lu et al., 1983; Wagner et al., 1984). The methylation on the N6 position of the adenine residue in the GATC sequences is carried out by the product on the dam gene (Marinus and Morris, 1975; Geier and Modrich, 1979) with a short delay relative to DNA synthesis (Lyons and Schendel, 1984).

Methyl-directed mismatch repair requires the mutH, mutL, mutS and mutU gene products (see Radman and Wagner, 1986). In addition, the mismatch repair in crude cell extracts requires ATP and the single-strand DNA binding (SSB) protein (Lu et al., 1984). The products of these mut genes have been identified (Lu et al., 1984; Pang et al., 1985) but the molecular mechanisms of mismatch repair remains largely obscure. The mutU (uvrE, uvrD) gene product has been identified as the DNA helicase II (Kumara and Sekiguchi, 1984). Helicase II translocates unidirectionally in the 3' to 5' direction along single-stranded DNA and the strand displacement is stimulated by the SSB protein (Matson, 1986; Matson and George, 1987). The purified MutS protein was shown to bind to DNA regions containing a single base pair mismatch (Su and Modrich, 1986). The purified MutH protein was shown to nick (very inefficiently) the unmethylated duplex DNA 5' to the GATC sequence (Welsh et al., 1986).

In vivo experiments with heteroduplex DNA of phage Φx174 containing 0, 1 or 2 GATC sequences gave a clue to the role of GATC sequences in mismatch repair (Längle-Rouault et al., 1986; 1987). "2 GATC" heteroduplexes were efficiently mismatch repaired in E.coli, whereas there was very little, if any, mismatch repair observed with "1 GATC" and "0 GATC" heteroduplexes. Methylation of "2 GATC" heteroduplexes abolished mismatch repair on the methylated strand. Thus, unmethylated GATC sequences are required for mismatch repair and they must be present in the substrate DNA. When GATC sequences are not present in Φx174

218

DNA, then a single persistent nick in the heteroduplex (transfection of a ligase-ts mutant at 40°C for 1 hour) can complement both for GATC sequences and for the MutH function (but not for other Mut-functions). This result led us to propose that the mechanism of mismatch repair in E.coli is the following: (i) MutS and MutL proteins recognize the mismatch and somehow (ii) allow MutU (helicase II) to enter the DNA at, or close to, the mismatch site, melts the DNA in both directions until the two flanking unmethylated GATC sequences are reached. The MutH protein cuts unmethylated GATC sequences, perhaps like a single-strand specific restriction endonuclease, leading to (iii) a large gap which is filled-in by polymerase and (iv) sealed by ligase (see figure 1).

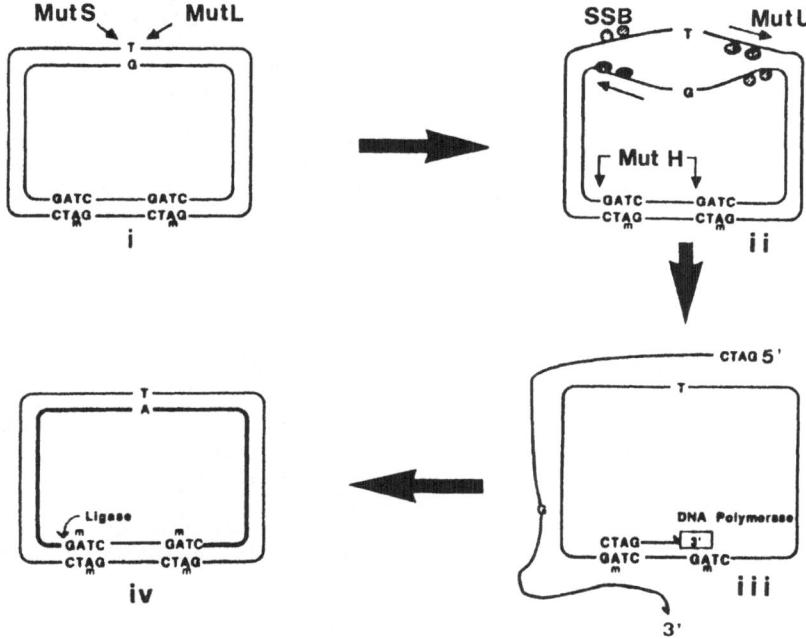

Figure 1: Molecular model for methyl-directed mismatch repair in Φx174 phage heteroduplex DNA as proposed by Längle-Rouault et al., (1987). See text for explanations.

Molecular specificity of methyl-directed mismatch repair

The E.coli mismatch repair system does not recognize and repair all mismatches with equal efficiency. The extent of repair depends on both the mismatch and the neighboring nucleotide sequence. In general, it appears that transition mismatches (G:T and A:C) are better repaired than transversion mismatches (G:A, C:T, A:A, G:G, T:T and C:C (Dohet et al., 1985; Kramer et al., 1984; Jones et al., 1987a) and, for a given mismatch, repair efficiency increases with increasing G:C content in the neighboring nucleotide sequence (Jones et al., 1987a). This correlation between G:C content and mismatch repair efficiency is strongest for transversion mismatches (G:A and C:T) and was observed for a region including four base pairs on either side of the mismatch. The correlation is less evident when the region considered is smaller than four base pairs on either side of the mismatch, but is still apparent when the region is as large as ten base pairs on either side of the mismatch.

The E.coli mismatch repair system can recognize and repair
frameshift : wild type heteroduplexes (Dohet et al., 1986). These
heteroduplexes do not contain a mismatch, but rather have one strand
in which there is an extra, and therefore unpaired, base. When such
heteroduplexes are unmethylated, mismatch repair works equally well
on either strand and the nature of the unpaired base does not influence
the extent of repair. The repair of frameshift : wild type heteroduplexes
is methyl directed. The E.coli mismatch repair system is able to correct
both addition and deletion frameshift mutations arising during the
course of replication.

Although heteroduplexes with a single unpaired base are
substrates for the mismatch repair enzymes, heteroduplexes with a large
single stranded loop are not (Kramer et al., 1982; Wagner et al., 1984).
However, there does not appear to be some activity in mismatch repair
deficient cells (mutH, mutL, mutS and mutU) which can act to repair
such looped structures even in fully GATC methylated DNA (Dohet et
al., 1987). Heteroduplex regions with such large non-homologies are
not likely to occur frequently as a result of replication errors, but
have been shown to occur in the course of bacteriophage lambda
recombination (Lichten and Fox, 1984). The repair of such heteroduplexes
during recombination may allow the insertion or deletion of large
non-homologies and thus contribute to the formation of recombinant
genotypes even under conditions where mismatch repair does not occur
(e.g., in fully GATC methylated DNA).

It appears that the E.coli mismatch repair system can recognize
and act upon some chemically induced lesions in DNA. Dam mutants are
more sensitive than wild type cells to base analogs (2-aminopurine
and 5-bromouracil) and small alkylating agents (Glickman and Radman,
1980; Glickman et al., 1978; Jones and Wagner, 1981; Karran and Marinus,
1982). These sensitivities depend on the operation of the mismatch
repair system, i.e. they can be alleviated by the addition of a mutH,
mutL, mutS or mutU mutation. In addition, mismatch repair deficient
cells are hypermutable by 5-bromouracil, non-covalent intercalating
agents (9-aminoacridine and ellipticines) and methylating agents
(Rydberg, 1978; Shanabruch et al., 1983; Skopek and Hutchinson, 1984
and B.René and C.Paoletti, pers.commun.).

The finding that increasing the number of G:C base pairs close
to a mismatch increases the extent of repair of that mismatch suggests
that mismatch repair operates most efficiently in regions of stable
double helix (Jones et al., 1987a). It may be that the common feature
of repairable heteroduplexes with mismatches or unpaired bases is the
ability of all bases in the heteroduplex to exist primarily in an
intrahelical configuration. Model building and NMR studies have shown
that single unpaired bases, as in frameshift: wild type heteroduplexes,
can assume an intrahelical configuration (Pardi et al., 1982). Further,
the results of an NMR study of synthetic oligonucleotide heteroduplexes
revealed that a poorly repaired mismatch can assume an extrahelical
configuration (Fazakerley et al., 1986). The sequences of the
oligonucleotides used in this study were taken from the phage lambda
heteroduplexes used in the study of Jones et al. 1987 and consisted
of the mismatch and five bases on either side on it. Three better repaired mismatches were also examined (two G:T and one G:A) and all were
found only in a helical form (Fazakerley et al., 1986).

The suggestion that the intrahelical mismatches may be
preferentially repaired was made earlier (Radman et al., 1985; Werntges
et al., 1986). Recent work on crystal structures of mismatch-containing
oligonucleotides (Hunter et al., 1986) indicates that the changes in
the glycosidic bond angles, in the base stacking and in the exposure

of functional groups into the major and minor grooves, could all contribute to the recognition of mismatches by the mismatch repair enzymes; such structural signals would be lost upon opening-up of the double-helix in the case of unrepaired mismatches (Fazakerley et al., 1986).

REPLICATION ERRORS AND MISMATCH REPAIR

The potent mutator effect of mutH, mutL, mutS and mutU mutants indicates that over 99% of all replication errors are repaired by the mismatch repair system in E.coli (Glickman and Radman, 1980). In other words, nonrepairable mismatches (e.g. transversion mismatches in A:T rich DNA) and replication errors not recognized by the mismatch repair system (e.g., large single stranded loops) comprise less than one percent of all replication errors. It may be that the polymerase simply does not attempt to incorporate incorrect bases in regions of DNA where mismatch repair functions poorly, i.e., A:T rich regions. However, it appears that the proofreading exonuclease activity associated with the T4 phage polymerase is most active in A:T rich regions (Petruska and Goodman, 1985). Thus the findings that poorly or unrepaired mismatches can exist in extrahelical configurations and that T4 phage DNA polymerase makes fewer mistakes in A:T rich regions than in G:C rich regions suggests that poorly repaired mismatches formed in the course of DNA replication (e.g., transversions in A:T rich regions) may disrupt base pairing between the newly synthesized and template strands such that the incorrect bases are most often removed by the single strand specific "proofreading" exonuclease associated with the DNA polymerase. It may be that the net effect of the differences in specificity between mismatch repair and polymerase proofreading is that all substitution mutations arise with similar frequencies throughout the E.coli genome. Support for these ideas comes from experiments examining the spontaneous reversion rates of a series of trpA mutations in wild type and mut E.coli (Choy and Fowler, 1985). In the wild type background, all sites revert at a similar rate (10^{-10}-10^{-9} mutants/base pair replicated). However, in a mutL or mutS background there is a wide range of reversion rates, from 10^{-6} for a site requiring a G:C to A:T transition, i.e., involving repairable mismatches, to 4 x 10^{-10} for a site requiring an A:T to C:G transversion, i.e., involving nonrepairable mismatches.

REPAIR OF DEAMINATED 5-METHYL-CYTOSINE

A series of transfection experiments by Jones et al., (1987b) provided evidence for the existence of very short patch repair proposed by Lieb (Lieb, 1983; Lieb, 1985). Using a series of heteroduplexes with individual mismatches in and surrounding a CCAGG sequence in the cI gene, they found that mismatch repair occured in fully GATC methylated DNA only when the mismatch was at the second position (from either end) of the sequence. No repair was observed when the mismatch was at the first base pair outside the CCAGG sequence or at the center base pair of the sequence. The repair acts on the G:T mismatch and not on the A:C mismatch and acts only to replace the thymine residue and thus restore the G:C base pair. A second site where repair of the thymine of a G:T mismatch gave a mutant CCAGG sequence revealed that the process does not repair only to wild type, but rather acts to restore the CCAGG sequence. Repair events did not extend ten base pairs to either side, but did frequently include a mismatch two base pairs away on the 5' side of the excised thymine.

Short patch repair also operates in GATC unmethylated or hemimethylated DNA, but the effects are less noticeable due to the operation of methyl directed mismatch repair. As is the case for methyl directed repair, short patch repair requires intact mutL and mutS genes. However, short patch repair is fully operational in mutH or mutU bacteria and requires a functional dcm gene, which codes for the E.coli cytosine methylase. Because transfections using DNA from phages grown in dcm bacteria were performed in dcm⁺ cells in order to see short patch repair, the results do not allow the conclusion that cytosine methylation is not required for short patch repair. However, the finding of Lieb (submitted) that mutants at the second cytosine of CCAG sequences are subject to short patch repair suggests that it is not. The requirement of the short patch repair system for functional mutL and mutS genes explains the hyporecombinogenic effect of these mutations on Pam3 x Pam80 crosses (Glickman and Radman, 1980). The fact that mutH and mutU have a slight hyperrecombinogenic effect on these same crosses suggests that it may have been uniform corepair rather than the absence of repair that allowed linear mapping of markers in the studies discussed above.

The specific action of the E.coli short patch repair system on cytosine methylation sequences suggests that it might function on the E.coli chromosome to reduce mutations arising from deamination of 5meC to form thymine. However, it has been reported that all three 5meC sites in the lacI gene are hotspots for 5meC to T transitions (Coulondre et al., 1978). It may be that these hot spots would be even "hotter" in the absence of mismatch repair. However, the lacI studies were conducted with F-lac episomes and it may be that the number of mismatch repair enzymes is sufficiently small that non-chromosomal DNA is not efficiently repaired. (An estimate from lacZ gene fusion experiments is that there are only 10-20 mismatch repair enzyme molecules per cell; A.Brandenburger, J.D. Franssen and M. Radman, unpublished). The idea is supported by the findings that the mutator effects of mutH and mutS are weaker on the F-lac episome than on the E.coli chromosome (Leong et al., 1986) and that sequencing of spontaneous chromosomal lacI mutants revealed no obvious 5meC mutational hot spots (Glickman et al., 1986).

MISMATCH REPAIR IN EUKARYOTES : SPECULATIONS

The similarities in mismatch repair specificities in E.coli (see above), S.pneumoniae (Claverys et al., 1983; Lacks et al., 1982) and S.cerevisiae (White et al., 1985) suggest that there may be similarities in mismatch repair processes among diverse organisms, including both prokaryotes and eukaryotes. Most organisms do not use GATC methylation to direct mismatch repair as GATC sequences are methylated only in a limited number of bacterial species (Barbeyron et al., 1984). It is unlikely that any methylation is involved in directing mismatch repair in eukaryotes, since the chromosomal DNA of yeast and Drosophila contains no detectable methylation and mammalian DNA contains only 5meC, which appears to be involved in gene regulation (Hattman, 1981; Proffitt et al., 1984). It has been suggested that mismatch repair in eukaryotes acts at nicks in the DNA and is therefore directed to the newly synthesized strands by the transient nicks in them (Hare and Taylor, 1985). Such a directed repair may occur in eukaryotes, although to date the experiments purporting to demonstrate it lack sufficient controls to eliminate other interpretations (Folger et al., 1985; Hare and Taylor, 1985).

The presence of considerable amounts of 5meC in mammalian cells (Hattman, 1981) and the high rate of spontaneous deamination of 5meC to form thymine (Lindahl, 1982) suggest the need for a repair system analogous to the short patch repair system in E.coli. There is currently no evidence for such a repair system in eukaryotes. It has been suggested that a short patch repair system, similar to that of E.coli, but perhaps less specific, could promote the diversification of genes by acting on heteroduplexes formed from different members of such repetitive gene families as the histocompatibility genes (Kourilsky, 1983) or the variable region of the immunoglobulin genes (Radman, 1983). Directed repair of such heteroduplexes would tend to homogenize divergent sequences. The results of transfections of mouse cells with complex heteroduplexes containing many mismatches and small unpaired regions indicate that cellular processing of such heteroduplexes can produce "patchwork" sequences (Abastado et al., 1984).

The best candidate for an eukaryotic mechanism for repairing replication errors may be recombination repair via double strand breaks as it requires no means of strand discrimination. However, the model does require that an additional copy of the genetic information be readily available for recombination. This requirement could be fullfilled in higher organisms by restricting the process of the chromatin-free region of the DNA immediately behind the replication fork. Recombination events in this region which terminate with a crossover could be delected as sister chromatid exchanges. Recombinational mismatch repair could explain the occurence of much higher levels of sister chromatid exchange than are predicted from the frequency of mitotic recombination. The stimulation of sister chromatid exchanges by double strand breaks produced _in vivo_ (Natarajan et al., 1985) and the occurence of homologous recombination as a result of double strand gap repair (Brenner et al., 1986) have recently been demonstrated in cultured mammalian cells.

REFERENCES

Abastado, J.P., Cami, B., Dinh, T.H., Igolen, J., and Kourilsky, P., 1984, Proc. Natl. Acad. Sci. USA, 81:5792-5796.
Barbeyron, T., Kean, K., and Forterre, P., 1984, J. Bact.,160:586-590.
Brenner, D.A., Smigocki, A.C., and Camerini-Otero, D.R., 1986, Proc. Natl. Acad. Sci. USA, 83:1762-1766.
Choy, H.E., and Fowler, R.G., 1985, Mut. Res., 142:93-97.
Claverys, J.P., Mejean, V., Gasc, A.M., and Sicard, A.M., 1983, Proc. Natl. Acad. Sci. USA, 80:5956-5960.
Coulondre, C., Miller, J.H., Farabaugh, P.J., and Gilbert, W., 1978, Nature 274:775-780.
Dohet, C., Wagner, R., and Radman, M., 1985, Proc. Natl. Acad. Sci. USA 82:503-505.
Dohet, C., Wagner, R., and Radman, M., 1986, Proc. Natl. Acad. Sci. USA, 83:3395-3397.
Dohet, C., Dzidic, S., Wagner, R., and Radman, M., 1987, Mol. Gen. Genet. 206:181-184.
Fazakerley, G.V., Quignard, E., Woisard, A., Guschlbauer, W., van der Marel, G.A., van Boom, J.H., Jones, M., and Radman, M., 1986, EMBO J., 5:3697-3703.
Folger, K.R., Thomas, K., and Capecchi, M.R., 1985, Mol. Cell. Biol. 5:70-74.
Geier, G.E., and Modrich, P.J., 1979, J. Biol. Chem., 254:1408-1413.
Glickman, B.W., Van der Elsen, P., and Radman, M., 1978, Mol. Gen. Genet. 163:307-312.
Glickman, B.W., and Radman, M., 1980, Proc. Natl. Acad. Sci. USA, 77: 1063-1067.
Glickman, B.W., Fix, D.E., Yatagai, F., Burns, P.A., and Schaaper, R.M., 1966, in : Mechanisms of DNA damage and repair. Simic, M., Grossman, L., and Upton, A., Eds. Plenum Press, N.Y. 425-438.
Hare, J.T., and Taylor, H., 1985, Proc. Natl. Acad. Sci. USA, 82: 7350-7354.
Hattman, S., 1981, The Enzymes, 14:517-548.

Hunter, W.N., Brownn, T., Anand, N.N., and Kennard, O., 1986, Nature,
320:552-555.
Jones, M., and Wagner, R., 1981, Mol. Gen. Genet., 184:562-563.
Jones, M., Wagner, R., and Radman, M., 1987a, Genetics, in press.
Jones, M., Wagner, R., and Radman, M., 1987b, J. Mol. Biol., in press.
Karran, P., and Marinus, M.G., 1982, Nature, 296:868-869.
Kourilsky, P., 1983, Biochimie, 65:85-93.
Kramer, W., Schughart, K., and Fritz, H.J., 1982, Nucl. Acids. Res.
10:6475-6485.
Kramer, B., Kramer, W., and Fritz, H.J., 1984, Cell, 38:879-881.
Kumura, K., and Sekiguchi, M., 1984, J. Biol. Chem., 259:1560-1565.
Lacks, S.A., Dunn, J.J., and Greenberg, B., 1982, Cell, 31:327-336.
Längle-Rouault, F., Maenhaut-Michel, G., and Radman, M., 1986, EMBO J.,
5:2009-2013.
Längle-Rouault, F., Maenhaut-Michel, G., and Radman, M., 1987, EMBO J.,
submitted.
Leong, P.M., Hsia, H.C., and Miller, J.H., 1986, J. Bact. 168:412-416.
Lichten, M.J., and Fox, M.S., 1984, Proc. Natl. Acad. Sci. USA, 81:
7180-7184.
Lieb, M., 1983, Mol. Gen. Genet., 191:118-125.
Lieb, M., 1985, Mol. Gen. Genet., 199:465-470.
Lindahl, T., 1982, Annu. Rev. Biochem., 51:61-87.
Lu, A.L., Clark, S., and Modrich, P., 1983, Proc. Natl. Acad. Sci. USA,
80:4639-4643.
Lu, A.L., Welsh, K., Su, S.S. and Modrich, P., 1984, Cold Spring Harbor
Symp. Quant. Biol., 49:589-596.
Lyons, S.M., and Schendel, P.F., 1984, J. Bact., 159:421-423.
Marinus, M.G., and Morris, N.R., 1975, Mutat. Res., 28:15-26.
Matson, S.W., 1986, J. Biol. Chem., 261:10169-10175.
Matson, S.W., and George, J.W., 1987, J. Biol. Chem., accepted.
Natarajan, A.T., Mullenders, L.H.F., Meijers, M., and Mukherjee, U.,
1985, Mut. Res. 144:33-39.
Pang, P.P., Lundberg, A.S., and Walker, G.C., 1985, J. Bact., 163:1007-1015.
Pardi, A., Morden, K.M., Patel, D.J., and Tinoco, I. Jr., 1982,
Biochemistry, 21:6567-6574.
Petruska, J., and Goodman, M.F., 1986, J. Biol. Chem., 260:7533-7539.
Proffitt, J.H., Davie, J.R., Swinton, D., and Hattman, S., 1984,
Mol. Cell. Biol., 4:985-988.
Pukkila, P.J., Peterson, J., Herman, G., Modrich, P., and Meselson, M.,
1983, Genetics, 104:571-582.
Radman, M., Wagner, R.E., Glickman, B.W., and Meselson, M., 1980, in :
Progress in Environmental Mutagenesis, Alacevic, M. ed., Elsevier,
Amsterdam, 121-130.
Radman, M., 1983, UCLA Symp. Mol. Cell. Biol., 11:287-298.
Radman, M., Dohet, C., Jones, M., Doutriaux, M.P., Längle-Rouault, F.,
Maenhaut-Michel, G., and Wagner, R., 1985, Biochimie, 66:745-752.
Radman, M., and Wagner, R., 1986, Annu. Rev. Genet., 20:523-538.
Rydberg, B., 1978, Mutat. Res., 52:11-24.
Shanabruch, W.G., Rein, R.P., Behlau, I., and Walker, G.C., 1983,
J. Bact. 153:33.
Skopek, T.R., and Hutchinson, F., 1984, Mol. Gen. Genet., 195:418-423.
Su, S.S., and Modrich, P., 1986, Proc. Natl. Acad. Sci. USA 83:5057-5061.
Wagner, R., Dohet, C., Jones, M., Doutriaux, M.P., Hutchinson, F. and
Radman, M., 1984, Cold Spring Harbor Symp. Quant. Biol., 49:611-615.
Welsh, K.M., Lu, A.L., and Modrich, P., 1986, Fed. Proc., 45: abst.1758.
Werntges, H., Steger, G., Riesner, D., and Fritz, H.F., 1986, Nucleic
Acids Res., 14:3773-3790.
White, J.H., Lusnak, K., and Foger, S., 1985, Nature, 315:350-352.

RECOGNITION OF DNA SEQUENCES BY RESTRICTION ENDONUCLEASES

Guenter Maass

Biophysikalische Chemie
Med. Hochschule Hannover
D-3000 Hannover 61

Introduction

Type II restriction endonucleases recognize DNA sequences 4 or 6 basepairs long with very high specificity in a vast excess of non-specific DNA binding sites. They cleave the DNA within or in the immediate vicinity of the recognition sequence. Because of the immense importance of these enzymes in analysis, preparation and in vitro recombination of DNA many investigations have been carried out to understand the structural and mechanistic features underlying their mode of recognition[1]. Among approximately 500 type II restriction endonucleases described, EcoRI is the so far best studied enzyme. EcoRI is a dimeric protein with two identical sub-units, the aminoacid sequence of which is known[2,3]. It recognizes the palindromic sequence

$$-G \!\!\downarrow\!\! A\ A\ T\ T\ C-$$
$$-C\ T\ T\ A\ A \!\!\uparrow\!\! G-$$

and cleaves the DNA double strand at the marked positions. For its enzymatic activity it needs Mg^{2+} as a co-factor. It was the first DNA-binding protein that was co-crystallized with its substrate TCGCGAATTCGCG by Frederick et al.[4] in the absence of Mg^{2+}. The 3Å-X-ray analysis revealed a complementary interface between the enzyme and the major groove of the DNA. The binding of the enzyme introduces distortions of the DNA, so called neo-kinks, and thereby widens the major groove. In a more recent though preliminary presentation, a more detailed interpretation of the 3Å-structure was given by J. Rosenberg[5], who suggested that EcoRI interacts with its canonical sequence via 12 hydrogen bonds between glutamic acid and arginine residues and the guanine and adenine residues of GAATTC. However, the data available at present are not yet sufficient to understand the molecular basis of the EcoRI action. A more detailed knowledge of the thermodynamic, kinetic and structural aspects of the EcoRI-DNA interaction is needed. This contribution intends to present physico-chemical studies on the binding and catalytic properties of EcoRI and to discuss structural aspects of the interaction.

Thermodynamics of EcoRI-DNA recognition

A very interesting aspect of the enzymology of restriction endonucleases concerns their high specificity. According to Halford[6] the reactivity of EcoRI towards its recognition sequence is by at least seven orders of magnitude higher than to any other hexanucleotide sequence. This value represents a lower limit, since small impurities of non-specific nucleases present in the enzyme preparation cannot be differentiated from a non-specific cleavage activity of the enzyme itself.

The enzymatic reaction has to occur in at least two steps:
- Binding to the recognition sequences
- Cleavage of the specific complex

Both steps can contribute to the specificity of the reaction. From detailed studies of operator-repressor systems[7,8] it is known that the repressor discriminates its operator site from non-specific sites by large differences in the free enthalpy of binding: non-specific binding arises preferentially from charge-charge interactions between phosphates and positively charged aminoacid residues, and specific binding is the consequence of a complementary pattern of hydrogen bond donors and acceptors on DNA and protein.

A similar motif has been expected to explain the high specificity of the DNA sequence recognition by restriction nucleases. Numerous studies[1] were performed to verify this assumption. Common to all these studies is the finding that EcoRI shows a strong non-specific binding to DNA[10,11]. The complex is characterized by 8 ionic interactions[9]. At least two of these ionic contacts are located within the canonical sequence. This was deduced from binding studies on oligonucleotides[19] and from studies in which the interference of ethylation of phosphates with specific binding was investigated[1]. These ion pairs do not contribute to the sequence specificity of the enzyme[8], they are, however, required to position the respective recognition elements on the protein and the major groove of the DNA.

Using nitrocellulose filter binding assays[8,12,13] it has been shown that in the absence of Mg^{2+}-ions binding to the canonical sequence is 4 - 5 orders of magnitude more stable than binding to non-specific DNA sequences. These data are at variance with observations from our own laboratory obtained on specific octanucleotides using CD-titrations which show no significant increase in the binding affinity due to specific interactions[11]. It has to be pointed out, however, that these results were obtained on oligonucleotides with the terminal 5'-phosphate missing. In recent nitrocellulose filter binding studies on the EcoRI-tridecamer d(TCGCGAATTCGCG) interaction we found a binding affinity of $4 \cdot 10^8$ M^{-1}. This value is still lower than the one determined by Jen-Jacobsen et al.[8] for the same complex under comparable conditions. Comparing binding constants obtained for an octamer ($5 \cdot 10^6$ M^{-1}) and a decamer ($1 \cdot 10^7$ M^{-1}) each containing the canonical sequence, it becomes evident that basepairs and phosphate groups adjacent to the hexameric recognition sequence contribute to the binding affinity. Competition cleavage experiments in the presence of Mg^{2+} show a specific binding of d(GGAATTCC), which is only two orders of magnitude stronger than non-specific binding[14].

The discrepancies between these two sets of data cannot be reconciled, unless one assumes that in the presence of Mg^{2+} the specific binding affinity differs from that determined in its absence. Another explanation may result from an inherent problem of the nitrocellulose filter binding-technique: A large number of binding sites with low binding affinity show qualitatively similar behaviour as one binding site with high affinity. Different from "classical" titrations, which determine a quantity directly proportional to the number of occu-

227

pied binding sites, in the nitrocellulose filter assay a DNA molecule is already retained on the filter, if only one of its binding sites is occupied by a protein[15].

Irrespective of the different affinities found for the specific interaction, even the highest value reported for the difference in binding of the EcoRI enzyme to its recognition site and non-specific alternative sequences is not sufficient to account for the minimum selectivity value of 10^7 estimated by Halford[6]. Consequently, a substantial amount of the specificity has to be contributed by the catalytic step, which can result from the complementary interface of EcoRI-enzyme and the DNA in the activated complex with the hydrogen bonds being in an optimal arrangement.

We have studied the accuracy of the EcoRI catalyzed cleavage of DNA under normal buffer conditions (20 mM Tris HCl, pH 7.2, 50 mM NaCl, 10 mM $MgCl_2$, 37°C) using the plasmid pUC8 as substrate which has one canonical site[16]. Under the buffer conditions applied at high enzyme concentrations and after long incubation times pUC8 DNA is not only cleaved at its canonical site, but with rates by a factor of 3000 and 9000 resp. lower also at the non-canonical sites -TAATTC- and -GAGTTC-. Two other non-canonical sites, -GAATAC- and -CAATTC-, are also cleaved, but considerably more slowly. The results show, that erroneous cleavage is only detectable at sites differing in one basepair from the canonical site. The digestion pattern obtained under normal buffer conditions is identical to that produced under EcoRI* conditions. The kinetics, however, are different. Cleavage at non-canonical sites is faster in the presence of $MnCl_2$, at low ionic strength (20 mM Tris, 1 mM $MgCl_2$) and pH 8.8, and in the presence of 40 % glycerol than at normal buffer conditions. From these results it is concluded that the mechanisms of recognition are essentially the same for EcoRI and EcoRI* activity, the difference being that under EcoRI* conditions the rate of cleavage at the canonical site is slowed down, while it is enhanced at non-canonical sites. Under all conditions the hydrolysis rates follow a hierarchical order: G >> T > C at position 1, A >> T at position 2, and A >> G at position 3. These results are in agreement with the observation of Goodman et al.[17] for the hierarchy at position 1 and the general rules derived for EcoRI* activity by Rosenberg and Greene [18].

Kinetics of the EcoRI-DNA interaction

From the above discussion it becomes quite obvious that information on the kinetic behaviour of the EcoRI-restriction endonuclease-DNA interaction is essential for its understanding. Most of the kinetic information available on this system has been derived from classical Michaelis-Menten type kinetics. For the cleavage of pBR322 under steady state conditions a K_M of 5 nM and a turnover number of 1.8 min^{-1} was obtained[14] (0.08 NaCl, 0.02 $MgCl_2$, pH 7.2, 37°C). Michaelis-Menten conditions, however, are not appropriate to measure the catalytic rate constant because of the considerable amount of non-specific binding. In pBR322 e.g., the nonspecific binding sites are in 700-fold excess over the specific site. Using the observed difference in binding between specific and non-specific sites of about two orders of magnitude and assuming a non-specific binding constant of 10^5 M^{-1} a 10 - 15 fold excess of enzyme over DNA is needed for an equivalent saturation of specific and non-specific sites.

Single turnover experiments using a quenched-flow apparatus with enzyme in excess over DNA result in catalytic constants for cleavage of the first and second phosphodiester bond of 0.35 sec^{-1} at 21°C. Similar results have been obtained by Jen-Jacobsen on the same enzyme and for the cleavage of pMB9 (k_{cat}=0.6 s^{-1})[19]. The absolute value of 0.35 s^{-1} appears to be rather low as compared to non-specific exonucleases. This may be taken as an indication that in case of EcoRI the catalytic reaction involves more steps than only the hydrolytic phosphodiester bond cleavage. The complexity of the cleavage mechanism has been extensively discussed by Halford[41].

The cleavage rate significantly depends on sequences adjacent to the recognition site. We have shown that A - T base pairs enhance and G - C base pairs slow down the hydrolytic activity of the EcoRI endonuclease[20]. The differences can amount to one order of magnitude. Similar effects have been

obtained for the isoschizomers Hae III, Bsp RI and BsuRI
which recognize the sequence $\begin{smallmatrix}-G & G\!\downarrow\!C & C-\\ -C & C\!\uparrow\!G & G-\end{smallmatrix}$ and cleave it at the
indicated position[21]. The preference is qualitatively the
same for all three enzymes: A G G C C T > T G G C C A >
G G G C C C ≈ C G G C C G.

Beyond the sequences immediately adjacent to the recognition
site sequences further away are of importance for the location of the target sequence. Non-specific binding presents a
basis for the protein to facilitate its search for the specific site by linear diffusion[7,8,22]. Since different
approaches to detect linear diffusion in the case of EcoRI
were at variance[23,24,25] we have re-investigated this problem
using a simple and direct approach[26]. Our results show that
depending on the reaction conditions EcoRI as well as Hind
III and Bam HI cleave long substrates faster than short ones.
For EcoRI the mean diffusion length is approximately 1000
base pairs at 1 mM $MgCl_2$. With a dissociation rate of $10 \ s^{-1}$
this corresponds to a linear diffusion coefficient of
$5 \cdot 10^{-14} \ m^2 s^{-1}$. At 10 mM $MgCl_2$, linear diffusion of EcoRI over
chainlengths beyond 100 base pairs could not be detected[24].
This may be due to a faster dissociation of the EcoRI from
non-specific DNA sequences. This assumption would also recon-
cile the differences found in the specific binding in the
absence[1] and presence of Mg^{2+}. The dependence of linear
diffusion on Mg^{2+}-concentration explains why EcoRI at 20 mM
$MgCl_2$ does not cleave plasmid DNA with two sites, 96 and 318
base pairs apart in a processive manner[24]. In the presence of
nonsaturating amounts of the prokaryotic histone-like protein
HU (NS2) small and large DNA substrates are cleaved with
identical rates by EcoRI indicating that proteins bound to
DNA constitute a barrier against linear diffusion. This finding argues against an in vivo role of linear diffusion,
however, recently it could be shown by electron-microscope
studies, that in the E. coli cell EcoRI is preferentially
localized in the periplasmatic space[27]. Here, due to the
absence of DNA-binding proteins, EcoRI very well may use
linear diffusion to find its target.

Structural aspects of the recognition process

As shown above using non-specific interactions linear diffusion is a means to guide the enzyme to its recognition site. The selection of the recognition site is based on the formation of complementary interactions between corresponding groups on the protein and the nucleic acid[28,29] as well as on kinetic principles, i. e. the individual rates of the various steps involved in the reaction pathway. In the EcoRI-system the major contribution to specific complex formation arises from 12 hydrogen bonds which have been suggested on the basis of EcoRI* cleavage data[18] and a preliminary 3Å X-ray analysis of the co-crystal. In addition, hydrophobic contributions have to be considered. Base modifications present a means to probe the DNA-determinants relevant for recognition, site directed mutagenesis has opened new perspectives to exchange individual aminoacids which are presumed to be involved in recognition and catalysis. In addition, this new technique of "protein engineering" allows to determine the energetics of individual intermolecular interactions in a way as introduced by Fersht[30] and will give more detailed insights into the mechanism of the catalytic step.

DNA-modification

Base analogue substitutions in oligodeoxyribonucleotides containing the EcoRI recognition site have consistently indicated that the enzyme recognizes its specific site via the major groove[1]. All these studies show that modifications of purine moieties pointing towards the major groove inhibit the EcoRI endonuclease more than corresponding modifications of the pyrimidine part of the canonical sequence. For example, methylation of the adenine bases blocks the hydrolytic reaction, substitution of the adenine moieties by 2-aminopurines in only one strand inhibits the cleavage of the phosphodiester bond in the modified strand completely and reduces the cleavage rate in the unmodified strand considerably[31]. Modification of the purines leads to perturbations of the hydrogen bond structure of the protein-DNA interface. Substitutions of the thymidines by deoxyuridine or bromodeoxyuridine lead to a reduction of the reactivity by at most 50 %. This observation is essentially independent of whether the substitution has been carried out in the fourth or fifth

position of the hexameric recognition sequence[32]. These data
were confirmed for the dT - dU exchange in the fourth position by Brennan et al.[33]. They report, however, a 2.2-fold
increase in the reactivity upon the dT-BrdU-exchange in the
same position and find no cleavage upon a dT-dU exchange in
the fifth position. With respect to these apparent discrepancies it has to be pointed out that Fliess et al. used undeca-
nucleotides and Brennan et al. octanucleotides as substrates.
Chainlength dependences have to be expected as long as the
oligonucleotides are too short in order to extend over the
entire surface of interaction between protein and DNA. It has
to be considered that base-modifications or substitutions may
also change the secondary structure of the DNA and that this
effect becomes more pronounced with shorter substrates. This
interpretation is supported by the finding that modifications
of bases adjacent to the recognition sequence[32,33] may have
substantial influence on the cleavage rate and that the
enzyme covers at least ten basepairs when complexed to DNA.
Substitution of the methyl group of the thymidines by the
considerably larger propyl group blocks the cleavage reaction
completely. The same behaviour was found substituting dT by
α-D-arabinosyluracil, which may introduce a distortion of the
regular geometry of the phosphodiester backbone.

Altogether the results suggest that in forming the specific
complex there is apparently little direct contact between
EcoRI and the thymidine residues, whereas all the hydrogen
bond contacts are confined to the purine bases via the major
groove. This assignment is in agreement with the 3Å structure
of the EcoRI-oligodeoxyribonucleotide complex, from which it
was concluded that there are twelve hydrogen bonds between
arginine and glutamic acid side chains of the protein and the
guanine and adenine bases of the DNA and no obvious direct
contacts to the thymidines. It has to be mentioned, however,
that the X-ray studies were performed in the absence of Mg^{2+},
the presence of which may introduce structural alterations.

Obviously, the thymidine bases are of minor importance for
EcoRI to recognize its specific sequence. This finding should
not be generalized to other restriction endonucleases. The

situation is different for the EcoRV endonuclease which recognizes the sequence -GATATC-[32]. Analogous studies as carried out with EcoRI have been done with EcoRV using the undecanucleotide AAAGATATCTT as the substrate. The substitution of either thymidine within the recognition sequence by uridine inhibits the cleavage by EcoRV completely. This indicates that hydrophobic interactions are essential for catalysis by EcoRV. The reduction of the rate of cleavage by 80 % and 100 % upon substitution of dT by BrdU or β-5-propyluridine indicates steric effects. Similar as with EcoRI the introduction of arabinosyluracil for either thymidine distorts the phosphodiester backbone and inhibits the cleavage reaction completely. This example shows that different restriction enzymes use different strategies to recognize their target site. This fact becomes also obvious for the isoschizomers Hae III, Bsp RI and Bsu RI which respond very differently to base modifications in their common recognition sequence -GGCC-[21].

Aminoacid modification of the EcoRI endonuclease

As compared to DNA modification studies only few investigations were concerned with the identification of the regions or even the residues of the protein involved in the binding and catalytic step of the EcoRI-substrate reaction. From chemical modification studies it was shown that lysine[10], glutamic acid[34] and histidine residues are essential for the activity of EcoRI; it has not been possible, however, to identify those modified residues in the protein sequence which are responsible for the binding and/or the catalytic step of the enzyme. We could show recently by an immunological study[35] and by crosslinking experiments[36] that the aminoacid sequence 137 - 157 of the EcoRI enzyme is involved in the binding of DNA. It is interesting to note that the substrate binding site identified by this approach has a high sequence homology with those helical parts of various DNA binding proteins which are involved in DNA recognition[37]. The region of protein-DNA interaction recently suggested by J. Rosenberg[5] coincides with the amino acid sequence

137-157. The catalytic site, however, is supposed to be located in the N-terminal half of the EcoRI molecule. Possible candidates to be involved in the catalysis are the Glu 103, Glu 111 and His 114 residues. In Staphylococcus nuclease[38] and DNAase I from bovine pancreas[39,40] a Glu or Asp and a His residue resp. have been suggested to be involved in the nucleophilic attack of a water molecule on the phospho- diester bond.

In order to identify amino acid residues involved in the recognition and cleavage process we have started with site directed mutagenesis experiments. The following single amino acid substitutions were obtained: Glu$\longrightarrow$Gln (111), Glu$\longrightarrow$Gln (144), Arg$\rightarrow$Lys (145). As was concluded from circu- lar dichroism spectroscopy and gel filtration experiments the secondary structure or gross conformation of the enzyme in- cluding its dimeric structure is essentially not affected by the single amino acid substitution. The mutants still bind to DNA, however, with reduced affinity, and their cleavage rates are reduced by a factor of 50 - 100 as compared to wildtype EcoRI. It is certainly too premature to interpret these preliminary observations in the sense of structure-function relationships. As to the role of the glutamic acid residues several possible contributions to the catalytic action of EcoRI can be discussed: the carboxylate group can be involved in the binding of the essential Mg^{2+}-cofactor, it can parti- cipate in a charge relay system similarly as proposed for the catalytic action of the DNAase I[40] besides its main role as proton acceptor in a hydrogen bond between the protein and the canonical sequence in the recognition complex.

Little is known about the role of Mg^{2+} in the catalytic process. From x-ray studies the position of Mg^{2+} could not yet be identified so far. Yet one of its functions may be to direct the complex into a position which renders it optimal for reaction by virtue of metal complex formation, and/or to provide a source of hydroxyl ions from metal bound water molecules at neutral pH, which may act as nucleophiles in the hydrolytic reaction. In the latter case the Mg^{2+}-ion would be the catalyst. The chances are great that these, as well as questions concerning the relationship between enzyme structure, energy of binding and catalytic activity can be addressed by introducing the techniques of protein engineering.

References

1.   Modrich, P., CRC Crit. Rev. Biochem. 13, 287 (1982)
2.   Newman, A. K., Rubin, R. A., Kine, S. H., and Modrich P., J. Biol. Chem. 256, 2131 (1981)
3.   Greene, P. J., Gupta, M., Boyer, H. W., Brown, W. E., and Rosenberg, J. M., J. Biol. Chem 256, 2143 (1981)
4.   Frederick, C., Grable, J., Melia, M., Samudzi, C., Jen-Jacobsen, L., Wang, B. C., Greene, P., Boyer, H. W., and Rosenberg, J. M., Nature 309, 327 (1984)
5.   Rosenberg, J. M., Proceedings of the 14th Aharon-Katzir-Conference, Bielefeld 1985
6.   Halford, S. E., Biochem. Soc. Transactions 8, 399 (1980)
7.   Berg, O. G., Winter, R. B., and von Hippel, P. H., Biochemistry 20, 6926 (1981)
8.   Winter, R. B., Berg, O. G., and von Hippel, P. H., Biochemistry 20, 6961 (1981)
9.   Jen-Jacobsen, L., Lesser, D., and Kurpiewski, M., Cell 45, 619 (1986)
10. Woodhead, J. L., and Malcolm, A. D. B., Nucl. Acids Res., 8, 389 (1980)
11. Goppelt, M., Pingoud, A., Maass, G., Mayer, H., Koester, H., and Frank, R., Eur. J. Biochem 104, 101 (1980)
12. Modrich, P., and Zabel, D., J. Biol. Chem., 251, 5866 (1976)
13. Halford, S. E., and Johnsen, N. P., Biochem. J., 191, 593 (1980)
14. Langowski, J., Pingoud, A., Goppelt, M., and Maass, G., Nucl. Acids Res. 8, 4727 (1980)
15. Clore, G. M., Gronenborn, A. M., and Davies, R. W., J. Mol. Biol. 155, 447 (1982)
16. Pingoud, A., Alves, J. Urbanke, C., and Zabeau, M., to be published
17. Goodman, H. M., Greene, P. J., Garfin, D. E., and Boyer, H. W., in Vogel, H. J. (Ed.): "Nucleic Acid-Protein Recognition", Academic Press, New York, pp 239 - 259 (1977)

18. Rosenberg, J. M., and Greene, P. J., DNA 1, 117 (1982)

19. Jen-Jacobsen, L., Kurpiewski, M., Lesser, D., Grable, J., Boyer, H. W., Rosenberg, J. M., and Greene, P. J., J. Biol. Chem. 258, 14638 (1983)

20. Alves, J., Pingoud, A., Haupt, W., Langowski, J., Peters, F., Maass, G., and Wolff, C., Eur. J. Biochem., 140, 83 (1984)

21. Wolfes, H., Fliess, A., and Pingoud, A., Eur. J. Biochem., 150, 105 (1985)

22. Richter, P.H., and Eigen, M., Biophys.Chem. 2, 255 (1974)

23. Jack, W. E., Terry, B. J., and Modrich, P., Proc. Natl. Acad. Sci. USA, 79, 4010 (1982)

24. Langowski, J., Alves, J., Pingoud, A., and Maass, G., Nucl. Acids Res. 11, 501 (1983)

25. Terry, B. J., Jack, W. E., and Modrich, P., J. Biol. Chem. 260, 13130 (1985)

26. Ehbrecht, H. J., Pingoud, A., Urbanke, C., Maass, G., and Gualerzi, C., J. Biol. Chem. 260, 6160 (1985)

27. Kohring, G. W., Mayer, F., and Mayer, H., Europ. J. Cell Biology 37, 1 (1985)

28. von Hippel, P. H., and Berg, O.G., Proc. Natl. Acad. Sci. USA, 83, 1608 (1986)

29. Modrich, P., and Roberts, R. J., in "Nucleases", Linn, S. M., and Roberts, R. J., ed., 109, Cold Spring Harbor Laboratory, New York (1982)

30. Fersht, A. R., Leatherbarrow, R. J., and Wells, T. N. C., TIBS 11, 321 (1986)

31. Haupt, W., Ph. D. thesis, University of Hannover (1983)

32. Fliess, A., Wolfes, H., Rosenthahl, A., Schwellnus, K., Bloecker, H., Frank, R., and Pingoud, A., Nucl. Acids Res., 14, 3463 (1986)

33. Brennan, C. A., von Cleve, M. D., and Gumport, R. I., J. Biol. Chem. 261, 7270 (1986)

34. Woodhead, J. L., and Malcolm, A. D. B., Eur. J. Biochem. 120, 125 (1981)

35. Scholtissek, S., Pingoud, A., Maass, G., and Zabeau, M., J. Biol. Chem. 261, 2228 (1986)

36. Wolfes, H., Fliess, A., Winkler, F., and Pingoud, A., Eur. J. Biochem., in press (1986)

37. Matthews, B. W., Ohlendorf, D. H., Anderson, W. F.,

Fisher, R. G., and Takeda, Y., Cold Spring Harbor Symp. Quant. Biol. 47, 427 (1983)

38. Cotton, F. A., Hazen, E. E.(Jr.), and Legg, M. J., Proc. Natl. Acad. Sci. USA, 76, 2551 (1979)

39. Price, P. A., Stein, W. H., and Moore, S., J. Biol. Chem. 244, 924 (1969)

40. Suck,D., Oefner,C., and Kabsch,W.,EMBO J., 3, 2423 (1984)

41. Halford, S. E., TIBS 8, 455 (1983)

MECHANISM AND SPECIFICITY OF TWO RESTRICTION ENZYMES, CauI and CauII,

THAT RECOGNIZE ASYMMETRICAL DNA SEQUENCES

S. Paul Bennett and Stephen E. Halford

Department of Biochemistry, Unit of Molecular Genetics
University of Bristol
University Walk, Bristol, BS8 1TD, U.K.

RECOGNITION SITES

Type II restriction endonucleases are enzymes that recognize specific nucleotide sequences on duplex DNA and cleave both strands of the duplex at fixed locations relative to their recognition sites (1). The only co-factor that they need is Mg^{2+}, while type I and Type III restriction enzymes also need ATP and S-adenosyl methionine for maximal activity (2). Type I and type III enzymes also differ from type II by cleaving DNA at variable distances from their recognition sites (2). Hence, the study of type II restriction enzymes was initially motivated by their unique applications in the analysis of DNA and in the construction of recombinant DNA molecules (3). However, these enzymes also provide examples of DNA-protein interactions whose mechanisms are amenable to molecular analysis.

Inspection of many different species of bacteria has revealed, to date, over 600 different type II restriction enzymes, and a list of their recognition sites covers a wide variety of DNA sequences (4). The current listing of restriction enzymes constitutes, in effect, a library for the study of DNA-protein interactions. One enzyme can be taken to study one aspect of the problem and another enzyme a different aspect due to its different recognition sequence.

The majority of type II restriction endonucleases each recognize a unique sequence, of 4,6 or 8 bp, that is symmetrical in that both strands of the duplex have the same 5' - 3' sequence. For example, both the EcoRI and EcoRV restriction enzymes (from two different strains of Escherichia coli) recognize such sequences (4).

```
EcoRI      5'-GAATTC-3'
           3'-CTTAAG-5'

EcoRV      5'-GATATC-3'
           3'-CTATAG-5'
```

(The phosphodiester bonds that are cleaved by these enzymes are marked by arrows). However, for other restriction enzymes, the recognition sites are asymmetric, either completely so as in the case of MboII (5) or partially asymmetric as illustrated by both CauI and CauII.

239

```
        ↓
MboII   5'-GAAGA(N)₈-3'
        3'-CTTCT(N)₇-5'
                  ↑

        ↓
CauI    5'-GGACC-3'
        3'-CCTGG-5'
            ↓↑
CauII   5'-CCGGG-3'
        3'-GGCCC-5'
            ↑
```

Thus, depending on orientation, the centre of the recognition site for CauI contains either an A.T or a T.A base pair while the equivalent position for CauII has either G.C or C.G base pairs.

At the present time, much more is known about EcoRI than any other restriction enzyme (1, 7-9). However, the EcoRI protein has an extremely low solubility, which eliminates the application of many biophysical techniques that could otherwise have been used to analyse its interaction with DNA (10). This disadvantage does not apply to the EcoRV nuclease and consequently this enzyme has particular advantages for studying how a protein interacts with a unique symmetrical sequence (11). Strains of E.coli that over-produce the EcoRV enzyme have been constructed so large amounts of the protein can be easily purified (12). In addition, EcoRV is the only restriction enzyme, apart from EcoRI (8), for which crystals of the protein have been described (13).

Most of the type II restriction endonucleases that recognize completely asymmetric sequences, such as MboII, cleave the DNA at fixed locations several nucleotides away from their recognition sites (4). But no studies on the mechanism of action of such enzymes have been reported to date. However, we have recently investigated the properties of two restriction enzymes, CauI and CauII, that recognize partially asymmetric sequences. We report here on the similarities and differences between these two enzymes and other restriction enzymes, such as EcoRI and EcoRV, that interact with unique symmetrical sequences. The reason for selecting CauI and CauII for study was that the source, Chloroflexus aurantiacus, is the only bacterium known to produce two such enzymes (14). In addition, C. aurantiacus is a thermophile so its enzymes should be more stable than the enzymes from many other sources.

SUBUNIT STRUCTURES

We have isolated and purified to homogeneity both the CauI and CauII restriction endonucleases. Neither enzyme is present in C. aurantiacus at high levels and preparations from 10L cultures of this strain typically yielded about 3μg CauI and 0.4μg CauII. During these purifications, we discovered that C. aurantiacus contains another restriction enzyme in addition to the two noted previously (14). The third enzyme, CauIII, cleaves DNA at the same site as PstI: apart from finding that it has a much higher thermo-stability than PstI, CauIII was not studied further.

M_r values for individual polypeptides were determined from their electrophoretic mobility through polyacrylamide in the presence of SDS (15): gels were stained with silver (16), for Coomassie Blue lacked sufficient sensitivity. M_r values for the native proteins were estimated from either their sedimentation velocities through sucrose gradients (17) or by gel filtration through Sephacryl S-200

(18): fractions from either the gradients or the filtrates were assayed for enzyme activity.

On SDS gels, purified samples of CauI contained a single polypeptide of M_r 27,500. The active form of CauI yielded M_r values of 56,000 and 64,000 by sedimentation velocity and gel-filtration respectively. This data (not shown) indicates that the CauI restriction endonuclease exists in solution as a dimer of two identical protein subunits. Nearly all restriction enzymes that recognize unique symmetrical sites, such as EcoRI and EcoRV, are also dimers of identical subunits (1,11,13). It was suggested many years ago (19) that these enzymes might interact with their targets on the DNA in symmetrical fashion, forming a complex in which the axes of symmetry of the DNA and of the dimeric protein duplicate each other. This arrangement could place the active site from one subunit against the scissile phosphodiester bond from one strand of the DNA and likewise the second subunit against the second strand. For EcoRI, direct visualization of such a complex has been obtained by X-ray crystallography (8). We describe later an adaptation of this model that can account for a dimer of identical subunits interacting with the asymmetrical recognition site for CauI.

In contrast, SDS gels of our most purified preparations of the CauII restriction enzyme always revealed two polypeptides, of M_r 31,500 and 29,000, present in about equal amounts. We do not believe that either of these proteins is an impurity in our preparations: both of them co-purified with the CauII activity on a large number of chromatography matrices. Moreover, CauII showed anomalous behaviour upon gel-filtration and, depending on the ionic strength and the pH of the buffer, eluted from Sephacryl S-200 at a range of different volumes. But the two polypeptides always eluted together. [These anomalies are due in part to interactions with acidic groups on the gel matrix, as also seen with EcoRV (11) and perhaps other DNA binding proteins, and in part to dissociation of the protein into subunits]. However, we cannot eliminate the possibility that the smaller peptide is a proteolytic product of the larger : our preparations of CauII yield insufficient protein to test for this.

The CauII enzyme was sedimented through sucrose in both the presence of 50mM NaCl (where the enzyme is largely inactive) and in the absence of NaCl (the conditions required for maximal activity of this enzyme). The former yielded an M_r value of 33,000 and the latter 54,000 (data not shown). Hence, we suggest that the CauII restriction enzyme is active as a heterologous dimer, made up of one subunit at M_r 31,500 and one at M_r 29,000. The loss of activity caused by NaCl could be due, at least partly, to the dissociation of this dimer into its constituent subunits. It therefore seems likely that both subunits interact with DNA but, if CauII is a heterologous dimer, each subunit could interact differently with the DNA. At the recognition site for CauII, the two strands of the DNA have different 5' - 3' sequences.

CATALYTIC MECHANISMS

In a complex between a dimeric restriction enzyme and a symmetrical recognition site, each subunit of the protein might be able to cleave one strand of the duplex within the life-time of the complex. This concerted pathway would not be available to a monomeric restriction enzyme unless the monomer contained two active sites within one polypeptide. Certain restriction enzymes appear to cleave DNA by concerted mechanisms. Examples include the EcoRV and SalGI enzymes

(11,20) and with these the intact duplex of the substrate is converted directly to the final product cleaved in both strands. The mechanism of action of the EcoRI enzyme is more complicated: this dimeric enzyme can dissociate from the DNA after cleaving its recognition site in just one strand, and a second independent reaction is then needed to cleave the second strand (1,7,21).

The distinction between these pathways can be made by using as the substrate for the restriction enzyme a covalently closed circle of DNA that contains one copy of the recognition sequence. If each strand of this DNA is cut in separate consecutive reactions, the covalently closed substrate will be converted initially into open-circle DNA which in turn will be cut to yield linear DNA as the final product. The covalently closed, open-circle and linear forms of the same DNA can be separated from each other by electrophoresis through agarose and, with radio-labelled DNA, the amount of each form at various time-points during the reactions can be measured by scintillation counting (20,22). However, if open-circle DNA is not observed, then either the enzyme cleaves both strands in a single concerted reaction or, alternatively, it cleaves the open-circle intermediate much faster than the covalently closed substrate. For SalGI, further experiments eliminated the latter possibility (20).

We have used this method on both the CauI and CauII restriction endonucleases. The substrate was the replicative DNA (RFI) from bacteriophage phiX174, whose sequence has a single recognition site for each of these enzymes (23). Reaction conditions were the optimal for these enzymes : 10mM Tris-HCl, 5mM MgCl$_2$, 10mM 2-mercaptoethanol, 0.02% (w/v) bovine serum albumin, pH 7.5. Experiments with CauI were done at both 55° and 70°C, with equivalent results, but only the lower temperature was used with CauII for this enzyme is unstable at 70°C. Both CauI and CauII cleaved the covalently closed DNA directly to the linear DNA : no open-circle DNA was generated during the time-courses of these reactions (data not shown, as similar reactions on other restriction enzymes are described elsewhere : 20,22).

For the CauII restriction enzyme, an alternative approach to determining the concertedness of its reactions is provided by the DNA sequence,

```
         ↓↓
5'-CCCGGG-3'
3'-GGGCCC-5'
       ↑↑
```

which is present on the replicative DNA from phage M13mp10. This sequence contains two recognition sites for CauII, one (CCCGG in the top strand) that would be cleaved as indicated by the small arrows and a second (CCGGG in the top strand) that would be cleaved at the large arrows. A concerted reaction at this site would cleave both strands at one or the other of these sites. But cleavage of each strand by independent reactions would yield a heterogeneous mixture of products with either zero, one or two bp 5'-extensions. In the latter case, the distribution of products may be biased if the enzyme were to cut preferentially at either the C-C or the C-G phosphodiester bond.

The method of Brown and Smith (24) was used to identify where CauII cut the above sequence in both the top and bottom strands (Fig. 1). The top (labelled) strand was cut almost exclusively at the CG dinucleotide marked by the large arrow : only a small fraction of the DNA was cut at the site of the small arrow (Fig. 1, lane R). In the bottom (template) strand, a strong bias was again observed

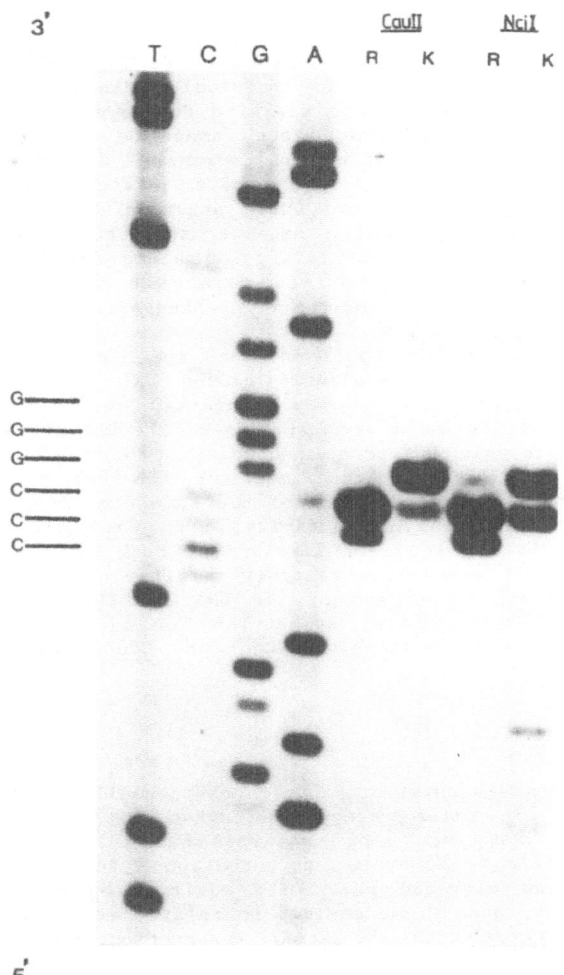

Fig. 1. Single stranded M13mp10 DNA was sequenced by the chain-termination method (25). Lanes T,C, G and A denote the respective di-deoxy channels. In a separate experiment, the same primer/template combination was used as a substrate for the Klenow fragment of DNA polymerase I in the presence of dTTP, dCTP and dGTP: labelling was by a pulse of [α-^{35}S]dATP followed by a cold chase of dATP (24). The DNA was washed with phenol and then ether, after which it was digested with either 3 units of CauII or 6 units of NciI. The DNA from each digest was divided into two fractions, R and K. The K samples were treated with 1 unit of Klenow (with all 4 dNTPs for 15 min at 37°C) while the R samples were left untreated. All samples were analysed by electrophoresis through an 8% polyacrylamide gel (25) alongside the sequencing ladder. After fixing the DNA, the gel was dried and then autoradiographed. The sequence of the labelled primed strand reads 5' to 3' from the bottom to the top of the gel. The phosphodiester bonds in the primed strand that were cleaved by either CauII or NciI are identified from the R lanes, while those in the template strand are identified from the K lanes. (24).

for the site marked by the large arrow but this is now at the C-C dinucleotide (Fig. 1, lane K). Hence, at this duplex, CauII prefers the right hand recognition site over the left hand, in both strands of the DNA, rather than any particular dinucleotide. This indicates a concerted mechanism for cleaving duplex DNA. The reason for the preference between the two sites cannot be due to the 6 bp sequence itself for it is a perfect palindrome. Sequences outside of the palindrome must influence the choice. The substrate used in Fig. 1 has a longer region of double stranded DNA on the right of the target than on the left, and perhaps CauII locates its recognition sequence by facilitated diffusion along duplex DNA (26). The NciI restriction enzyme, which recognizes the same sequence as CauII, shows a slight preference for the right hand site (Fig. 1).

For both the CauI and the CauII enzymes, the kinetic experiments described earlier, with the covalently closed DNA from phiX174 as the substrate, are consistent with a concerted mechanism for the cleavage of duplex DNA, in which both strands of the DNA are cut within the life-time of a single DNA-protein complex. For CauII, this was confirmed by the site-location experiment on the DNA from M13mp10 that is shown in Fig. 1. In buffers of the same composition as used for experiments on their enzymic activity, both CauI and CauII are dimeric proteins. Hence, it seems likely that both of these enzymes form complexes at their respective recognition sites where one subunit is in place to cleave one strand of the DNA and the other subunit the other strand. The potential symmetries for such complexes, at the asymmetrical recognition sites for CauI and CauII, are discussed below.

DNA SEQUENCE SPECIFICITIES

Restriction enzymes such as EcoRI and EcoRV normally show extremely high specificities for their recognition sites on DNA, and cleave DNA at these sites very much more readily than at any alternative DNA sequence (7,27). However, perturbations to the reaction environments, such as elevated pH or the addition of water-miscible organic solvents, can cause these enzymes to relax their specificities. This was observed first with the EcoRI restriction enzyme and the altered specificity, known as EcoRI* (28), results in DNA cleavage at a number of sequences that differ from the recognition site by one or two base pairs (29). The EcoRV enzymes similarly shows a relaxed specificity and, in reactions at pH 8.5 in the presence of DMSO (Dimethyl sulphoxide), EcoRV* cleaves DNA at most but not all sequences that differ by one base pair from the recognition site (27).

The recognition site for CauI has at its centre either an A.T or a T.A base pair, while the equivalent position for CauII is either C.G or G.C. Hence, one might expect the degree of specificity that the CauI and the CauII enzymes show for their respective recognition sites to differ from enzymes which recognize unique symmetrical sites. However, in the standard reaction conditions for these enzymes, we were unable to detect any cutting of phiX174 DNA other than at their recognition sites. But when the reactions were carried out at high pH and in the presence of DMSO, both CauI and CauII cut the DNA at a number of additional sites. We shall describe their altered specificites as CauI* and CauII*.

To determine the DNA sequence specificity of CauI* and CauII* we used the plasmid pAO3 as a substrate on account of its small size (1,683bp) and the availability of its complete DNA sequence (30).

Table 1.   Relaxed specificities of <u>Cau</u>I and <u>Cau</u>II

In the reactions given below, <u>Cau</u>I cleaved the plasmid pAO3 at many
locations in addition to the canonical <u>Cau</u>I sites, while <u>Cau</u>II cleaved
the DNA at five additional sites.   The locations of a sample of the
<u>Cau</u>I* sites and all five <u>Cau</u>II* sites were mapped: co-ordinates for
the map positions are from the sequence of pAO3 in Oka et al (30). In
the vicinity of each mapped site, sequences related to the canonical
site were found at the positions noted below.   These sequences are
listed from the "top" strand of pAO3 (30) with numbering from the
first base.   Bases that deviate from the canonical sites are
underlined.

A)   <u>Cau</u>I* : reactions typically contained 60 units <u>Cau</u>I and 1µg DNA
        in 20µl of 20mM Tris, 5mM MgCl$_2$, 10mM 2-mercaptoethanol,
        0.02% (w/v) bovine serum albumin, 20% (v/v) DMSO, pH
        9.0, for 2 h at 65°C.

Map Positions	Related Sequences
65	GGA<u>G</u>C at 52
285	GGTC<u>G</u> 283
305	GGTA<u>C</u> 304
500	GGAC<u>T</u> 492
1660	G<u>TT</u>CC 1652

B)   <u>Cau</u>II* : reactions typically contained 100 units <u>Cau</u>II and 0.5µg
        DNA in 20µl of 20mM Tris, 5mM MgCl$_2$, 10mM
        2-mercaptoethanol, 0.02% (w/v) bovine serum albumin,
        30% (v/v) DMSO, pH 8.5, for 2 h at 55°C.

Map Positions	Related Sequences
235	CCAGG at 240
385	CC<u>A</u>GG 385
505	CC<u>A</u>GG 506
535	CC<u>T</u>GG 519
585	<u>T</u>CGGG 591

The recognition sequence for CauI appears twice on this DNA and that for CauII at four locations. But when treated with CauI under the perturbed reaction conditions described in Table IA, pAO3 was cut at a large number of additional sites. In contrast, perturbation to the reaction environment for CauII (Table IB) caused this enzyme to cleave pAO3 at only five further sites in addition to the canonical sequences. The locations of some of the CauI* sites and all five of the CauII* sites were mapped, relative to the positions of other known restriction sites on pAO3, to an accuracy of about +/- 10 bp (Table I). The mapping was done primarily by the method of Smith and Bernstiel (31) using small restriction fragments from pAO3, each labelled at one 5'terminus. The DNA sequences in the vicinity of each of the mapped sites for CauI* and CauII* were then inspected to see if they contained any sequences related to the canonical sites for CauI and CauII. In every case, a sequence was found that differed from the canonical site by one base pair (Table I). We propose that the CauI* and CauII* activities cleave DNA at these related sequences.

From the limited sample of CauI* sites in Table IA, it appears that the CauI restriction enzyme can relax its specificity so as to cleave virtually any sequence with 4/5 homology to the canonical site for CauI, except that no examples were found where the A.T (or T.A) base pairs at the centre were replaced by either G.C or C.G base pairs. The A.T/T.A degeneracy at the centre of the CauI site seems not to be accompanied by a lower discrimination against other base pairs. In contrast to the specificity of CauI*, all bar one of the CauII* sites on pAO3 had single substitutions in which the G.C (or C.G) base pair at the centre of the CauII site had been replaced by either A.T or T.A base pairs. Hence, with CauII, the G.C/C.G degeneracy at the centre of its recognition site is accompanied by a lower discrimination against other base pairs.

DNA-PROTEIN CONTACTS

There is no evidence, with any restriction enzyme to date, that the recognition site on duplex DNA adopts a conformation that is radically different from B-DNA, such as left-handed DNA or a cruciform. Only discrete variations, such as that observed in the complex between EcoRI and its substrate (8), can play a role in the recognition process. Both the CauI and CauII restriction enzymes were equally active at their respective recognition sites on either the negatively supercoiled or the relaxed forms of the replicative DNA from phiX174 (data not shown). For CauI, this was observed even in reactions at 70°C, where the supercoiled DNA could have readily adopted alternative conformations (32). Consequently, these enzymes must recognize their DNA sequences from the functional groups that each base pair exposes to either the major or minor grooves on B-DNA (33). The types of functional groups are illustrated in Fig. 2.

In the major groove, each of the four possible base pairs has a unique arrangement of hydrogen bonding groups that is distinct from the arrangements seen with the other three base pairs (33). However, for a protein to unambiguously distinguish each of the base pairs, it will in general be necessary for that protein to form at least two hydrogen bonds with each base pair (29). In contrast, the arrangements of hydrogen bonding groups exposed to the minor groove are not unique to each base pair (33,34). For example, an A.T base pair has two hydrogen bond acceptors (and no donors) in the minor groove, N3 of adenine and O2 of thymine (Fig. 2). But upon

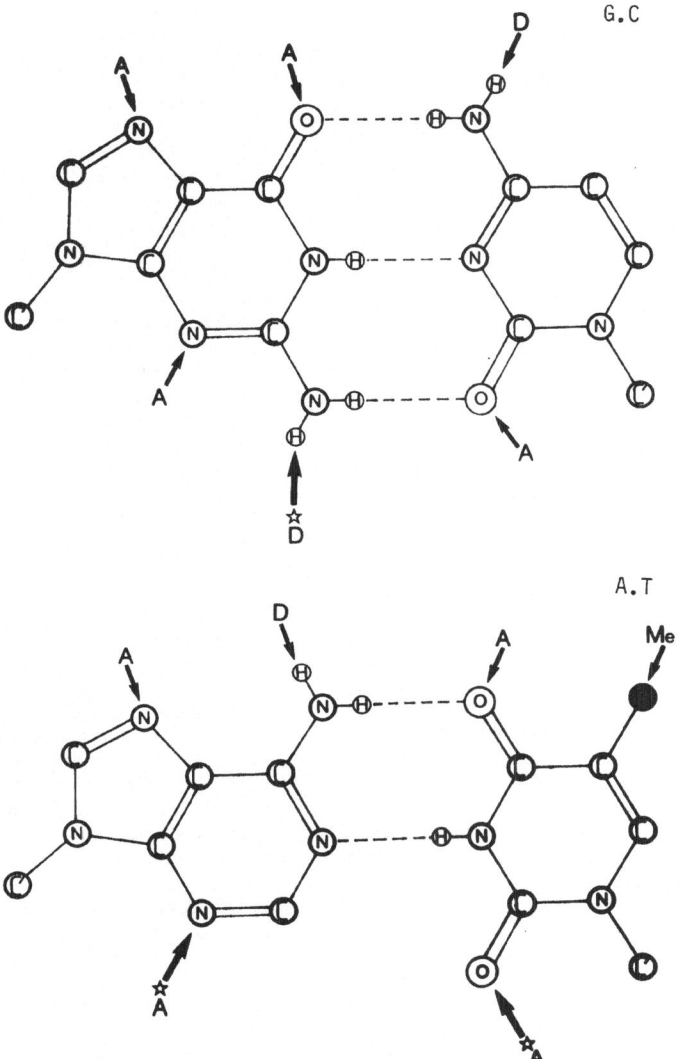

Fig. 2.   Watson-Crick base pairs for G.C and A.T are represented.
The edges of the base pairs that would be exposed to the major groove
of B-DNA are those above each structure : edges exposed to the minor
groove are below each structure.   The functional groups that would
be accessible from either major or minor grooves (33) are identified
by arrows labelled either A for hydrogen bond acceptor or D for hydrogen
bond donor.   Bold arrows marked with asteriks denote the functional
groups that may be involved in DNA sequence recognition by the CauI
and CauII restriction enzymes.

inversion of A.T to yield a T.A base pair, the N3 of adenine will occupy a virtually identical position to that of the thymine O2 in A.T. The inversion will also move the O2 of thymine to a position that is very close to that previously occupied by the N3 of adenine. Hence, from the minor groove, A.T and T.A base pairs will be virtually indistinguishable. Similarly, G.C and C.G base pairs are virtually indistinguishable from the minor groove but these are distinct from A.T/T.A on account of the N2 amino group of guanine which could act as a hydrogen bond donor (Fig. 2).

Rosenberg et al (34) have used the similarities between certain base pairs to derive models that explain how some restriction enzymes recognize degenerate DNA sequences. Our data on CauI and CauII correlate strongly with their models. Consider first the CauI restriction enzyme. Four out of the five base pairs at the CauI recognition site are defined unambiguously and these four are related by symmetry. The CauI enzyme is a dimer of identical protein subunits, and it cleaves both strands of the DNA within a single DNA-protein complex. Hence, we suggest that this complex contains two protein subunits bound symmetrically to the four unique base pairs, which presumably, are recognised from the major groove. At the centre of its recognition site, CauI interacts with either an A.T or a T.A base pair. This implies recognition from the minor groove (34). The two hydrogen bond acceptors in the minor groove of an A.T. (or T.A) base pair are in equivalent positions either side of the inversion axis, so this recognition process maintains the symmetry of the DNA-protein complex. Yet the specificity of CauI demonstrates an effective discrimination against G.C or C.G base pairs at this central position. This could be due to steric hindrance within the minor groove (34), on account of the N2 amino group of guanine in place of the proton on C2 of adenine (Fig. 2).

The above model cannot, however, be applied directly to the CauII restriction enzyme. In the minor groove of a G.C base pair, the inversion to C.G exchanges the positions of the two hydrogen bond acceptors (N3 of guanine and O2 of cytosine) but the hydrogen bond donor (N2 of guanine) is located very close to the inversion axis so its position remains unaltered. The closeness of the N2 amino group to the inversion axis creates a problem for a symmetrical restriction enzyme: either one or the other of the two protein subunits could interact with this group, but not both together, and thus the symmetry of the DNA-protein complex would be lost (34). The CauII restriction enzyme appears to have solved this problem by being a heterologous dimer made up of different protein subunits. The CauII enzyme normally recognizes either G.C or C.G at the centre of its target site, but the specificity of CauII[*] indicates that it can fail to discriminate against either A.T or T.A at this position. From the minor groove, the only difference between G.C/C.G and A.T/T.A, in terms of hydrogen bonding groups, is the N2 amino group of guanine (Fig. 2). Hence, if the active site of CauII can accommodate the minor groove of either G.C or C.G, it must also be able to accommodate either A.T or T.A with the loss of only one hydrogen bond.

ACKNOWLEDGEMENTS

We thank Nigel Brown and Joe Winnie for advice and discussions, and the Science and Engineering Research Council for financial support.

REFERENCES

1. P. Modrich and R.J. Roberts, Type II Restriction and Modification Enzymes,in : "Nucleases", S.M. Linn and R.J. Roberts, eds., Cold Spring Harbor, New York (1982).
2. T.A. Bickle, ATP-dependent Restriction Endonucleases, in : "Nucleases", S.M. Linn and R.J. Roberts, eds., Cold Spring Harbor, New York (1982).
3. S.N. Cohen, A.C.Y. Chang, H.W. Boyer and R.B. Helling, Construction of biologically functional bacterial plasmid in vitro, Proc Natl. Acad. Sci. U.S.A. 70: 3240 (1973).
4. C. Kessler, T.S. Neumaier, and W. Wolfe, Recognition sequences of restriction endonucleases and methylases, Gene 33 : 1 (1985).
5. N.L. Brown, C.A. Hutchinson, and M. Smith, The specific non-symmetrical sequence recognized by restriction endonuclease MboII, J.Mol. Biol. 140 : 143 (1980).
6. E. Molemans, J. van Emmeloo, and W. Fiers, The sequence specificity of endonucleases CauI and CauII, Gene 18 : 93 (1982).
7. S.E. Halford, How does EcoRI cleave its recognition site on DNA? Trends Biochem. Sci. 8 : 455 (1983).
8. C.A. Frederick, J. Grable, M. Melia, C. Samudzi, L. Jen-Jacobson, B.C. Wang, P. Greene, H.W. Boyer, and J. Rosenberg, Kinked DNA in crystalline complex with EcoRI endonuclease, Nature 309 : 327 (1984).
9. C.A. Brennan, M.D. Van Cleve, and R.I. Gumport, Effects of Base Analogue Substitutions on Cleavage by EcoRI Restriction Endonuclease, J. Biol. Chem. 261 : 7270 (1986).
10. P.A. Luke, and S.E. Halford, Solubility of the EcoRI restriction endonuclease and its purification from an over-producing strain, Gene 37 : 241 (1985).
11. P.A. Luke, S.A. McCallum, and S.E. Halford, The EcoRV Restriction Endonuclease, in : "Gene Amplification and Analysis, vol 6", J. Chirikjian, ed., Elsevier, New York. (in press).
12. L. Bougueleret, M.L. Tenchini, J. Botterman, and M. Zabeau, Overproduction of EcoRV endonuclease and methylase, Nucl. Acids. Res. 13 : 3823 (1985).
13. A. D'Arcy, R.S. Brown, M. Zabeau, R.W. van Resandt, and F. Winkler, Purification and crystallization of the EcoRV restriction endonuclease, J. Biol. Chem. 260 : 1987 (1985).
14. A.H.A. Bingham, and J. Darbyshire, Isolation of two restriction endonucleases from Chloroflexus aurantiacus, Gene 18: 87 (1982).
15. U.K. Laemmli, Cleavage of structural proteins during the assembly of the head of bacteriophage T4, Nature 227 : 680 (1970).
16. C.R. Merril, D. Goldman, and M.L. Van Keuren, Silver staining methods for polyacrylamide gel electrophoresis, Methods in Enzymology, 96 : 230 (1983).
17. R.G. Martin and B.N. Ames, A Method for determining the sedimentation behaviour of enzymes : applications to protein mixtures, J. Biol. Chem. 236 : 1372 (1961).
18. P. Andrews, The gel-filtration behaviour of proteins related to their molecular weights over a wide range, Biochem. J. 96 : 595 (1965).
19. T.J. Kelly, and H.O. Smith, A restriction enzyme from Hemophilus influenza J. Mol. Biol. 51 : 393 (1970).
20. A. Maxwell, and S.E. Halford, The SalGI restriction endonuclease: mechanism of DNA cleavage, Biochem. J. 211 : 402 (1983).
21. S.E. Halford, and N.P. Johnson, Single turnover of the EcoRI restriction endonucleases, Biochem. J. 211 : 405 (1983).

22. S.E. Halford, N.P. Johnson, and J. Grinsted, The reactions of the EcoRI and other restriction endonucleases, Biochem. J. 179 : 353 (1979).

23. F. Sanger, A.R. Coulson, T. Friedman, G.N. Air, B.G. Barrell, N.L. Brown, J.C. Fiddes, C.A. Hutchinson, P.M. Slocombe, and M. Smith, The nucleotide sequence of bacteriophage φX174. J. Mol. Biol. 125 : 225 (1978).

24. N.L. Brown, and M. Smith, A general method for defining restriction enzyme cleavage and recognition sites, Methods in Enzymology 65 : 391 (1980).

25. F. Sanger, S.Nicklen, and A.R. Coulson, DNA sequencing with chain terminating inhibitors, Proc. Natl. Acad. Sci. U.S.A. 74, 5463 (1977).

26. O.G. Berg, R.B,. Winter, and P.H. von Hippel, Diffusion-driven mechanism of protein translocation on nucleic acids, Biochemistry 20 : 6929 (1981).

27. S.E. Halford, B.M. Lovelady, and S.A. McCallum, Relaxed specificity of the EcoRV restriction endonuclease, Gene 41 : 173 (1986).

28. B. Polisky, P. Greene, D.E. Garfin, B.J. McCarty, H.M. Goodman, and H.W. Boyer, Specificity of substrate recognition by the EcoRI restriction endonuclease, Proc. Natl. Acad. Sci. U.S.A 72 : 3310 (1975).

29. J.M. Rosenberg, and P. Greene, EcoRI* specificity and hydrogen bonding, DNA 1 : 117 (1982).

30. A. Oka, N. Nomura, M. Morita, H. Sugisaka, K. Sugimoto, and N. Takanami, Nucleotide sequence of small ColEI derivatives, Mol. Gen. Genet. 172 : 151 (1979).

31. H.O. Smith, and M.L. Bernstiel, A simple method for DNA restriction site mapping, Nucl. Acids. Res. 3 : 2387 (1976).

32. D.M.J. Lilley, DNA supercoiling and DNA structure, Biochem. Soc. Trans, 14 : 211 (1986).

33. N.C. Seeman, J.M. Rosenberg, and A. Rich, Sequence-specific Recognition of double helical nucleic acids by proteins, Proc. Natl. Acad. Sci. U.S.A. 73 : 804 (1976).

34. J.M. Rosenberg, H.W. Boyer and P Green, The structure and function of the EcoRI restriction endonuclease, in : "Gene Amplification and Analysis, Vol 1", J. Chirikjian, Ed., Elsevier, New York (1981).

STRUCTURE OF THE DNA-ECORI ENDONUCLEASE RECOGNITION COMPLEX

John M. Rosenberg, Judith A. McClarin, Christin A. Frederick,
Bi-Cheng Wang and John Grable

Department of Biological Sciences, University of Pittsburgh
Pittsburgh, PA 15260

Herbert W. Boyer and Patricia Greene

Department of Biochemistry and Biophysics, University of
California, SF, San Francisco, CA 94143

The 3 Å structure of the co-crystalline recognition complex between
EcoRI endonuclease and the cognate oligonucleotide TCGCGAATTCGCG has been
solved by the ISIR method using a platinum isomorphous derivative [1,2,3].
Refinement is in progress. The endonuclease-DNA recognition complex consists of a distorted double helix and a protein dimer composed of identical
subunits related by a two-fold axis of rotational symmetry (see Fig. 1).
The distortions of the DNA are not seen in the crystal structure of the
dodecamer, CGCGAATTCGCG [4]; hence they are induced by the binding of the
protein. The distortions are concentrated into separate features which are
localized disruptions of the double helical symmetry. These disruptions

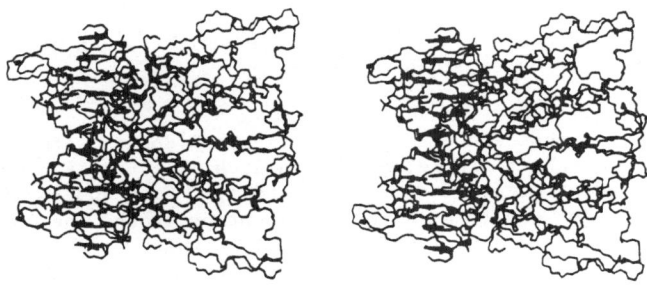

Fig. 1. A stereo drawing of the main chain atoms of the protein and the
non-hydrogen atoms of the DNA, which is towards the left of the
complex. The two-fold axis of rotational symmetry is horizontal
and in the plane of the figure.

appear to have structural consequences which propagate over long distances through the DNA via twisting and perhaps bending effects. They are therefore referred to as neo-kinks (see Figs. 2 and 3). The type I neo-kink spans the central two-fold symmetry axis of the complex and it introduces a net unwinding of 25° into the DNA. This increases the separation of the DNA backbones across the major groove thereby facilitating access by the protein to the base edges, which are at the floor of the groove. The type-I neo-kink also re-aligns adjacent adenine residues within the central AATT tetranucleotide so as to create the detailed geometry necessary for amino acid side chains to bridge across these purines (see below).

0 1 2   3   4   5   6   7 8   9 10 11 12

T pC pG pC pG pA pA pT pT pC pG pC pG

| | |

Type II   Type I   Type II
neo-kink   neo-kink   neo-kink

Fig. 2. The sequence of the tridecameric oligonucleotide used to make the DNA-protein complex. Also shown is the location of the kinks and the base numbering scheme, which was chosen to be consistent with the numbering system used by Dickerson and co-workers for their similar dodecamer [4].

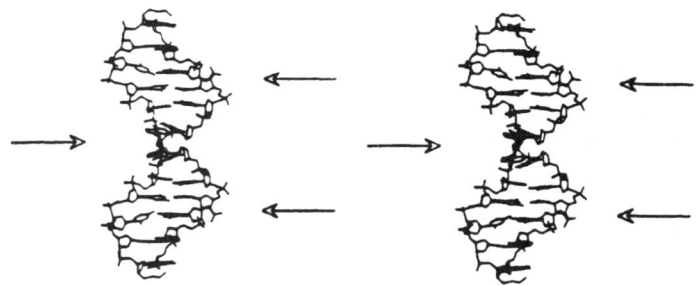

Fig. 3. A stereo figure of the DNA indicating the type I and type II neo-kinks. The single arrow on the right points to the center of the type I neo-kink. The two-fold symmetry axis passes through both the arrow and the center of the type I neo-kink; hence the type I neo-kink possesses two-fold symmetry. The two arrows on the left point to the centers of the type II neo-kinks which are identical by virtue of the two-fold symmetry.

Each subunit is composed of a single principal domain with a central 5-stranded wall of β-sheet bracketed by α-helices, i.e. it is organized according to α/β architecture (see Figs. 4 and 5). Each domain also possesses an extension called an "arm", which wraps around the DNA. The β-sheet can be subdivided into topological motifs which have identifiable functional roles. The first three strands are anti-parallel (β1, β2 and β3 in Fig. 4); they form a 3-stranded anti-parallel motif which is associated with phosphodiester bond cleavage. Similarly, β3, β4 and β5 form a 3-stranded parallel motif which is associated with sequence recognition and the subunit interface. The two segments overlap to form the whole 5-stranded β-sheet.

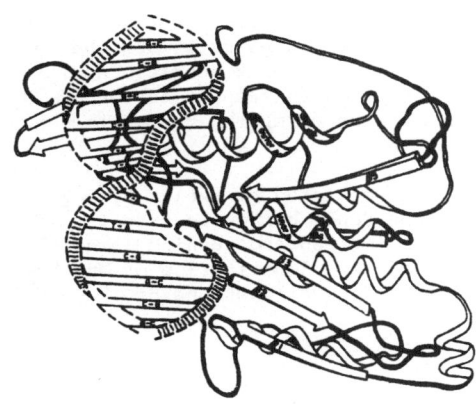

Fig. 4.   Schematic backbone drawing of one subunit of (dimeric) EcoRI endo-
          nuclease and both strands of the DNA in the complex. The arrows
          represent β-strands, the coils represent α-helices and the ribbons
          represent the DNA backbone. The figure has been oriented as in
          Fig. 1 and the protein subunit shown would be the one which is
          towards the rear of that figure. The helices in the foreground
          of the diagram are the inner and outer recognition helices. They
          connect the third β-strand to the fourth and the fourth β-strand
          to the fifth. The two helices also form the central interface
          with the other subunit. The amino-terminus of the polypeptide
          chain is in the arm near the DNA.

The surface of the protein is involuted to form two symmetry related
clefts which contain segments of the DNA backbone including the scissile
bond. The co-crystals were grown in the absence of Mg^{+2}, which prevents
DNA cleavage, but the DNA can be cleaved in situ by diffusing Mg^{+2} into the
co-crystals. The Mg^{+2}-treated co-crystals continue to diffract X-rays;
structure determination of the resulting enzyme-product co-crystal is in
progress. The amino acid residues responsible for forming the cleavage site
are somewhat displaced from the scissile bond. That is, the structure reported here appears to be a specifically bound, inactive conformer which
can isomerize to an active enzyme upon addition of Mg^{+2}. We suggest that
this represents a functional intermediate on the catalytic pathway.
According to this model, which we term "allosteric activation", the protein-
base interactions at the DNA recognition site have a strong allosteric effect on the equilibrium between the inactive and active forms such that the
active form is favored only when the cognate sequence is bound (under physiological conditions). Allosteric activation would enhance the specificity
of EcoRI endonuclease by inhibiting cleavage at non-cognate sites. The
allosteric activation model also could account for the relaxation of specificity under EcoRI* conditions by invoking a solvent mediated shift of the
conformational equilibrium towards the active form even when EcoRI* sites
were bound to the enzyme.

The interactions which determine the recognition specificity depend
critically on the relative positioning of the bases and amino acid side
chains at the DNA-protein interface. Unitary α-helices position the key
amino acid residues with respect to the DNA (see Fig. 6). These α-helices
are organized into modules with a spatial division of labor across the recognition site. The outer G-C base pairs are recognized by identical,
symmetry-related outer modules. Each outer module consists of a single
α-helix. The inner tetranucleotide, AATT, is recognized by the inner mod-

ule which consists of two symmetry-related α-helices, one from each subunit.
Amino acid side chains from the modules establish the relative position of
the α-helices so as to form a four helix bundle. Additional amino acid
side chains position the bundle with respect to the DNA by interacting
with the DNA backbone and by anchoring the recognition bundle within secondary structure of the complex.

Sequence specificity is mediated by twelve hydrogen bonds between the
protein and bases within the EcoRI hexanucleotide (see Fig. 7). Bidentate
hydrogen bonds between arginine 200 and guanine (arg::G) determine the base
specificity of the outer module. Substitution of any base other than
guanine would lead to rupture of at least one of these hydrogen bonds. The

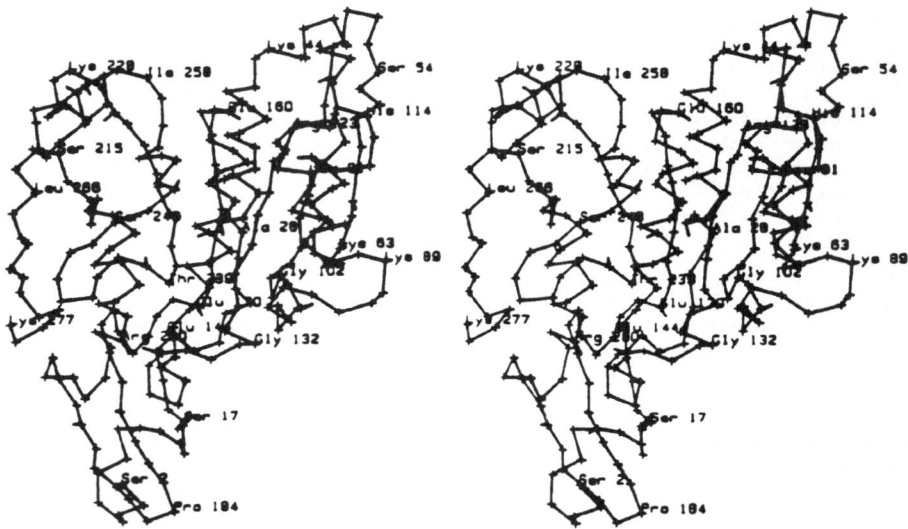

Fig. 5.   A stereo drawing of the α-carbon trace of one subunit of EcoRI
          endonuclease.

inner module also utilizes bidentate hydrogen bonds, but in a bridging tet-
rad arrangement with glutamic acid 144 and arginine 145 forming four hydrogen bonds to adjacent adenine residues (glu-arg::AA). Substitution of any
other base for either adenine residue would also result in rupture of at
least one hydrogen bond. No hydrogen bonds are formed with the pyrimidine
residues, however they are recognized by hydrogen bonds to the purines on
the complementary strand. The twelve hydrogen bonds are therefore unique
to the canonical EcoRI hexanucleotide. These interactions are also consistent with the spectrum of EcoRI* cleavage rates because the observed hier-
archies of cleavage rates can be predicted simply by counting the maximal
number of hydrogen bonds possible between the protein and relevant EcoRI*
sites [3,5].

254

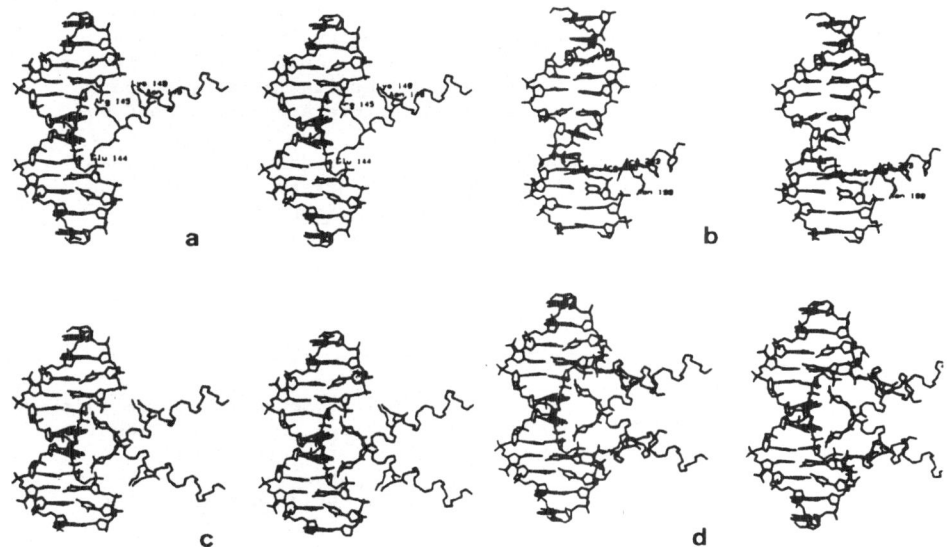

Fig. 6. Stereo drawings showing the recognition α-helices and modules.
a) The "inner" α-helix, which is part of the inner recognition
module. This helix is also a crossover helix, connecting the third
and fourth strands of the β-sheet. Glutamic acid 144 interacts
with adenine residues in the lower half of the DNA and arginine
interacts with adenine residues in the upper half. Lysine 148 and
asparagine 149 interact with the phosphate moiety from guanine 4.
b) The "outer" α-helix, which is also one of two, identical outer
recognition modules. Each outer module determines the specificity
for one of the two G-C base pairs at the ends of the EcoRI hexanucleotide via an interaction between arginine 200 and guanine.
The outer α-helix also connects the fourth and fifth β-strands.
Asparagine 199 interacts with the phosphate moiety from cytosine
3' (C3 on the opposite strand), while arginine 203 interacts with
phosphate moieties from cytosine 3' and guanine 4'. In the other
views in this figure, the two-fold symmetry axis is in the plane
of the figure, however, this view has been rotated approximately
20° for clarity. c) The inner recognition module, consisting of
the inner α-helices from both subunits. This module determines
the specificity in the inner tetranucleotide, AATT via interactions
between glutamic acid 144 and arginine 145 from both subunits and
all four adenine residues. d) The four helix bundle consisting
of the inner and outer α-helices from both subunits.

The recognition interactions are stabilized by electrostatic interactions between oppositely charged amino acid side chains at the recognition site. This suggests that the DNA-protein interaction energy is not a simple additive sum over the individual interactions. The formation of some correct protein-base interactions probably facilitates formation of additional correct interactions while incorrect interactions with non-cognate bases probably have an inhibitory effect. We refer to this mechanism as "cooperative enhancement" because it sharpens the differentiation between cognate and non-cognate sites. Glutamic acid 144 side chains from both subunits are centrally located in the electrostatic array. EcoRI methylase mediated modification of either central adenine residue results in methylation of the N6 amino group. This modification would rupture a hydrogen bond between the N6 amino group and glutamic acid 144. It would also physically displace one or both of these negative charges. The charge displacement should perturb the entire recognition interface, thereby sharpening the discrimination between the modified and unmodified EcoRI sites.

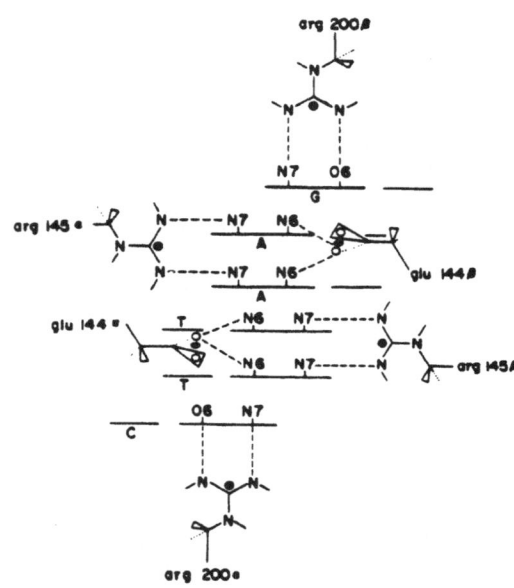

Fig.7. A schematic representation of the recognition interactions and the twelve hydrogen bonds which determine the specificity of EcoRI endonuclease. α and β in this figure refer to the bases and amino acid side chains have been shifted from the current model as shown in Fig.6 in the interests of clarity.

REFERENCES

1. C. A. Frederick, J. Grable, M. Melia, C. Samudzi, L. Jen-Jacobson, B.-C. Wang, P. J. Greene, H. W. Boyer, and J. M. Rosenberg, Kinked DNA in crystalline complex with EcoRI endonuclease, Nature 309:327-331 (1984).

2. J. M. Rosenberg, J. A. McClarin, C. A. Frederick, B.-C. Wang, H. B. Boyer, and P. Greene, The 3 Å structure of a DNA-EcoRI endonuclease recognition complex, Chemica Scripta 26: (1985) in press. Paper presented by John M. Rosenberg at the Conference on "Molecular Evolution of Life", Lidingo, Sweden, 8-12 September 1985.

3. J. A. McClarin, C. A. Frederick, B.-C. Wang, P. Greene, H. W. Boyer, J. Grable, and J. M. Rosenberg, Structure of the DNA-EcoRI endonuclease recognition complex at 3 Å resolution, Science 234:1526-1541 (1986).

4. R. E. Dickerson and H. R. Drew, Structure of a B-DNA dodecamer: II. Influence of base sequence on helix structure, J. Mol. Biol. 149: 761-786 (1981).

5. J. M. Rosenberg and P. J. Greene, EcoRI* specificity and hydrogen bonding, DNA, 1:117-124 (1982).

ANTIBODIES TO NUCLEIC ACIDS

Marc Leng

Centre de Biophysique Moléculaire, C.N.R.S.
1A, avenue de la Recherche Scientifique
45071 Orléans   edex 2, France

The ability of vertebrate species to cope with foreign substances depends in part on their capacity to synthesize antibodies which recognize specifically and bind to the foreign substances called antigens. An antibody is a serum protein belonging to the family called immunoglobulins. The immunoglobulins are biologically active glycoproteins capable of carrying out several different functions. They can bind specifically to antigens, activate the complement systems, mediate cytotropic reactions and act as antigen receptors on lymphocyte membranes.

The antigenic functions of nucleic acids were recognized much later than those of proteins and polysaccharides. The discovery that the sera of patients with systemic Lupus erythematosus, a disease of the immune system, contained antibodies reacting with natural DNA gave an impetus to studies for experimental inductions of antibodies to nucleic acids. It is now well-established that the synthesis of antibodies specifically recognizing bases, nucleosides, nucleotides, single-stranded and multi-stranded polynucleotides, can be induced in animals immunized with the right antigen (general reviews [1-7]). These antibodies are largely used in many laboratories for at least three main reasons,

- They have many applications as biochemical reagents.

- The study of the antigen-antibody complexes can help to the elucidation of the mechanism of recognition between nucleic acids and proteins. The ability to get experimentally induced homogeneous monoclonal antibodies make these studies very promising.

- They are interesting models for the understanding of the autoimmune phenomena.

We here report some results illustrating the first two points. The first part of the paper deals with some characteristics of immunoglobulins and the procedures of immunization, the second part with some properties of the antibodies to nucleosides(tides) and of the antibodies to double-stranded polynucleotides.

Immunoglobulins

An antibody is a serum protein belonging to the family called immuno-globulins. There are five classes of immunoglobulins (IgG, IgM, IgA, IgD and IgE) which are built of heavy and light chains. All the immunoglobu-lins can be regarded as being derived from a basic structure of two light polypeptide chains of molecular weight 22000-23000 (about 220 amino acid residues) and two heavy polypeptide chains of molecular weight 50000-70000, linked together by the disulphide bridges of cystine residues. The sequence analyses of numerous immunoglobulins have demonstrated the exis-tence of distinct regions of sequence homology, called domains. Each light chain has two domains, one variable domain (V_L) and one constant (C_L). Each heavy chain has one variable domain (V_H) and several constant do-mains. The constant domains are common to immunoglobulins of the same group. The variable domains vary from antibody to antibody. They have sections of variable amino acid sequence that are involved in the antigen binding site (an IgG has two antigen binding sites) and determine the antigen binding specificity.

Under suitable conditions, limited proteolysis with papain results in three fragments of about equal molecular weight, two Fab fragments and one Fc fragment. The antigen binding site is localized on the Fab fragment while the Fc fragment is involved in complement fixation and other biolo-gical functions. A schematic representation of the basic immunoglobulin structure is shown in figure 1.

The Y-shaped molecule is not rigid. The Fab fragments can change their shape and their relative orientations because some flexible segments of the polypeptide chains are located at the hinge regions and between the V_L and C_L domains (general reviews [8-9]).

Antigen

An antigen is any substance which when injected into an animal in-duces the formation of antibodies capable of reacting specifically with the antigen. The substance has to be of large molecular weight. Low mole-cular substances (approximately less than 1500) do not induce the forma-

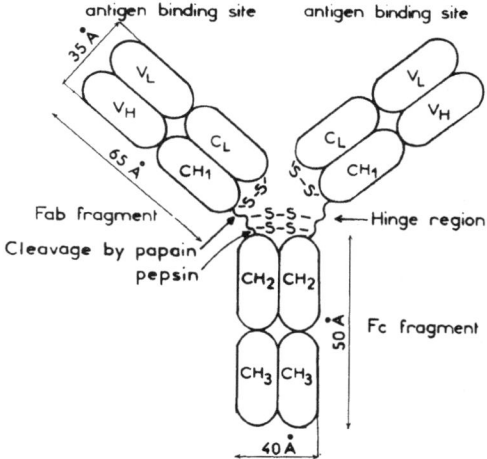

Fig. 1 - Schematic representation of the basic immunoglobulin structure.

tion of antibodies. However, antibodies can be raised to small molecules by immunization with conjugates made up of low molecular weight substances called haptens, covalently linked to a macromolecule (generally a protein).

Antigen-antibody complexes

The reaction of formation of a complex between an antibody and the corresponding antigen is reversible. Both products are not chemically modified. The specificity of the interaction is a function of the stability of the complex formed between the antigen and the antibody. The intermolecular forces involved in the antibody-antigen complexes are similar to those engaged in any type of molecular complex (electrostatic, hydrogen bonding, stacking and hydrophobic interactions). The thermodynamic parameters for numerous antibody-hapten complexes have been determined. The values of the association constants vary in a large range ($\simeq 10^4$-10^{11} M^{-1}). The thermodynamic parameters for the complexes between an antibody and an antigen bearing several antigenic determinants are more difficult to estimate. Depending upon their relative concentrations, the immunoglobulins (IgG have two antigen binding sites, IgM have ten...) can bind to the same macromolecule, or bind to several macromolecules leading to a large molecular weight soluble complex, or cross-link several macromolecules leading to the formation of a precipitate (general reviews [8]).

Immunochemical procedures

The three major steps involved in the production of antibodies to nucleic acids are : the synthesis of the antigen, the immunization of animals, the characterization of the antibodies.

Antigen. Low molecular weight compounds (haptens) are not antigenic and thus have to be covalently linked to a carrier. Bovine serum albumin has often been used but some other proteins might be better carriers. The choice of the covalent linkage depends upon the nature of the hapten and several chemical reactions have been described[10]. When possible, the periodate oxidation reaction, as proposed by Erlanger and Beiser, is one of the most convenient[11] and is summarized in figure 2.

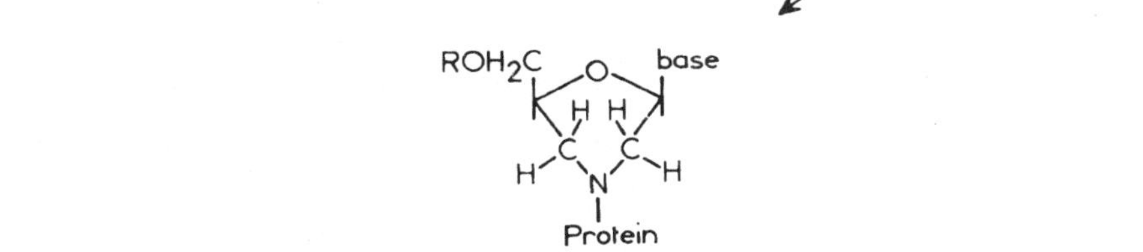

Fig. 2 - Conjugation of a ribonucleoside, ribonucleotide or oligoribonucleotides to a protein according to Erlanger and Beiser[11] (From M. Leng, Biochimie, 67:309, 1985 ; with permission).

Immunization. There are many immunization protocols[1-3,12]. In rabbits, large molecular weight nucleic acids are often mixed with positively charged methylated bovine serum albumin and form insoluble complexes. These complexes or the hapten-protein conjugates are prepared for injection in the form of a stable emulsion by mixing with complete Freund's adjuvant. Several inoculations are done at various time intervals. Rabbits possessing antibodies are boosted with antigen emulsified in incomplete Freund's adjuvant. The amount of antigen for each inoculation is of the order of 50-200 μg.

Large amounts of homogeneous antibodies (monoclonal antibodies) can be obtained after hybridization of lymphoid cells from an immunized animal (mouse for example) with cells from a mouse myeloma adapted to growth in culture[13].

Specificity. The specificity of the antibodies can be characterized by several physico-chemical methods. Radioimmunoassays (RIA) are now largely used because of their sensitivity and reliability. After determination of conditions giving 50 % binding of the radioactive hapten or polymer (tracer), competition experiments are performed with unlabeled inhibitors. This enables to prove the specificity of the antibodies, to titrate the amount of unlabeled antigen in the samples to be analysed, and to get some knowledge on groups or atoms of the antigen involved in the recognition by the antibodies. Another technique, also largely used, is the enzyme-linked immunosorbent assay (ELISA) which does not need a radioactive tracer[14].

ANTIBODIES TO NUCLEOSIDES(TIDES)

In order to obtain antibodies against nucleosides(tides), the nucleoside(tide) has to be covalently linked to a carrier. This modifies the nucleoside(tide) and consequently the specificity of the antibodies. In other words, the specificity of the polyclonal antibodies to a nucleotide will depend on whether the nucleotide is linked to the carrier via the sugar ring, the phosphate group or a group of the base. Nevertheless, some general features can be given and the antibodies to adenosine (or to adenosine 5'-monophosphate) are taken as an example (general reviews[3,5,6,15]).

The base plays an important role in the recognition by the antibodies. The antibodies have a much lower affinity for Guo, Cyd or Thd than for adenosine (the ratio of the association constants K(adenosine) over K(nucleoside) is larger than 1000). Inosine is recognized but less than adenosine. In general, the antibodies bind almost equally well adenosine and d-adenosine. The base donor and acceptor groups for H bonds are important in the formation of the hapten-antibody complex. The antibodies bind to oligo(A), poly(A), denatured DNA but do not bind to native DNA. These results show the potential of the antibodies to adenosine. They can be used to titrate the percentages of adenine residues in a nucleic acid (after hydrolysis of the sample, if necessary) or to detect defects (denatured regions) in a nucleic acid. Several examples illustrate the latter point. The covalent binding of chemical compounds to DNA can distort the DNA conformation. This distortion which can interfere with biological events has to be characterized. The covalent binding of chemical carcinogens to DNA is thought to play an important role in the first step of cancer process. The mechanism of action of drugs is related to their bindings to DNA. The comparison of the affinities of the antinucleoside antibodies to modified and unmodified DNAs has brought some knowledge on the local denaturation of the double-helix (see for example [16-18]). An interesting study on metaphase chromosomes has compared the patterns pro-

260

duced by quinacrine or by the Giemsa G-banding technique to the patterns produced by the antinucleoside antibodies after chemical modifications of the chromosomes[19].

The association constants for the binding of Fab fragments or IgG and several haptens deduced from dialysis equilibrium, fluorescence quenching and competition experiments[20] are of the order of 10^6 M^{-1}. Much higher values were obtained with antibodies against several modified nucleosides. These high values (and a very small binding to unmodified nucleoside) have allowed the titration of the modified bases (1 modified base per 10^{-4} - 10^{-6} nucleotide residue) after in vivo modification of DNA[3,21].

More generally, it seems better to use monoclonal antibodies than polyclonal antibodies because the specificity of monoclonal antibodies is better characterized than that of polyclonal antibodies. On the other hand, it is much easier to get polyclonal than monoclonal antibodies. Nevertheless, it is necessary to emphasize that even with monoclonal antibodies, some unexpected cross-reaction can occur and thus, as with many other probes, control experiments have to be carried out.

ANTIBODIES TO DOUBLE-STRANDED POLYNUCLEOTIDES

Up to now, it does not seem possible to induce the synthesis of antibodies to B-DNA by immunization with B-DNA (denatured DNA is a good immunogen). On the other hand, several double-stranded polynucleotides are immunogens (reviews[1,2,5,15]). We will present some results obtained with the antibodies to poly(I).poly(C) and with the antibodies to Z-DNA.

Specificity

Animals were immunized with poly(I).poly(C) or with Z-DNA complexed with methylated bovine serum albumin. To induce the synthesis of the antibodies to Z-DNA, chemically or enzymatically modified poly(dG-dC). poly(dG-dC) or poly(dA-dC).poly(dG-dT) were used as antigens. Several studies have shown that these modified polymers are in the Z conformation under physiological conditions (reviews[22,23]).

The specificity of the antibodies has been studied by several physico-chemical and immunological techniques.

The antibodies to poly(I).poly(C) bind to poly(I).poly(C), poly(A). poly(U) but do not bind to poly(G).poly(C), poly(X).ploly(U), poly(dG-dC).poly(dG-dC) and native DNA. Several chemical modifications of the bases do not lead to large changes in the association constants. These results strongly suggest that the antibodies recognize mainly the phosphodiester backbone of the double helix[1,2,5,15].

Several modified poly(dG-dC) or poly(dA-dC).poly(dG-dT) (Z-form) have been used as antigens. All the results agree that the antibodies recognize Z-DNA but do not recognize linear B-DNA, denatured DNA, DNA-RNA hybrids, single or double-stranded RNA, poly(dG).poly(dC) or poly(dA-dT). poly(dA-dT). Several sites seem to be recognized by the antibodies[2,22,24-29]. This has been confirmed by the study of monoclonal antibodies to Z-DNA. For example, some antibodies are sensitive to the base composition of Z-DNA, some are sensitive to ionic strength... which demonstrate the existence of several antigenic domains in Z-DNA such as the helical convex surface or the phosphodiester backbone of the double helix[28,30-32].

Association constants

The association constants for the binding of anti poly(I).poly(C) Fab fragments (purified by affinity chromatography) and poly(I).poly(br5C) have been deduced from fluorescence experiments[33]. In the presence of poly(I).poly(br5C), the fluorescence of Fab fragments is quenched. Assuming the same fluorescence quenching for all the bound Fab fragments, r and r/c can be calculated ; r is the molar ratio bound Fab fragments per nucleotide and c is the concentration of free Fab fragments. The results (Fig. 3) have been analysed using the following equation :

$$\frac{r}{c} = K(1-nr)\left(\frac{1-nr}{1-(n-1)r}\right)^{n-1}$$

in which K is the intrinsic association constant and n is the number of nucleotide residues covered by each ligand. This equation assumes no-cooperative effect ; the non-linearity of the variation of r/c as a function of r is due to the fact that the ligand covers more than one lattice residue[34]. One deduces that n is equal to 7 (3-4 base pairs) and K are equal to 3.2×10^6 M^{-1}, 1.5×10^6 M^{-1} and 0.3×10^6 M^{-1} in 20 mM, 50 mM and 155 mM Na^+, respectively. From the linear variation[35] of log K as a function of log Na^+ , 1.5 phosphate groups among the seven phosphate groups covered by the Fab fragment binding site are found to be involved in electrostatic interactions.

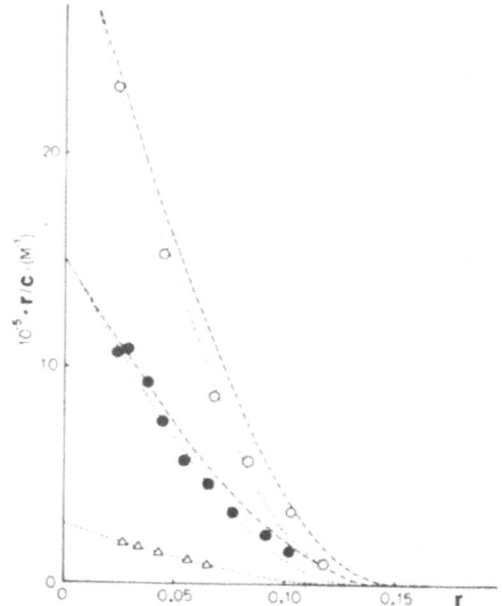

Fig. 3 - Binding isotherms. Variation of r/c as a function of r. (O) 15 mM NaCl, (●) 50 mM NaCl, (△) 0.15 M NaCl. (All the solutions contain 5 mM Tris-HCl, pH 7.5, 0.1 mM EDTA). Temperature 15°C. The dotted lines are calculated assuming n = 7 (----) and n = 6 (_ _ _) (from M. Guigues and M. Leng, Eur. J. Biochem., 69:615, 1976 ; with permission).

Size of the binding site

Each anti poly(I).poly(C) Fab fragment binding site covers about 7 nucleotides, i.e., 3-4 base pairs. This does not mean that all these nucleotide residues interact with the aminoacids of the binding site but they are no longer available for the binding of another Fab fragment. Another estimate of the number of residues covered by each binding site can be deduced from the analysis of the antigen-antibody precipitate, formed in the presence of a large excess of antibodies relative to the amount of antigen to be precipitated. By this technique[26,33] about 5 base pairs of poly(I).poly(C) are covered by each binding site of the antibodies to poly(I).poly(C) and about 4 base pairs of modified poly(dG-dC) (Z-DNA) by each binding site of the antibodies to Z-DNA. An approximate size of the antibody binding site is 14 Å x 20 Å x 10 Å.

Recently, a new approach[32] has been developed to investigate at the nucleotide level the patterns of Z-DNA recognition by the antibodies to Z-DNA. The N(7) of guanine residues in Z-DNA react with the chemical compound diethylpyrocarbonate[36,37] (they are much less reactive in B-DNA). On the other hand, it is well-established that Z-DNA is efficiently stabilized by free energy of torsional stress within a negatively supercoiled DNA (review[22]). In physiological conditions, $(dG-dC)_n$ sequences inserted in closed circular DNA are in B or in Z conformation depending upon the superhelical density of DNA[22]. The study of the reactivity of diethylpyrocarbonate and $(dG-dC)_n$ sequences inserted in plasmids-antibodies to Z-DNA complexes shows that only a few guanines are protected from modification. Each antibody binding site covers about 6 base pairs. The number of guanine residues protected from reaction with diethylpyrocarbonate depends upon the monoclonal antibody which reflects different interactions between Z-DNA and the antibodies.

Stability of the double helices in the presence of specific antibodies

As already pointed out, antibodies are useful reagents in the titration and location of the corresponding antigens. However, since the antibodies bind to a given form and if this form is in equilibrium with another form, the addition of the antibodies will shift the equilibrium towards the form they recognized. This can be illustrated by the following example. In the right salt conditions, the mixture of poly(A) and poly(U) leads to the formation of a triple-stranded helix poly(A).2 poly(U). On heating, a biphasic melting is observed, the triple-stranded helix is first transformed in the double-stranded helix poly(A).poly(U) and then at higher temperature, poly(A).poly(U) melts[38].

$$poly(A).2\ poly(U) \rightleftharpoons poly(A).poly(U) + poly(U)$$

$$poly(A).poly(U) \rightleftharpoons poly(A) + poly(U)$$

The antibodies to poly(I).poly(C) bind to poly(A).poly(U) and to poly(A).2 poly(U) but they have a higher affinity[39] to poly(A).poly(U) than to poly(A).2 poly(U) (they do not bind to poly(A) or to poly(U)). Thus, in the presence of the antibodies, one expects the conversion poly(A).2 poly(U) - poly(A).poly(U) to be shifted towards the right and the conversion poly(A).poly(U) - poly(A) + poly(U) to be shifted towards the left. In other words, the presence of the antibodies to poly(I).poly(C) should decrease the Tm of poly(A).2 poly(U) and should increase the Tm of poly(A).poly(U). As shown in figure 4, this has been verified, the differences in Tm depending upon the antibodies concentration (theoretical calculations on the helix-coil transition of DNA in the presence of large ligands have been performed by McGhee[40]).

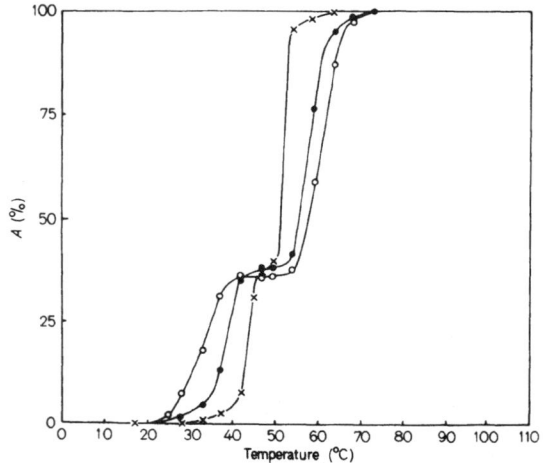

Fig. 4 - Thermal stability of poly(A).2
poly(U). Variation of the absorbance
at 260 nm as a function of temperature. Polynucleotide (×), polynucleotide + anti poly(I).poly(C) Fab
fragments (●) at P/Fab = 46, (o)
P/Fab = 26. Poly(A).2 poly(U) concentration 0.1 M. Solvent 40 mM
Nacl, 10 mM sodium cacodylate, pH 7,
0.1 mM EDTA (from M. Guigues and M.
Leng, Eur. J. Biochem., 69:615,
1976 ; with permission).

Several experiments have shown that the antibodies to Z-DNA shift the
B-DNA - Z-DNA transition towards the right[25,32,41-44]. More generally, any
equilibrium A⟷B can be shifted toward the left or the right in the
presence of antibodies which have more affinity for A or for B, respectively.

Detection of Z-DNA

Antibodies to Z-DNA have been largely used to detect sequences of Z-
DNA in free DNA molecules or in complex structures such as chromosomes or
nuclei (general review [7,22,23]). An example is the visualization of Z-DNA
in form V-DNA by electron microscopy. Form V-DNA has been prepared from
pBR 322 DNA by annealing covalently closed complementary single strands
according to the procedure of Stettler et al.[45]. The antibodies were first
reacted with form V-DNA and then the complexes were reacted with ferritin
labeled goat immunoglobulins anti rabbit immunoglobulins[46]. Figure 5 shows
that the antibodies to Z-DNA bind to form V-DNA and that they are distributed all along the DNA molecules. There are several sequences in the Z-
form and not one long sequence. From inspection of pBR 322 base sequence,
it has been deduced that the regions recognized by the antibodies contain
(A.T) and (G.C) base pairs.

Antibodies to Z-DNA react with fixed chromosomes and nuclei. Since
the fixatives (in general acetic acid) can remove the proteins and thus
change the torsional strain, these experiments do not prove the presence
of Z-DNA in chromosomes or nuclei but show that many DNA regions have the
potential for forming Z-DNA structure (reviews[7,22,23]). In one case, the
nucleotide sequence recognized by the antibodies has been determined[47].

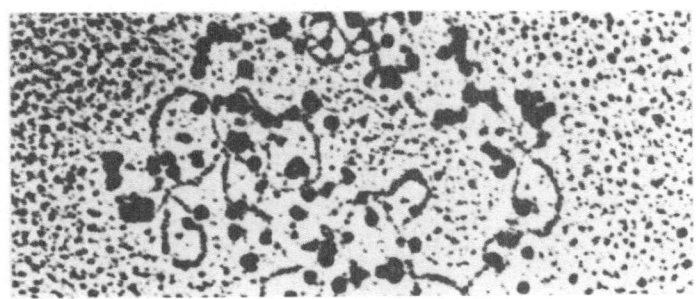

Fig. 5 - Electron micrograph of ferritin-labeled goat antibodies bound to the form V-DNA - antibodies to Z-DNA complexes.

Antibodies to Z-DNA interact strongly with R-band positive heterochromatic segments of fixed metaphase chromosomes of Cebus[48]. These segments are constituted of a satellite DNA, the repeat unit of which is about 1520 base pairs long. This segment contains a $(AC)_{15}$ region which, _in vitro_, adopts the Z conformation under topological constraints. Moreover, an adjacent sequence which is $(AC)_n$ rich but with several bases out of alternation is also in the Z conformation (Vogt _et al._, to be published).

In conclusion, it seems likely that in the near future, antibodies to nucleic acids will be largely used in many laboratories and in particular to study the polymorphism of DNA in complex structures.

REFERENCES

1. F. Lacour, E. Nahon-Merlin, and M. Michelson, Immunological recognition of polynucleotide structure, Curr. Top. Microbiol. Immunol. 62:1 (1973).
2. B.D. Stollar, The specificity and applications of antibodies to helical nucleic acid antigens, CRC Crit. Rev. Biochem. 3:45 (1975).
3. T.W. Mums, and M.K. Liszewski, Antibodies specific for modified nucleosides : an immunochemical approach for the isolation and characterization of nucleic acids, Prog. Nucl. Acid Res. Mol. Biol. 24:109 (1980).
4. P.T. Strickland, and J.M. Boyle, Immunoassay of carcinogen-modified DNA, Prog. Nucl. Acid Res. Mol. Biol. 31:1 (1984).
5. T.M. Jacob, and C. Srikumar, Nucleic acid reactive antibodies-specificity and applications, J. Biosci. 7:61 (1985).
6. M. Leng, Immunological detections of lesions in DNA, Biochimie 67:309 (1985).
7. B.D. Stollar, Antibodies to DNA, CRC Crit. Rev. Biochem. 20:1 (1986).
8. E.A. Kabat, Basic principle of antigen-antibody reactions, in : "Methods in Enzymology" H. Van Vunakis and J.J. Langone, ed., vol. 70, part A, Academic Press, New-York (1980).
9. M. Marquart, and J. Deisenhofer, The three-dimensional structure of antibodies, Immunology Today 3:160 (1982).
10. B.F. Erlanger, The preparation of antigenic hapten-carrier conjugates : a survey, in "Methods in Enzymology", H. Van Vunakis and J.J. Langone, ed., vol. 70, part A, Academic Press, New-York (1980).

11. B.F. Erlanger and S.M. Beiser, Antibodies specific for ribonucleosides and ribonucleotides and their reaction with DNA, Proc. Natl. Acad. Sci., 52:68 (1964).

12. B.A.L. Hurn, and C.M. Chantley, Production of reagent antibodies, in "Methods in Enzymology", H. Van Vunakis and J.J. Langone, ed., vol. 70, part A, Academic Press, New-York (1980).

13. C. Milstein, From antibody structure to immunological diversification of immune response, Science 231:1261 (1986).

14. E. Enguall, Enzyme immunoassay ELISA and Emit, in "Methods in Enzymology", H. Van Vunakis and J.J. Langone, ed., vol. 70, part A, Academic Press, New-York (1980).

15. D.B. Stollar, The experimental induction of antibodies to nucleic acids, in "Methods in Enzymology", H. Van Vunakis and J.J. Langone, ed., vol. 70, part A, Academic Press, New-York (1980).

16. E. Sage, M. Spodheim-Maurizot, P. Rio, M. Leng, and R.P.P. Fuchs, Discrimination by antibodies between local defects in DNA induced by 2-aminofluorene derivatives, FEBS Lett. 108:66 (1979).

17. R.M. Santella, D. Grunberger, S. Broyde, and B.E. Hingerty, Z-DNA conformation of N-2 acetylaminofluorene modified poly(dG-dC).poly(dG-dC) determined by reactivity with anticytidine antibodies and minimized potential energy calculations, Nucleic Acids Res. 9:5459 (1981).

18. W.I. Sundquist, S.J. Lippard, and D.B. Stollar, Binding of cis- and trans-diamminedichloroplatinum(II) to deoxyribonucleic acid exposes nucleosides as measured immunochemically with anti-nucleosides antibodies, Biochemistry 235:1520 (1986).

19. R.R. Schreck, D. Warburton, O.J. Miller, S.M. Beiser, and B.F. Erlanger, Chromosome structure as revealed by a combined chemical and immunochemical procedure, Proc. Natl. Acad. Sci. 70:804 (1973).

20. J. Lavayre, and M. Leng, Purification and specificity of antibodies to adenosine, Biochimie 59:33 (1977).

21. M.C. Poirier, Antibodies to carcinogen-DNA adducts, J. Natl. Cancer Inst. 67:515 (1981).

22. A. Rich, A. Nordheim, and A.H.J. Wang, The chemistry and biology of left-handed Z-DNA, Ann. Rev. Biochem. 53:791 (1984).

23. M. Leng, Left-handed Z-DNA, Biochim. Biophys. Acta 825:339 (1985).

24. E.M. Lafer, A. Mollar, A. Nordheim, B.D. Stollar, and A. Rich, Antibodies specific for left-handed Z-DNA, Proc. Natl. Acad. Sci. USA 78:3546 (1981).

25. B. Malfoy, and M. Leng, Antiserum to Z-DNA, FEBS Letters 132:45 (1981).

26. B. Malfoy, N. Rousseau, and M. Leng, Interaction between antibodies to Z-form deoxyribonucleic acid and double-stranded polynucleotides, Biochemistry 23:54 (1984).

27. D.A. Zarling, D.J. Arndt-Jovin, M. Robert-Nicoud, L.P. McIntosh, R. Thomae, and T.M. Jovin, Immunoglobulin recognition of synthetic and natural left-handed Z-DNA conformations and sequences, J. Mol. Biol. 176:369 (1984).

28. D.A. Zarling, D.J. Arndt-Jovin, L.P. McIntosh, M. Robert-Nicoud, and T.M. Jovin, Interactions of anti-poly d(G-br^{5}C) IgG with synthetic, viral and cellular Z-DNA's, J. Biomol. Struct. Dyn. 1:1081 (1984).

29. L.H. Hanau, R.M. Santella, D. Grunberger, and B.F. Erlanger, An immunochemical examination of acetylaminofluorene-modified poly(dG-dC).poly(dG-dC) in the Z-conformation. J. Biol. Chem. 259:173 (1984).

30. A. Möller, J.E. Gabriels, E.M. Lafer, A. Nordheim, A. Rich, and B.D. Stollar, Monoclonal antibodies recognize different parts of Z-DNA, J. Biol. Chem. 257:12081 (1982).

31. J.S. Lee, M.L. Woodsworth, and J.P. Latimer, Functional groups on "Z" DNA recognized by monoclonal antibodies, FEBS Lett. 168:303 (1984).

32. L. Runkel, and A. Nordheim, Chemical footprinting of the interaction between left-handed Z-DNA and anti Z-DNA antibodies by diethylpyrocarbonate carbethoxylation, J. Mol. Biol. 189:487 (1986).

33. M. Guigues, and M. Leng, Antibodies to poly(I).poly(C) : purification and interaction with polynucleotides, Eur. J. Biochem. 69:615 (1976).

34. J.D. McGhee, and P.H. von Hippel, Theoretical aspects of DNA-protein interactions : cooperative and non-cooperative binding of large ligands to one-dimensional homogeneous lattice, J. Mol. Biol. 86:469 (1974).

35. T.M. Record, Jr., T.M. Lohman, and P.L. de Haseth, Ion effects on ligand-nucleic acids interactions, J. Mol. Biol. 107:145 (1976).

36. W. Herr, Diethylpyrocarbonate : a chemical probe for secondary structure in negatively supercoiled DNA, Proc. Natl. Acad. Sci. USA 82:8009 (1985).

37. B.H. Johnston, and A. Rich, Chemical probes of DNA conformation : detection of Z-DNA at nucleotide resolution, Cell 452:713 (1985).

38. C.L. Stevens, and G. Felsenfeld, The conversion of two-stranded poly(A+U) to three-stranded poly(A+2U) and poly A by heat, Biopolymers, 2:293 (1964).

39. M. Guigues, and M. Leng, Recognition of polynucleotides by antibodies to poly(I).poly(C), Nucleic Acids Res. 3:3337 (1976).

40. J.D. McGhee, Theoretical calculations of the helix-coil transition of DNA in the presence of large, cooperatively binding ligands, Biopolymers 15:1345 (1976).

41. T.M. Jovin, L. McIntosh, D.J. Arndt-Jovin, D.A. Zarling, M. Robert-Nicoud, J.H. Vande Sande, K.F. Jorgensen, and F. Eckstein, Left-handed DNA : from synthetic polymers to chromosomes, J. Biomol. Struct. Dyn. 1:21 (1983).

42. B. Revet, D.A. Zarling, T.M. Jovin, and E. Delain, Different Z-DNA forming sequences are revealed in ØX174 RFI by high resolution darkfield immuno-electron microscopy, EMBO J. 3:3353 (1984).

43. E.M. Lafer, R. Sousa, R. Ali, A. Rich, and B.D. Stollar, The effect of anti-Z-DNA antibodies on the B-DNA-Z-DNA equilibrium, J. Biol. Chem. 261:6438 (1986).

44. F.M. Pohl, Dynamics of the B- to Z- transition in supercoiled DNA, Proc. Natl. Acad. Sci. USA, in press.

45. U.H. Stettler, H. Weber, Th. Koller, and Ch. Weissmann, Preparation and characterization of form V DNA, the duplex DNA resulting from association of complementary, circular single-stranded DNA, J. Mol. Biol. 31:21 (1979).

46. M.C. Lang, B. Malfoy, A.M. Freund, M. Daune, and M. Leng, Visualization of Z sequences in form V of pBR 322 by immuno-electron microscopy, EMBO J. 1:1149 (1982).

47. B. Malfoy, N. Rousseau, N. Vogt, E. Viegas-Pequignot, B. Dutrillaux, and M. Leng, Nucleotide sequence of an heterochromatic segment recognized by the antibodies to Z-DNA in fixed metaphase chromosomes, Nucleic Acids Res. 14:3197 (1986).

48. E. Viegas-Pequignot, C. Derbin, B. malfoy, E. Taillandier, M. Leng, and B. Dutrillaux, Z-DNA immunoreactivity in fixed metaphase chromosomes of primates, Proc. natl. Acad. Sci. USA 80:5890 (1983).

SUMMARY AND CONCLUSIONS

Wolfram Saenger

Institut für Kristallographie, Freie Universität

Berlin, Takustr. 6, D-1000 Berlin 33, FRG

After almost two weeks of constant assimilation of science either in the form of 19 lectures, of 18 research seminars, of 48 posters, and of 3 vivid round table discussions, or in discussions after lectures, during meals, and in the free afternoons, I try to write a summary to draw conclusions referring to the lectures which we have heard during the school on "DNA-Ligand Interactions: From Drugs to Proteins". I also include my view of some research seminars which were of particular interest in connection with the lectures. The reader should be aware that these are my personal impressions resulting from recollections and from the notes I took during the lectures and seminars, and that somebody else would have and will have another view of the scientific contents which were presented. In the following I shall order my thoughts in three categories of increasing complexity: DNA structure and small ligands, DNA-drug interaction, DNA-protein interaction.

Structure of DNA and its Complexes with Small Ligands

SUMMARY. The discovery of Z-DNA and the crystal structure analysis of oligonucleotides confirmed solution studies which suggested that DNA is not a uniform, stiff rod. Depending on nucleotide sequences, it displays structural variations which, at the atomic level, were described by O. Kennard with emphasis on base pair mismatches, and by R.E. Dickerson with a detailed treatment of the binding to DNA of the drugs netropsin, distamycin, and related substances. In general, the geometry of specific and non-specific binding of drugs to DNA begins to unravel. We also have some insight into the binding of water to DNA. Water binds to phosphate groups and exerts a structural specificity for B, A, Z form, whereas structural and sequence specificity was observed in the binding of water to the minor groove of B-DNA (sequence AATT) and of Z-DNA (sequence CGCG...), where it forms spines of hydration. The major groove of A-DNA hosts water pentagons in the sequence TATA. In Z-DNA the binding of $Mg^{++}(H_2O)_6$

and $Co^{++}(NH_3)_6$ is well defined (seminar R. Gessner) whereas in crystals of A- and B-DNA, metal ions could not be located. The theme of oligo- and polynucleotide structure in solution was further illustrated by D.R. Kearns who developped the use of 1D and 2D NMR techniques which also provide insight into the DNA double helix. G.V. Fazakerley reported in a seminar on NMR studies on oligonucleotides with mismatch base pairs which destabilize the DNA double helix depending on flanking regions. Contrasting this finding a stabilizing effect of amino acid amides on the B-DNA double helix was observed by J. Subirana using X-ray fiber diffraction methods.

Concering the occurrence of Z-DNA, M. Leng reviewed the use of antibodies raised against Z-DNA to very sensitively monitor the existence of the left-handed form. R.D. Wells reported on experiments suggesting that $d(CG)_n$ and related inserts in plasmids give rise to Z-DNA stretches, B/Z junctions and yet unidentified "anisotropic" DNA structures. These studies suggest that we still have to go a long way until we can claim that we really understand "DNA structure".

CONCLUSIONS. We have some structural information on DNA hydration, but the binding of metal ions is still poorly understood, at least for A- and B-DNA. As to base pair mismatches, their geometry is well described and it appears that flanking sequences play a role in their formation. The fine structure of DNA and the rules governing sequence specific effects on the local geometry are emerging, but the transition from B- to A- and Z-DNA, cruciform structures and the bending of DNA in A/T tracts which were dealt with in round table discussions, are still a matter of debate. It is clear, however, that these "superstructures" will have profound implications on DNA structure and function.

DNA-Drug Interactions

SUMMARY. A general understanding emerges from crystallographic, UV, fluorescence and 2D NMR spectroscopic studies of DNA-drug complexes, supported by foot printing methods employing gelelectrophoresis. These were reviewed by J.B. LePecq, R.E. Dickerson and M. Waring. The bisintercalator (quinoxaline) drugs give rise to Hoogsteen base pairing at intercalation sites and they appear to be specific for G/C sequences (M. Waring). C. Hélène has linked oligonucleotides with intercalators and these molecules bind to specific DNA sequences. Concerning DNA-drug interaction and function, a caveat was expressed by J.B. LePecq. Obviously, drugs often do not act on the level of DNA binding. They interfere with other proteins such as topoisomerase II and with the SOS system, and these interactions are frequently mistaken as occurring with DNA.

CONCLUSIONS. We have some understanding of DNA associating with several drugs by "outside" minor groove binding or base pair intercalation. In some cases these interactions provide an explanation for drug action and serve to design new, better drugs. Much more detailed investigations are neces-

sary, however, to really understand the biological target of
the different drugs which are of current interest.

DNA-PROTEIN INTERACTIONS

Repressor-operator interactions

SUMMARY. The physico-chemical aspects of proteins binding to
DNA were reviewed by P.von Hippel and by D. Porschke. In
organelles, there is probably cascade binding where one
protein induces weak binding which is followed by strong,
specific annealing. New insight into DNA-protein interaction
can be gained by direct cross-linking using UV laser pulses
in the nsec range (P.von Hippel), and studies on lac repres-
sor/ operator binding suggest that the repressor slides
along DNA until it finds the operator. This contrasts tet repressor which hops and jumps until it finds the specific
target (D. Porschke). It does not distort DNA in contrast to
the catabolite activator protein (CAP) which forces DNA to
bend by about 180°. The melting temperature of DNA is reduced
by amides of aromatic amino acids Trp, Tyr, Phe, an effect
also seen with single strand binding protein (SSB) which contains large amounts of these amino acids (D. Porschke). In
another series of experiments on SSB protein, G. Maass showed
that the binding of the tetrameric protein to DNA is cooperative depending on the length of the DNA.
    The complex formation and structure of the lac repressor/
operator system was investigated by B. Müller-Hill in a
series of genetic experiments. In his lecture (but not in
his notes), Müller-Hill reported on gel-electrophoretic
behavior of lac repressor/DNA complexes which suggests that
at low lac repressor concentrations, DNA forms loops of
about 6 helical turns, which are closed by one lac repressor
bound in bidentate mode to two operators simultaneously. If
the concentration of repressor is increased, the binding is
relaxed and two separate strands of DNA are bound. The three
dimensional structure of the lac head piece was established
as helix-turn-helix motif by R. Kaptein using 2D NMR techniques, and studies on the complex with operator are under
way supported by CIDNP methods to circumvent the problems
inherent in line broadening.
    In a comparable study reported as research seminar, F.
Buck assigned NMR signals for $O_R 1$, $O_R 3$ operators, studied
the complex formation with cro Repressor, and observed structural changes. NMR was also used in a study of Lex A repressor, which with Rec A protein is involved in the SOS system
(seminars Schnarr, diCapua). Lex A is cleaved by Rec A in
two domains, of which the amino terminal (68 amino acids)
cover 8 base-pairs in the complex with DNA (M. Schnarr). The
Rec A protein of 38 kDa binds to single and double stranded
DNA to form helical filaments which were investigated and
characterised by electron microscopy (E. diCapua).
    Very detailed structural information on repressors was
provided by the crystal study of the E. coli trp repressor
and aporepressor, and a model for interaction with the
operator DNA was proposed by P. Sigler. Trp repressor displays a helix-turn-helix motif in the binding region compa-

rable to lambda-cI, but is different from CAP and cro proteins. Since crystals of the repressor/operator complex were
grown, more information will be available shortly. The structure of the catabolite activator protein CAP and its model-
built interaction with DNA was described by T.A. Steitz who
also provided details of the structure of the Klenow fragment
of DNA polymerase I. A complex of the latter with the nucleotide 5'-TMP suggests several hypotheses concerning the
functional mechanism which are scrutinized by mutagenesis
experiments. For another DNA processing enzyme, resolvase,
crystals but no heavy atom derivatives are available. It is
hoped that a Ser-Cys point mutation will provide a useful
site for Hg^{++} binding (T.A. Steitz).

Methylases and Restriction Endonucleases

Because methylases and restriction endonucleases recognise the same DNA sequences, their structural, functional,
and evolutionary relationships were discussed at some length.
Concerning methylases, M. Radman provided a general review
on DNA methylation and its involvement in the control of
mismatch repair. Transitions (G/T and A/C) are well, transversions (purine/purine and pyrimidine/pyrimidine) are
less-well repaired. The number of G/C pairs flanking the
mismatch site is important, a feature that sheds light on
the influence of sequence on DNA structure, as was also
suggested by the studies of G.V. Fazakerley and G. Maass. In
a survey on endonucleases, S.E. Halford limited himself to
type II enzymes because those of types I and III are far
more complicated and less-well studied at present. Depending
on the Mg^{++} concentration, EcoRI as prototype of these
enzymes cleaves only one strand of double helical closed
circle DNA in presence of low Mg^{++} concentration and then,
or directly with high Mg^{++} concentration, the second strand
is cut. If dimethylsulfoxide or excess Mg^{++} are added, the
DNA cleavage becomes less specific. The crystal structure of
EcoRI in a complex with a tridecamer containing the specific
GAATTC sequence (in the absence of the activator Mg^{++})
provides details of protein-nucleic acid interaction and
induced DNA bending in remarkable detail (J. Rosenberg). The
sequence GAA is recognized by Arg200(G), Glu144, and
Arg145(AA). The latter two amino acids were also traced in
an independent cyanobromide cleavage study, and kinetic
experiments suggested that the number of base pairs flanking
the cutting site have a marked influence on the time course
of enzyme action (G. Maass).
    Finally, J. Subirana presented latest results on DNA-
histone complexes and discussed in detail the different
models which have been proposed thus far. Obviously, a final
decision cannot be reached at present, and histone H1 appears
to have a key role in the assembly of nucleosome core into
higher order (solenoid) structures.

CONCLUSIONS. Although many details are presently known
concerning protein-DNA interactions and the enzymatic
processing of DNA, it is clear that more insight has to be
gained by crystal studies of DNA-protein complexes, supported
by physico-chemical data. The new techniques of site direc-

272

ted mutagenesis of strategic amino acids and the synthesis of longer oligonucleotides with different base pair sequences offer powerful tools to study all kinds of interactions once the three dimensional structures of the partners are known.
The influence of DNA sequence and flanking regions on enzymatic processes are not understood. They appear to be of importance as the studies on mismatch repair methylases and endonucleases have shown. Of major interest is also the sequence of events occurring in DNA replication/transcription and in translation processes, which we are just beginning to investigate and where detailed knowledge is lacking.

* * *

Thus far my own recollections. I have the impression that all the contributions have demonstrated the actuality and the vigorous interest which the field of DNA-ligand interactions is experiencing. Particularly encouraging to the organizers and the senior lectures was the active participation of young colleagues, their constant, demanding interest and eagerness to learn about the subject of this ASI. It appears to me that the interdisciplinary approach which we and others in this area of research have applied in recent years has borne fruit and that in the years to come there will be considerable insight into and understanding of the interactions governing associations of the genetic material, DNA, with water, metal ions, drugs, small and multi-subunit proteins. This knowledge will not only be of academic interest and we are all convinced that it will help understanding and directing future research in medical sciences, in the new, growing field of biotechnology and, of course, in molecular biology, the mother science of this NATO-ASI.

RESEARCH SEMINARS

F. Buck, Univ. Frankfurt (GFR): NMR studies of protein – DNA recognition: phage lambda
    cro protein and lexA protein of E. coli and their operators.

R. M. Burger, Albert Einstein, New York (USA): Mechanism of DNA cleavage by
    iron.bleomycin.

Fu-Ming Chen, Tennessee SU (USA): Thermal denaturation of DNAs covalently modified by
    (+)-trans-7,8- dihydroxy- anti-9,10-epoxy- 7,8,9,10-tetrahydro-benzopyrene.

E. DiCapua, ETH Zürich (CH): The interaction of recA protein with DNA.

G. V. Fazakerley, CEN Saclay (F): Structures of mismatched base pairs in DNA and their
    recognition by the E.coli mismatch repair system.

R. V. Gessner, MIT (USA): Z-DNA crystal structures at high resolution.

F. Grosse, MPI Göttingen, (GFR): Interaction of DNA polymerase alpha, RNase H and
    topoisomerase II on the in vitro replication of natural single stranded DNA.

B. J. M. Harmsen, Univ. Nijmegen (NL): Thermodynamics and kinetics of gene-5
    protein-DNA interaction.

U. Heinemann, FU Berlin (GFR): Crystal structure analysis of DNA octamers.

R. Lauster, MPI Berlin (GFR): Structural and fonctional studies on procaryotic DNA
    methylases.

W. Leupin, ETH Zürich (CH): NMR studies of complexes between drugs and DNA.

J. Liquier, Univ. Paris XIII, Bobigny (F): Heteronomous double stranded polynucleotide
    structures studied by I.R. spectroscopy.

M. J. Modak, Med. School, Newark (USA): Structure and function relationship in E.coli DNA
    polymerase: substrate and template primer binding domains.

J. Paoletti & L. Mir, Inst. Gustave Roussy, Villejuif (F): Synthesis, physico-chemical and
    biological properties of oligonucleotides covalently linked to 2-methyl
    9-hydroxyellipticinium.

G. Posfai, Academy of Science, Szeged (H): Studies on structure-function relationships in
    GGCC-specific modification methylases.

M. Schnarr, IBMC – CNRS, Strasbourg (F): LexA, the repressor of the SOS system.

J. Sponar, Academy of Science, Prague (CSSR): DNA conformational changes under
    dehydrating conditions caused by basic peptide ligands.

W. D. Wilson, Georgia SU, Atlanta (USA): Unusual structural and interaction properties of
    nonaltenating A.T sequences in DNA.

POSTERS

L. A. Allison, Univ. Washington, Seattle (USA): Ribosomal gene regulation during development in <u>Xenopus laevis</u>

H. Aoyama, UNICAMP, Campinas (Brasil): Inhibitors of DNA synthesis by DNA- and RNA-dependent DNA polymerases.

T. S. Balganesh, MPI Berlin (GFR): Characterization of the functional domains of <u>B. subtilis</u> phage coded methyltransferases.

A. S. Bourinbaiar, King's College London (UK): An application perspective for dicyclohexylammonium (DCHA): a polyamine inhibitor (PI) in improvement of chromosome resolution.

D. Burnouf, IBMC - CNRS, Strasbourg (F): Cis-DDP induced mutagenesis in plasmid pBR322 in <u>E.coli</u>.

J. L. Butour, CNRS, Toulouse (F): Digestion of DNA-Platinum(II) complexes by S1 nuclease: a kinetic study.

S. E Castell, Medical School Univ. Bristol (UK): Site specific recombination and relaxation by Tn 21 resolvase.

L. X.-Q. Chen, Univ. Chicago (USA): Photophysics of tryptophan residue in RNAase T1 and RNAase T1-2'GMP complex.

C. Cera, Univ. Padova (I): Interaction of photochemotherapeutic furocoumarins with structurally organised DNA.

R. Cini, Univ. Siena (I): Metal-nucleotide complexes. Cooperativity between Mg(II) and other divalent cations.

N. Dalay & B. Kirdar, Istanbul Med. Faculty (TC): Characterization of salt solubilized chromatin particles.

E. A. M. de Jong, Univ. Nijmegen (NL): The three-dimensional structure of the DNA binding wing of IKE gene5 protein as determined by 2D ^{1}NMR.

N. J. de Mol, SU Utrecht (NL): Interaction of potential quinone cytostatic drugs with DNA.

N. J. de Mol, SU Utrecht (NL): Effect of singlet oxygen diagnostic aids D_2O and DABCO on the photobinding of psoralens to biological macromolecules and induction of genetic effects.

E. de Vries, SU Utrecht (NL): Interaction between the adenovirus origin of replication and sequence specific DNA-binding proteins.

M. Erikson, Chalmers Univ. Technology, Gothenburg (S): Characterization of the covalent binding of benzo(a)pyrenediol-epoxide to DNA.

B. G. Feuerstein, UCSF (USA): Molecular mechanics calculations of polyamine – oligodeoxyribonucleotide interactions.

G. Flöser, Univ. Bayreuth (GFR): Spectrofluorometric studies on the daunomycin – DNA intercalation complex.

V. Garzino, Univ. Marseilles (F): Stage and tissue specific nuclear proteins present during _Drosophila_ embryogenesis.

M. W. Germann, Univ. Calgary (CND): Conformational transitions in DNA hairpins.

M. Gniazdowski, Lodz Med. Acad. (PL): Inhibition of DNA dependent RNA synthesis _in vitro_ by 8-methoxypsoralen.

E. Guittet, ICSN – CNRS, Gif (F): NMR studies on the influence of cis-platinum on DNA duplex formation.

U. Heinemann, FU Berlin (GFR): HB protein from _B. globigii_: primary structure and mode of DNA binding.

U. Heinemann, FU Berlin (GFR): Crystallization of the core fragment of _S. typhimurium_ araC protein.

P. Hantzopoulos, Mount Sinai Hospital, New York (USA): Expression in _E. coli_ of a cDNA clone encoding human alpha-galactosidase A.

M. M. G. Koning, Univ. Groningen (NL): Structure of the deoxyribonucleotide duplex d(GCGTTGCG).d(CGCAACGC) with and without thymine photodimer.

D. Kostrewa & R. V. Gessner, FU Berlin (GFR): Comparative 500 MHz NMR studies on the 5S RNA of two thermophile bacteria.

T. Kriebardis, CEN Saclay (F): _dam_ methylase from _E. coli_: secondary structure and methylation of pBR322 plasmid.

R. M. J. N. Lamerichs, Univ. Groningen (NL): Two-deminsional NMR studies on the complex of _lac_-headpiece and a 14 base pair _lac_- operator fragment.

S. R. LaPlante, Syracuse Univ. (USA): Hydrogen bonding effects on the ^{13}C-NMR spectra of the bases in DNA oligomer duplexes.

P. Laugaa & P. Leon, CNRS-INSERM, Paris (F): DNA bis-intercalating antitumor agents: importance of the length, rigidity and symmetry of the linking chain.

F. LeHégarat, Univ. Paris-Sud, Orsay (F): Purification of a novel histone-like protein from _B. subtilis._

R. B. Macgregor, MPI Göttingen (GFR): Viscosity dependence of ethidium-DNA intercalation kinetics.

C. Malvy, Inst. G. Roussy, Villejuif (F): Apurinic sites are potential targets for some DNA binding agents.

G. Manzini, Univ. Trieste (I): The interaction of actinomycin-D with cytosine-guanine containing oligodeoxynucleotides of different sequences in aqueous solution.

R. Marquet, Univ. Liège (B): Influence of the DNA length on the spermidine-induced condensation.

K. Mazeau, CBM – CNRS, Orléans, (F): Modeling of the binding of _cis_PtCl$_2$(NH$_3$)$_2$ to a d(GpCpG) sequence.

J. J. McDonald, Roswell Park Mem. Inst. Buffalo (USA): A molecular model of the relative spatial adjacency of aminoacyl and peptidyl tranfer RNAs, when occupying the A and P sites of the ribosome.

G. Otting, ETH Zürich (CH): 2D NMR studies of the DNA-binding domain 1-76 of the P22 c2 repressor.

E. Quignard, CEN Saclay (F): Effects of methylation on the amino group of adenine: study of oligonucleotides containing the Gm^{6}ATC site of dam methylase.

A. Rahmouni, CBM – CNRS, Orleans (F): Sequence specificity of the interstrand cross-links in the reaction of the antitumor drug cis-diamminedichloroplatinum(II) and DNA.

M. Rajagopalan, Indian Inst. Science, Bangalore (IND): Interaction of non-intercalative drugs with DNA: distamycin analogues.

A. Rodriguez, Univ. Complutense, Madrid (SP): Specific and non-specific interactions of C1 proteins (a class of HMG proteins from the fruit fly Ceratitis Capitata).

M. Takahashi, IBMC – CNRS, Strasbourg (F): Effect of association and dissociation of protein subunit on the DNA binding parameters: the case of recA-DNA interaction.

M. Takahashi, IBMC – CNRS, Strasbourg (F): Quantitative analysis of the DNA binding properties of cAMP receptor protein from E.coli: modulation of the interactions (specific and non-specific) upon variation of cAMP concentration.

A. Wissmann, Univ. Erlangen-Nürnberg (GFR): Effect of all possible base pair substitutions ot Tn10 tet operator O1 on tet repressor binding.

A. Woisard, CEN Saclay (F): Induction of left-handed Z-DNA in different polynucleotides.

N. Zein, Univ. Houston (USA): Studies of the reaction of mitomycin C with 2',3'-isopropylidene guanosine.

Participants at the NATO-ASI – FEBS-Course "DNA-Ligand Interactions: from drugs to proteins" (from left to right):

top row: Burnouf, Laugaa, Malvy, Guittet, Froystein, de Jong, Mazeau, Otting.

second row: Rahmouni, Zein, Bell, Böttcher, F.M. Chen, Posfai, de Vries, Alabert Cordoba, Diez Ibanez, Bourinbaiar, Feuerstein, Cera, Rajagopalan, Heinemann, Kirdar, Dalay, Flöser, MacGregor.

third row: Lamerichs, de Mol, Fazakerley, Leupin, Marquet, LaPlante, Sugiyama, Aoyama, Grosse, Takahashi, Burger, McDonald, Sponar, Gniadzdowski, Manzini, Prat, Germann.

Forth row: Rizzo, Balganesh, Lauster, Manoharan, Rodriguez, Belhadj, Kaptein, Müller-Hill, Kearns, Rosenberg, Dickerson, Kennard, von Hippel, Sigler, Kubista, Saenger, Wells, Steitz, Cini, Wilson.

Front row (standing): Morales, Garcia-Moll, Sanderson, Lippert, Parge, Kompis, Wissmann, Maas, Garzino, Leon, LeHégarat, Castell, Pörschke, Subirana, Erikson, Koning, L.C.-X. Chen, Gdaniec, Kulinska, Abdulwajid, Buck, Pearl, C. Paoletti, Allison, Sowers, Butour, LePecq.

Front row (kneeling): Bergerat, Halford, Quignard, Kriebardis, Waring, Gessner, Kostrewa, J. Paoletti, Woisard, Guschlbauer, diCapua, Hantzopoulos, Schnarr, Mir, Harmsen, Bamberger, Leng.

Missing: Frey, Hélène, Modak, Radman.

280

PARTICIPANTS (L: Lecturers, D: Director, CD: Co-Director)

Dr. A. W.   ABDULWAJID, Biology Division, Brookhaven National Laboratory, Upton NY 11973
      USA
Joao ALABART CORDOBA, Quimica Inorganica, Faculdad de Quimica de Tarragona, Pl. Imperial
      Tarraco s/n, E-43005 Tarragona, Spain
Lizabeth A. ALLISON, Department of Zoology, NJ-15, University of Washington, Seattle, WA
      98195 USA
Prof. Hiroshi AOYAMA, Depto. de Bioquimica, Istituto de Biologia, UNICAMP, CP 6109,
      BR-13081 Campinas SP, Brasil
Dr. Tanjore S. BALGANESH, Abtl. Trautner, M.P.I. für molekulare Genetik, Ihnestrasse 63-73,
      D-1000 Berlin 33 G.F.R.
Prof. Elchanan S. BAMBERGER, Department of Biology, Haifa University, P.O. Tivon, Oranim
      36910, Israel
Prof. Omran BELHADJ, Laboratoire de Biochimie Faculté de Sciences de Tunis, Tunis, Tunisia
Prof. Arthur BELL, Department of Pharmaceutical Chemistry, Sunderland Polytechnic
      Sunderland SR1 35D, U.K.
Agnès BERGERAT-COULAUD, Service Biochimie, Bat. 142, Department de Biologie, C.E.N.
      Saclay, F-91191-Gif-sur-Yvette Cédex, France
Dr. Artur BOETTCHER, Zaklad Biofizyki, Akademia Medyczina, Karlowicza 24, Bygdoszcz,
      Poland
Dr. Aldar   S. BOURINBAIAR, Haematology Department, King's College Hospital, London SE5
      9RS, U.K.
Dr. Friedrich BUCK, Institut für Biophysik. Chemie, Universität Frankfurt, Theodor Stern Kai
      7 D-6000 Frankfurt 70 G.F.R.
Dr. Richard M. BURGER, Laboratory of Chromosome Biology, Public Health Research Institute
      of the City of New York, 455 First Avenue, New York NY 10016 USA
Dr. Dominique BURNOUF, Institut de Biologie moléculaire et cellulaire, CNRS, 19 Rue
      Descartes, F-67084 Strasbourg Cédex France
Dr. Jean-Luc BUTOUR, Labor. Pharmacologie et Toxicologie Fondamentales, CNRS, 205 Route
      de Narbonne, F-31400 Toulouse, France
Sophie Elissa CASTELL, Department of Biochemistry, University of Bristol Medical School,
      University Walk Bristol BS8 1TD, U.K.
Cinzia CERA, Dipartimento di Chimica organica, Università di Padova, Via Marzolo 1, I-35100
      Padova, Italy
Prof. Fu-Ming CHEN, Department of Chemistry, Tennessee State University, Nashville TN
      37203 USA
Lin Xiang-Qun CHEN, Department of Chemistry, University of Chicago, 5735 S. Ellis Ave.,
      Chicago IL 60637, USA
Prof. Renzo CINI, Dipartimento di Chimica, Università di Siena, Pian de Mantellini 44, I-53100
      Siena, Italy
Prof. Nejat DALAY, Department of Biophysics, Istanbul Medical Faculty, Temel Bilimer Binasi
      Capa Istanbul, Turkey
Dr. Evert   A.M. de JONG, Department Biophys. Chemistry, University of Nijmegen,
      Toernooiveld, NL-6525 ED Nijmegen, Netherlands
Dr. Nicolas J. de MOL, Department Pharmaceutical Chemistry, Faculty of Pharmacy,
      Catharijnesingel 60 NL-3511 GH Utrecht, Netherlands
Erik de VRIES, Laboratory for physiol. Chemistry, University of Utrecht, Vondellaan 24
      NL-3511 Utrecht, Netherlands

Dr. Elisabeth DiCAPUA, Institut für Zellbiologie der E.T.H., Hönggerberg, CH-8093 Zürich,
Switzerland

Prof. Richard E. DICKERSON (L), Molecular Biology Institute, University of California at Los
Angeles, Los Angeles CA 90024 USA

Miguel DIEZ IBANEZ, Laboratoire de Biochimie Médicale, Faculté de Médicine, 7 Blvd. Jeanne
d'Arc, F-21033 Dijon, France

Magdalena ERIKSON, Department of Physical Chemistry, Chalmers University of Technology,
S-41296 Gotenburg, Sweden

Dr. G. Victor FAZAKERLEY, Service Biochimie, Bat. 142, Department de Biologie, C.E.N.
Saclay, F-91191-Gif-sur-Yvette Cédex, France

Dr. Burt FEUERSTEIN, Brain Tumor Res. Center, University of California, 783 HSW, San
Francisco CA 94143 USA

Götz FLOSER, Institut für Physik, Universität Bayreuth, Universitäts Strasse 30, D-8580
Bayreuth, G.F.R.

Prof. Harvey S. FREY, Herschman Laboratory, University of California at Los Angeles, Warren
Hall, Los Angeles CA 90024 USA

Niels Age FROYSTEIN, Department of Chemistry, University of Bergen, N-5000 Bergen,
Norway

Maria GARCIA-MOLL, Duke University Medical Center, P.O.Box 3945, Durham, NC 27710 USA

Veronique GARZINO, LGBC case 907, Faculté de Luminy, 70 Route Léon Lachamp, F-13288
Marseilles Cédex, France

Dr. Zofia GDANIEC, Institute of Bioorganic Chemistry, Polish Academy of Sciences,
Noskowskiego 12/14, PL-61-704 Poznan, Poland

Markus V. GERMAN, Department of Med. Biochemistry, University of Calgary, Health Science
Centre, 3330 Hospital Drive NW, Calgary Alberta T2N 4N1 Canada

Dr. Reinhard GESSNER, Institut für Biochemie, Freie Universität Berlin, Thielallee 63,
D-1000 Berlin 33 G.F.R.

Prof. Marek GNIAZDOWSKI, Department of General Chemistry, Medical Academy of Lodz, ul.
Lindleya, PL-90-131 Lodz, Poland

Dr. Frank GROSSE, M.P.I. für exptl. Medizin, Hermann-Rein-Strasse 3, D-3400 Göttingen
G.F.R.

Dr. Eric GUITTET, Laboratoire RMN, Institut des Substances Naturelles du CNRS, F-91190
Gif-Sur-Yvette, France

Dr. Wilhelm GUSCHLBAUER (D), Service Biochimie, Bat. 142, Department de Biologie, C.E.N.
Saclay, F-91191-Gif-sur-Yvette Cédex, France

Prof. Stephen E. HALFORD (L), Department of Biochemistry, University of Bristol, University
Walk, Bristol BS8 1TD, U.K.

Petros HANTZOPOULOS, Department of Microbiology, Mount Sinai School of Medicine, 1
Gustave Levy Plaza, New York NY 10029 USA

Dr. Ben J. M. HARMSEN, Department of Biophys. Chemistry, Faculty of Science, Toernooiveld,
NL-6525 ED Nijmegen, Netherlands

Dr. Udo HEINEMANN, Institut für Kristallographie, Freie Universität Berlin, Takustrasse 6,
D-1000 Berlin 33 G.F.R.

Prof. Claude HELENE (L), Laboratoire de Biophysique, Muséum National d'Histoire Naturelle,
61 Rue Buffon, F-75005 Paris, France

Prof. Robert KAPTEIN (L), Department of Physical Chemistry, University of Groningen,
Nijenborgh 16, NL-9747 AG Groningen, Netherlands

Prof. David R. KEARNS (L), Department of Chemistry B-014, University of California at San
Diego, La Jolla CA 92093 USA

Prof. Olga KENNARD (L), University Chemical Laboratory, University of Cambridge, Lensfield
Road, Cambridge CB2 1EW U.K.

Prof. Betul KIRDAR, Biyoloji Bölömü, Fen Edebiyat Fakultesi, Bogazics Universitesi,
TC-80815 Bebek, Istanbul, Turkey

Dr. Ivan KOMPIS, Department PF/ID, Bldg. 15, F. Hoffmann-La Roche & Co., Ltd., CH-4002
Basle, Switzerland

Thea KONING, Department of Physical Chemistry, University of Groningen, Nijenborgh 16,
NL-9747 AG Groningen, Netherlands

Dirk KOSTREWA, Institut für Biochemie, Freie Universität Berlin, Thielallee 63, D-1000
Berlin 33 G.F.R.

Dr. Tasos KRIEBARDIS, Service Biochimie, Bat. 142, Department de Biologie, C.E.N. Saclay,
F-91191-Gif-sur-Yvette Cédex, France

Mikael KUBISTA, Department of Physical Chemistry, Chalmers University of Technology,
    S-41296 Gotenburg, Sweden
Dr. Katarzyna KULINSKA, Institute of Bioorganic Chemistry, Polish Academy of Sciences,
    Noskowskiego 12/14, PL-61-704 Poznan, Poland
R.M.J.N. LAMERICHS, Department of Physical Chemistry, University of Groningen, Nijenborgh
    16, NL-9747 AG, Groningen, Netherlands
Steven R. LAPLANTE, Department of Chemistry, Syracuse University, 305 Bowne Hall,
    Syracuse NY 13244-1200 USA
Dr. Philippe LAUGAA, Department de Chimie Organique, Faculté de Pharmacie, 4 Ave de
    l'Observatoire, F-75006 Paris, France
Roland LAUSTER, Abtl. Trautner, M.P.I. für molekulare Genetik, Ihnestrasse 63-73, D-1000
    Berlin 33, G.F.R.
Prof. Françoise LE HEGARAT, Institut de Microbiologie, Université Paris Sud, Bat. 409,
    F-91405 Orsay, France
Dr. Marc LENG (L), Centre Biophysique Moléculaire, CNRS, 1A Ave. Recherche Scientifique,
    F-45071 Orléans Cédex 02, France
Pascale LEON, Department de Chimie Organique, Faculté de Pharmacie, 4 Ave de
    l'Observatoire, F-75006 Paris, France
Prof. Jean Bernard LEPECQ (L), Department Physicochimie Macromoléculaire, Institut Gustave
    Roussy, Rue Camille Desmoulins, F-94805 Villejuif Cédex, France
Dr. Werner LEUPIN, Institut für Molekularbiologie und Biophysik der ETH, Hönggerberg,
    CH-8093 Zürich, Switzerland
Prof. Bernhard LIPPERT, Institut für Anorg. & Analyt. Chemie, Universität Freiburg,
    Albertstrasse 21, D-7800 Freiburg im Breisgau, G.F.R.
Dr. Jean LIQUIER, UER de Medecine et de Biologie, Université Paris XIII, 74 Rue Marcel
    Cachin, F-93012 Bobigny Cédex, France
Prof. Günther MAAS (L), Abtl. Biophysikalische Chemie, OE 4350, Zentrum Biochemie,
    Medizinische Hochschule, Postfach 610180, D-3000 Hannover 61, G.F.R.
Dr. Robert B. MAC GREGOR, Abtl. Molekulare Biologie, M.P.I. für Biophys. Chemie, Postfach
    2841, D-3400 Göttingen - Niklausberg, G.F.R.
Dr. Claude MALVY, Department Physicochimie Macromoléculaire, Institut Gustave Roussy, Rue
    Camille Desmoulins, F-94805 Villejuif Cédex, France
Dr. Muthiah MANOHARAN, Department of Chem. & Biochemistry, University of Maryland, Box
    254, College Park MD 20742, USA
Prof. Giorgio MANZINI, Dipto. di Biochimica & Biofisica, Università di Trieste, Piazzale
    Europa 1, Trieste, Italy
Roland MARQUET, Laboratoire de Chimie macromoléculaire, Université de Liège, Sart-Timan,
    B-4000 Liège, Belgium
Karim MAZEAU, Centre Biophysique Moléculaire du CNRS, 1A Ave. Recherche Scientifique,
    F-45071 Orléans Cédex 02, France
Joseph J. MC DONALD, Unit of Theoretical Biology, Roswell Park Memorial Institute, 666
    Elm Street, Buffalo NY 14263 USA
Dr. Lluis MIR, Oncologie Moléculaire, Institut Gustave Roussy, Rue Camille Desmoulins,
    F-94805 Villejuif Cédex, France
Prof. Mukund J. MODAK, Department of Biochemistry, New Jersey Medical School, 100 Bergen
    Street, Newark NJ 07103 USA
Nydia M. MORALES, Department of Microbiology, School of Medicine, East Carolina
    University, Greenville NC 27834 USA
Prof. Benno MULLER-HILL (L), Institut für Genetik, Universität Köln, Weyertal 121, D-5000
    Köln 41, G.F.R.
Gottfried OTTING, Institut für Molekularbiologie und Biophysik der ETH, Hönggerberg,
    CH-8093 Zürich, Switzerland
Prof. Claude PAOLETTI, Institut Gustave Roussy, Rue Camille Desmoulins, F-94805 Villejuif
    Cédex, France
Dr. Jacques PAOLETTI, Institut Gustave Roussy, Les Hautes Bryuères, F-94805 Villejuif
    Cédex, France
Dr. Hans E. PARGE, Department of Molecular Biology MB3, Research Institute of Scripps
    Clinic, 10666 N. Torrey Pines Road, La Jolla CA 92037 USA
Dr. Laurence PEARL, Biomolecular Structure Unit, Institute of Cancer Research, Clifton
    Avenue, Sutton, Surrey SM2 5PX, U.K.

Dr. Dietmar POERSCHKE (L), M.P.I. für Biophysikalische Chemie, Postfach 2841, D-3400
    Göttingen - Niklausberg G.F.R.
Dr. György POSFAI, Institute of Biochemistry, Biological Research Center of the Hungarian
    Academy of Sciences P.O.Box 521, H-6701 Szeged, Hungary
Dr. Odette PRAT, Compagnie ORIS Industrie, Centre Marcoule LAPAM, BP N° 171, F-30205
    Bagnols sur Cèze Cédex, France
Etienne QUIGNARD, Service Biochimie, Bat. 142, Department de Biologie, C.E.N. Saclay,
    F-91191-Gif-sur-Yvette Cédex, France
Dr. Miroslav Radman (L), Institut Jacques Monod, C.N.R.S., Université Paris 7, Tour 43, 2,
    place Jussieu, F-75251 Paris Cédex 05, France
Abderrahim RAHMOUNI, Centre Biophysique Moléculaire du CNRS, 1A Ave. Recherche
    Scientifique, F-45071 Orléans Cédex 02, France
Malini RAJAGOPALAN, Biophysics Unit, Indian Institute of Science, Bangalore 560012, India
Dr. Vincenzo RIZZO, Ricerca & Sviluppo Chimico, Farmitalia Carlo Erba SpA, Via dei Gracchi
    35, I-20146 Milano MI, Italy
Ana RODRIGUEZ, Depto. de Bioquimica, Facultad de Quimicas, Universidad Complutense,
    E-28040 Madrid, Spain
Prof. John M. ROSENBERG (L), Department of Biological Sciences, University of Pittsburgh,
    Pittsburgh PA 15260 USA
Prof. Wolfram SAENGER (CD), Institut für Kristallographie, Freie Universität Berlin,
    Takustrasse 6, D-1000 Berlin 33 G.F.R.
Dr. Marc SANDERSON, Department of Molecular Biophysics, Yale University, P.O.Box 6666,
    New Haven CT 06511 USA
Dr. Manfred SCHNARR, Institut de Biologie moléculaire et cellulaire, CNRS, 19 Rue
    Descartes, F-67084 Strasbourg Cédex, France
Prof. Paul SIGLER (L), Department of Biochemistry & Molecular Biology, University of
    Chicago, 920 E. 58th Street, Chicago IL 60637 USA
Dr. Lawrence SOWERS, Molecular Biology Division, Universitry of Southern California, ACR
    406, Los Angeles CA 90089-1481 USA
Dr. Jaroslav SPONAR, Ustav org. Chemie ve Biochemie CSAV, Flemingovo nam. 2, CSSR-166 10
    Praha 6, Czechoslovakia
Prof. Thomas STEITZ (L), Department of Molecular Biophysics, Yale University, P.O.Box 6666,
    New Haven CT 06511 USA
Prof. Juan SUBIRANA (L), Unidad de Quimica Macromolecular, Escuela Técnica Superior de
    Ingenerios Industriales de Barcelona, Avda. Diagonal 647, E-08028 Barcelona,
    Spain
Dr. Hiroshi SUGIYAMA, Department of Synthetic Chemistry, Kyoto University, Sakyo-Ku, Kyoto
    606, Japan
Dr. Masayuki TAKAHASHI, Institut de Biologie moléculaire et cellulaire du CNRS, 19 Rue
    Descartes, F-67084 Strasbourg Cédex France
Prof. Peter von HIPPEL (L), Institute of Molecular Biology, University of Oregon, Eugene OR
    97403 USA
Prof. Michael J. WARING (L), Department of Pharmacology, Cambridge University, Hills Road,
    Cambridge CB2 2QD U.K.
Prof. Robert D. WELLS (L), Department of Biochemistry, University of Alabama, Schools of
    Medicine / Dentistry, Birmingham AL 35294 USA
Prof. W. David WILSON, Department of Chemistry, Georgia State University, Atlanta GA 30303
    USA
Andreas WISSMANN, Institut für Mikrobiologie & Biochemie, Universität Erlangen,
    Staudtstrasse 5, D-8520 Erlangen G.F.R.
Anne WOISARD, Service Biochimie, Bat. 142, Department de Biologie, C.E.N. Saclay,
    F-91191-Gif-sur-Yvette Cédex, France
Nada ZEIN, Department of Chemistry, University of Houston, Houston TX 77004 USA

Accuracy of recognition, 226, 228
Acridine, 142, 143, 220
Actinomycin D, 113, 114, 119-121, 124, 142, 144
Activator protein, 160
Adenovirus, 78
A-DNA, see DNA
Adriamycin, 113, 142, 143, 148
Affinity scale,
    for   amino acids, 87
Alkylating agents, 152, 221
Allosteric activation, 253
Alpha helix, 108-110, 252, 254, 255
    unitary, 254
Amino-acids,
    basic, 86-92, 106
    contributions of side chains
        to interactions, 86-90
    substitution, 233-235
2-Aminopurine, 220
AMSA, 142, 143, 147-149
Anthracycline, 142, 143, 149
Antibiotic-DNA interaction,
    binding sites, 114
    interaction, 113-122
        kinetics of, 119
        shuffling of, 119
Antibiotics, 113-124, 141-150
Antibodies, 257-267
    against adenosine, 257
    against poly(I).poly(C), 261, 262, 264
    against Z-DNA, 261
    association constants, 259, 262
    monoclonal, 259
Anticancer drugs, 113-124, 141-145
Antigen, 257-259
Anti-sense RNA, 127, 128, 130
Antitumor activity, 113, 143, 144
Arabinosyluracil, alpha-D-, 232
Arginine, 107
    and   EcoRI specificity, 234,254-256
Asparagine,
    and   EcoRI specificity, 255
Assignment,
    of NMR spectra, 194, 196, 200

Backbone conformation, 18
Bacteriophage,
    M13mp10, 242-244
    ϕX 174, 218, 219, 242, 244, 250
Base mismatches, 2, 3, 10, 15, 217-223
Base pairs, 2-4, 27, 28, 48, 162, 247
    Hoogsteen, 5, 29, 153
    mismatched, 2, 3
    stick representation of, 6, 163
    wobble, 2, 3
B-DNA, see DNA
Berenil, 125, 142, 145
Beta sheet, 252, 253, 255, 109, 110
Beta-galactosidase, 175
Beta-globin, 133, 134
Beta-lactamase, 132, 134
Binding,
    kinetics, 97
    probabilities, 161
    site, 50-55, 97-102, 113, 141-145,
        159-166, 174-179, 225-234,
        240-242, 245, 252-255, 260-264
    specification, 112- 118, 160, 162, 227
Bis-intercalation, 116-119, 123, 142, 144, 151
Bleomycin, 113, 142, 144, 152, 153
Bromoacetaldehyde, 67
5-Bromodeoxyuridine, 229
5-Bromouracil, 220

Catabolite activator protein (CAP), 48, 180, 185-190
CCAGG sites, 221
CCCGGG sites, 242
Circular dichroism (CD), 131
Cell death, 147, 153
Cell toxicity, 146
Cellular localization
    of DNA, 229
    of EcoRI, 234
Charge potential,
    electrostatic, 186
Chemotherapeutic agent, 113-124, 135,
    140-150

Chloroquine, 113, 142, 143
Chromatin, 109
Chromomycin, 113, 142, 145
Chromosomes,
  metaphase, 260, 261, 265
Cleavage,
  of nucleic acids, 135, 137, 225-234,
      239-248, 251-256
  kinetics,
    role of flanking sequences, 229, 230
    linear diffusion, 232
    Michaelis-Menten condition, 229
    single turnover experiment, 229
Chloroflexus aurantiacus, 240, 249
Crithidia fasciculata, 79-81
Co-crystallization, 188, 190
Code of recognition, 173-175, 179-181
Cooperative binding, 165
cro repressor, 48
Cruciforms in DNA, 46, 64
Cyclic AMP receptor, 93, 94
Cytotoxicity, 148, 150-153

Daunomycin, 113, 142, 143
Deamination,
  of cytosine, 217, 212
Deoxyuridine, 233
Dihydrofolate reductase, 146
Dimethyl sulfoxide, 244
Distamycin A, 124, 142, 145
Ditercalinium, 142, 144, 151, 152
DNA,
  A-form, 1, 3, 4, 9, 10, 12, 15-17, 105-107
  adenovirus, 80, 81
  anisomorphic, 64, 73, 76, 77
  B-form, 1, 3, 4, 10, 13, 15-17, 65, 105-107
    stabilization, 55
  B-Z   junction, 66, 69
  bent, 65, 79, 80
  binding,
    amino acids, 105
    Hoechst 33258, 50, 55
    netropsin, 50, 55
    proteins, 49
  breaks, 153
  C-form, 105, 107
  chemical synthesis, 173, 175
  conformation, 1-18
  counterions, 105
  cruciforms, 46, 64
  destabilization, 106
  drug binding, 45, 49
  form V, 264
  Herpes simplex, 64, 73-75
  information readout,
    extrinsic, 45
    intrinsic, 45, 59
  kinetoplast, 65, 79-81
  kinking, 186, 187
  left-handed, 17, 34, 64-69
  major groove, 10-12, 47

DNA, (continued)
  metal ions, 105
  methylated, 50
  minor groove, 47
    width, 58
  mitochondrial, 153
  modification, 137, 229
  nuclear, 153
  oligomer structures, 45
  pitch, 106
  reading by drugs, 59
  right handed, 17
  sequencing, 23
  slipped structures, 65
  strand breaks, 151-153
  structure, 1-18, 25-27, 38
  supercoiled, 76
  synthesis, 147
    inhibition of, 147-149
  Z-form, left handed, 1, 3-5, 10, 11, 14-17,
      23, 24, 26, 28, 35, 64-69
DNA-antibiotic interaction,
  dynamics of, 113-125
  effect on nucleosome structure, 121-125
  kinetics of, 119-121
  sequence recognition,
    molecular basis of, 116-119
  sequence selectivity, 113-116
DNAaseI, 150, 233, 234, 113, 115-116, 121
  cleavage by, 114, 116, 120, 121, 123
DNA-binding drugs, 141-153
DNA-drug complexes, 4, 5
DNA polymerase, 147, 185, 219
  Klenow fragment of, 188-190
DNA-protein interactions, 4, 5, 159-170
  affinity, 164-167
  binding selectivity in vivo, 170
  conformational change,
    by regulatory proteins, 167
  distortion of DNA or proteins, 167
  equilibrium selection, 167, 168
  evolutionary design of regulatory
      proteins, 169, 170
  free energy,
    of non-specific binding, 166, 167
    of specific binding, 166
  information content, 168
  regulatory proteins, 164, 167, 169, 170
Donor-acceptor groups, 6
Double helix,
  bending by cAMP receptor, 93, 94
  recognition of, 260
Drug binding,
  electrostatic forces, 55
  hydrogen bonding, 53, 54
  thermodynamics, 58
Drugs,
  antiparasitic, 142, 146, 151, 152
  antitumor, 142, 146, 150, 151
  DNA   binding, 45, 49, 113, 141-152

Dynamics,
  of excluded site binding, 94-96
  of oligonucleotide-oligopeptide complexes,
    91
  of repressor-operator recognition, 96-99

Echinomycin, 113-125, 142, 144, 153
Electric field jump, 93
Electro-optical methods, 85, 93
Electron microscopy, 230
Electrostatics,
  and EcoRI specificity, 254
Ellipticine, 142, 144, 147-149, 220
Enzyme linked immunosorbent assay (ELISA),
    260
Epipodophillotoxin, 150
Error correction, 218-221
  in replication, 217-223
Escherichia coli (E.coli), 127, 173-175, 179,
    180, 217-222, 239-240
Ethidium, 142, 143, 149, 150
Evolution,
  of genetic code, 91
  of melting proteins, 91
  of regulatory proteins, 169
Exonuclease,
  proofreading, 221

Fluorescence, 92, 131
Footprinting, 23, 114-122
  DNA-echinomycin, 113, 115, 116, 121, 122

GAATTC site, 46-56, 225-227, 239, 251-253
GATC sites, 218-212
Gel electrophoresis,
  two-dimensional, 68
Gene regulation, 63-72, 113-122, 127, 132,
    133, 159, 183
Glutamic acid,
  and EcoRI specificity, 237, 255, 256
Glycoprotein, 262
Glycosidic torsion angle, 8, 27
  syn-anti equilibrium, 27, 30-33
Grooves,
  in double helix, 9, 11

Hapten, 259, 260
Headpiece-operator complex, 193, 195, 196,
    204, 210, 211
Helical parameters, 5, 10, 16
Helical sense, 3, 27, 29, 33, 68
Helicase II, 218, 219
Helix-turn-helix motif, 183, 188
Heteroduplex DNA, 217, 218, 220
Histidine, 107
Histone, 121, 127, 109
Hoechst 33258, 50, 53, 55-57
Hoogsteen base pairing, 5, 29, 153
Hopping of proteins along DNA, 100-101
Hycanthone, 142, 143

Hydration, 51, 105, 106
  spine of, 51
Hydrogen bonds, 87, 117, 164, 246-248, 250
  bidentate, 53-54, 254
  in drug binding, 51-54
  and EcoRI specificity, 254
Hydrophobic interactions, 85-91, 179
Hydrophobicity
  effects on ligand binding, 88-91
Hydroxystilamidine, 142, 145

I-mutants, 212
Immunization, 257, 259, 260
Immunoglobulins, 257, 258
  Fab fragments, 257, 258
  structure of, 262, 263
Influenza virus, 133, 134
Information content
  of DNA grooves, 47
Intercalating agents, 116-123, 127-133, 137,
    138, 141-143, 148-150

Joint regions, 73
Jumping of proteins along DNA, 100-101

Klenow fragment, 185-188
Kinetics,
  of antibiotic-DNA interaction, 119
  of DNA binding to EcoRI, 229-231
  of ligand binding 95-98

lac, see also operator, repressor
  operator, 159, 160, 167, 173-175, 179, 180,
    191-214
  repressor, 99, 159, 160, 167, 173-180,
    191-214
   headpiece, 48, 180-210
lambda, see also operator, repressor
  operator, 173-179
  cro repressor, 48, 175-180
  repressor, 48, 175-180
Lexitropsins, 60
Linear diffusion, 230
Lucanthone, 142, 143
Lupus erythematosus, 257
Lysine, 107

Macromolecules,
  informational, 45
Magnesium,
  and EcoRI specificity, 225-227, 234,
    235, 244, 253
Melting of DNA, 67, 88, 89
Melting proteins, 93-96
6-Methyladenine, 217-219, 222
Methylation,
  of adenine, 217-219, 222
  of cytosine, 67, 217, 221-223
  of DNA, 217-221
5-Methylcytosine, 67, 217, 221-223
Methyl-propyl-EDTA-FeII, 114, 116

Mismatch, 2, 3, 15, 217-222
  base pair, 2, 3, 15
  repair, 217-222
   short patch, 222
Mitamycine, 113
Mitoxanthrone, 142, 143
Model building, 39, 49-54, 209, 210, 254-256
Modification,
  of DNA, 229-231
  of EcoRI, 233
  site-directed, 137
Molecular mechanics, 59
Monte Carlo simulation, 95, 97
Motion mechanisms, 100, 101
Mutagenesis,
  site directed, 189, 190, 234
Mutation,
  frameshift, 218
  transition, 217, 219-221
  transversion, 221, 223
Mutator genes, 218-220
Mutators, 217-222

Neo-kink, 225, 252
  type I, 252
  type II, 252
Netropsin, 50-61, 113, 124, 142, 145
Nitrocellulose filters,
  binding assay, 227
Nogalomycin, 113, 122
Nuclear magnetic resonance (NMR), 23-38,
      131, 191-216
  photo-CIDNP, 211
  two-dimensional, 24-40, 191-213
Nuclear Overhauser effect (NOE), 192-194,
     196, 200, 204, 206
  two-dimensional (NOESY), 25, 28, 30-32,
     34-36, 193, 195, 196, 200-202
Nucleases (see also restriction enzymes),
  Bal31, 49
  P1,   67, 75
  S1,   67, 75, 233
Nucleic acids,
  cleavage of, 135, 137
Nucleohistones, 109
Nucleosome, 109, 121-125
  particle, 121, 123, 124

Oligonucleotides, 1-18, 57-59, 116, 119,
     127-129, 132, 133, 137, 192, 196,
     231, 233
  crystallography of, 1-18, 50-59, 254
  NMR of, 51, 59, 220
Oncogenes, 136, 150-153
Operator DNA, 192-195
  gal, 174, 175
  lac, 173-178
  lambda, 175, 178
  trp, 183
Operator-repressor system, 226

Periplasmic space, 230
Photo-CIDNP, 211
Photosensitation, 137, 138
Plasmids, 67-70, 73, 173, 174, 176-178
  compatible, 173, 175
  pAO3, 244-246
  pBR322, 78, 177
  pPK201, 80
  pRW, 69
  pWB, 163
Plasmodium, 150, 151
poly(A), 260
poly(A).poly(U), 89, 90, 261, 263, 264
poly(dA).poly(dT), 56-58
poly(dA-dC).poly(dG-dT), 67, 261
poly(dA-dT), 58, 59, 121, 164, 261
poly(dG).poly(dC), 116, 261
poly(dG-dC), 57, 67, 261, 263
poly(dI-dbr^{5}C), 57
poly(dm^{5}C-dG), 58
poly(I).poly(br^{5}C), 261
poly(I).poly(C), 88, 90
poly(rA).poly(dT), 59
Polymer lattice, 94
Polynucleotides, 257-266
Polypeptides,
  basic, 106, 109, 110
Primary target,
  in drug action, 146
Proflavin, 142, 143
Promotor,
  tyrT, 115-117, 121-123
  major late, 78
Propeller twist,
  of base pairs, 6, 7
Propyluridine, 233
Protamines, 108
Protein,
  histone-like, 230
  single strand binding (ssb), 91, 94
  synthesis, 132, 133, 147
Protein-DNA recognition, 191
Protein-nucleic acid interaction,
  mechanism of, 85-87, 93-100, 159-168,
     173-180, 183, 185-189, 191-210,
     225-234, 239-249, 251-255, 260, 264
Proteinase K, 149
Protons,
  exchangeable, 29-39
  non-exchangeable, 29-39

Quinacrin, 142, 143, 261
Quinoxaline, 113-116

Radioimmunoassay (RIA), 260
Recognition,
  chemical basis of, 96-99, 162-164
  complex, EcoRI, 251
  primary sequence of, 162-164, 173-176,
     178
  helix, 173, 175, 179, 180, 191, 210

Recognition, (continued)
  interactions, 164, 256
  module,
    EcoRI, 254
    outer, 254
    inner, 254
  process, 230, 231
  sequence,
    of EcoRI endonuclease, 256
  secondary sequence of, 164
Recombination, 217
Regulatory proteins, 159-169
Relaxation,
  cross, 25
  proton, 24-40
  spin-lattice, 24, 25, 39
Repeats in DNA
  direct, 55-57
  inverted, 55-57
Repression, 173-175, 179
Repressor,
  gal, 174, 175
  homology of, 173, 175, 177
  lac, 173-175, 177-180
  lambda CI gene, 175, 178, 179
  lambda cro, 48, 173, 175, 178, 179
  tet, 104
  trp, 183
Repressor-operator complex, 207-210
  dynamics of, 96-99
  recognition, 94-97
Resolvase, 186-190
Resonances,
  assignment of, 30
Restriction enzyme (restriction
      endonuclease), 225-234, 239-248,
      251-256
  AatII, 177
  AccI, 82
  BamHI, 68, 70
  BspRI, 229
  BssHII, 68
  BsuRI, 229
  CauI, 239-242, 246-249
  CauII, 239-249
  CauIII, 240
  EcoRI, 49, 72, 77, 147, 225-234, 239-242,
      244, 246, 249, 250, 251-257
    cleavage and hydrogen bonding, 251-257
    DNA interaction, 226, 229, 251-257
    recognition sequence, 226, 256
    star activity, 229, 254, 256
  EcoRV, 177, 233, 239-241, 244
  HaeIII, 229
  HhaI, 68
  HincII, 73
  HindIII, 177
  MboII, 239, 240
  NciI, 243
  NcoI, 177

Restriction enzyme (restriction
      endonuclease), (continued)
  PstI, 73, 240
  SalGI, 241, 242, 249
  SssI, 177
  StuI, 80
  XbaI, 176
  XhoI, 177
  XmaI, 177
RNA,
  anti-sense, 130, 134
  messenger, 132
  synthesis, 130-133, 147
    inhibition of, 147, 148
RNA polymerase, 147, 160, 185
Roll of base pairs, 7-9

Sequence specific binding, 114, 16
Sequences,
  oligo-purine.oligo-pyrimidine, 73-76
  recognition, 226, 256
Slide of base pairs, 8, 9
Sliding of proteins along DNA, 100-101
Solvent accessibility, 211
SOS system, 152
Specificity of interactions,
  for individual amino acids, 86-90
  "window" of, 165
Spin diffusion, 30
Spine of hydration, 51-58
Stacking interactions, 87
Steady state kinetics, 229
Strand cleavage, 149, 150
Structure-activity relationship, 142, 147
Structures,
  slipped, 64, 75
Sugar conformation, 7, 8, 10, 17
Sugar pucker, 8, 29
Supercoiling, 69, 76
Symmetry,
  two-fold, 251-255
syn-anti conformation, 7-11, 14-17, 27,
      31-34

Targeting, 128, 137
Telestability, 76
tet repressor, 99, 104
Thermodynamics,
  of DNA binding to EcoRI, 226-228
  of drug binding, 58, 129-132
Tilorone, 113, 142, 143
Tilt of base pairs, 6-9, 17
Topoisomerase II, 146, 148-152
Transcription, 113, 127, 132
Translation, 127, 132-134, 136
Triostin, 113
Tris-intercalation, 151
Trypanosome, 79, 80, 135-137, 150, 151
Tryptophane, 183
Twisting of DNA, 252
tyrT promotor DNA, 115-117, 121-123

trp repressor, 48, 183

Uv sensitivity, 152

Viruses, 133, 134, 136
  adenovirus, 78, 79
  Herpes simplex, 73-75
  influenza virus, 133, 134
  SV40 virus, 64, 133-136, 150

X-ray diffraction,
  fiber, 105
  single crystal, 1-18, 45-56, 183, 185-189,
    251-256
    methods of 4, 5, 9

Z-DNA, see DNA